Advanced Power Cable Technology

Volume II
Present and Future

Authors

Toshikatsu Tanaka, Ph.D.
Central Research Institute
of Electric Power Industry
Tokyo, Japan

Allan Greenwood, Ph.D.
Philip Sporn Professor
Director
Center for Electric Power Engineering
Rensselaer Polytechnic Institute
Troy, New York

CRC Press, Inc.
Boca Raton, Florida

Library of Congress Cataloging in Publication Data

Tanaka, Toshikatsu.
 Advanced power cable technology.

 Bibliography: p.
 Includes index.
 Contents: v. 1. Basic concepts and testing —
v. 2. Present and future.
 1. Electric cables. 2. Electric lines—Underground.
I. Greenwood, Allan, 1923—   . II. Title.

TK3351.T36   1983   621.319′2   82-14597
ISBN 0-8493-5165-0 (v. 1)
ISBN 0-8493-5166-9 (v. 2)

International Standard Book Number 0-8493-5165-0 (v. 1)
International Standard Book Number 0-8493-5166-9 (v. 2)

Library of Congress Card Number 82-14597
Printed in the United States

PREFACE

We have learned much of past and present cable technology in the course of writing this book. An attempt has been made to describe the most recent information, but at the same time set out the fundamentals of cable technology in some depth. No details of "cable products" by individual manufacturers are described, instead, a focus is directed on general features of various kinds of cables. Although the expected prime readership will be cable engineers and those interested in cable research, the book will also be useful to university undergraduate and graduate students and management personnel who are interested in electric power engineering and underground power transmission in particular.

In the preparation of these two volumes abundant help was given by many distinguished cable research engineers in the U.S., Europe, and Japan. Mr. K. Masui of the International Electric Research Exchange helped in collecting related information internationally. We therefore accord him our sincere appreciation. Grateful thanks are also due to the following persons for their kind cooperation in obtaining necessary information and data: Dr. A. Lacoste of Electricite de France, Dr. P. Gazzana Prioroggia of Industrie Pirelli, S.P.A., Professor G. Wanser of Vorstansmitgelied der Kabel-und-Metallwerke Gute-hoffnungshutte Aktien-gesellschaft, Dr. R. Patsch of AEG-Telefunken, Dr. J. D. Endacott of BICC Research and Engineering Ltd., Mr. R. Jocteur of SILEC, Mr. I. Eyraud of Les Cables de Lyon, Mr. R. W. Samm of EPRI, Mr. E. E. McIlveen of the Okonite Company, Mr. R. B. Blodget of the Anaconda Company, Mr. L. D. Blais of Kaiser Aluminum and Chemical Corporation, Dr. H. C. Doepkin, Jr. of Phelps Dodge Cable and Wire Company, the late Dr. Z. Croitoru of Electricite de France, Dr. C. M. Cooke of Massachusetts Institute of Technology, Dr. A. Cookson of Westinghouse Electric Corporation, and so many friends in the Japanese cable industry. In addition we wish to acknowledge the CRC staff for their incessant encouragement and elaborate editorial work. Finally, indebtedness is expressed to Hiroko Tanaka for her patience and encouragement and for her help in typing the manuscript. In this, at different stages, she was assisted by Amelia Stewart, Janice Daigle, and Jane Burhans, who must also share thanks.

Toshikatsu Tanaka
Allan Greenwood

THE AUTHORS

Dr. Toshikatsu Tanaka is a fellow research scientist at the Central Research Institute of Electric Power Industry in Tokyo, Japan. Dr. Tanaka received his B.A., M.S., and Ph.D. degrees in materials science from Osaka University in Osaka, Japan. He was Visiting Lecturer at the University of Salford, Salford, England from 1970 to 1972, and Associate Professor at Rensselaer Polytechnic Institute, Troy, New York, and Visiting Research Scientist at the General Electric R:D Center, Schenectady, New York from 1975 to 1976. Dr. Tanaka is a member of several engineering and scientific organizations and has been the recipient of an IEEJ (the Institute of Electrical Engineers of Japan) award for his scientific paper.

Dr. Tanaka has been engaged in electrical insulation studies for more than 15 years. He has published about 100 research papers including several reviews. He is also involved in computer application to the electrical insulation field and in new areas such as research projects of superconducting energy storage. He wrote one book in Japanese on the subject of high voltage and electrical insulation and is currently in charge of the electrical insulating materials section of an encyclopedia of electrical engineering to be published in the near future. He was formerly a contributer to one chapter of *Digest of Literatures on Dielectrics* by the National Research Council in the U.S. from 1971 to 1973.

Dr. Tanaka is also active in the IEEJ, IEEE, CIGRE, and IEC on electrical insulating materials, serving as chairman and secretary of certain committees of the IEEJ and as Convenor of CIGRE and IEC. He has served as a correspondent to *EI News in Japan* of the Electrical Insulating Society (EIS) Newsletter of IEEE since 1974 and was an AdCom Member of EIS from 1977 to 1979. Dr. Tanaka holds five Japanese patents.

Dr. Allan Greenwood is Philip Sporn Professor of Engineering at Rensselaer Polytechnic Institute, Troy, N.Y. where he is the Director of the Center for Electric Power Engineering. Dr. Greenwood received his early engineering education at Cambridge University (B.A. degree 1943, M.A. degree 1948) and his Ph.D. degree in 1952 from the University of Leeds. He has had an active career in industry, notably with the General Electric Co. His areas of expertise are power switching equipment, power system transients problems and power cables. He holds 16 U.S. patents and about 70 foreign patents, is the author of more than 50 scientific and technical papers (three of which received prizes), and the author or co-author of two books. He is a Fellow of IEEE and serves on the Fellow Committee of the Power Engineering Society. He is a member of CIGRE, acting as Convenor of its Working Group 13.03, and is a member of the honor society Sigma Xi. Dr. Greenwood consults for national and state governments and private industry.

TABLE OF CONTENTS

Volume I

Volume II

Chapter 1
Present Cables and Their Improvement ... 1

Chapter 1

PRESENT CABLES AND THEIR IMPROVEMENT

1.1. EXTRUDED POWER CABLES

1.1.1. Development of Extruded Cables

There are a variety of extruded power cables available such as butyl rubber-insulated cable, ethylene-propylene-rubber insulated cable, polyethylene-insulated cable, and cross-linked polyethylene-insulated cable. They have superceded oil-filled type cable of low- and medium- voltage class and are replacing high-voltage and even extra-high voltage OF and POF cables.

Butyl rubber-insulated cable comprises butyl rubber insulation covered with chloroprene or vinyl protective sheath. Because of its relative resistance to heat, water, ozone, and so on, butyl rubber is considered suitable as high-voltage cable insulation. It replaced paper and/or natural rubber-insulated cables in the 600 V to 77 kV classes because of the reputation it earned for use in wet, high temperature and cold installation conditions. It has been widely used as submarine cable below 44 kV for more than 15 years. But it often faulted due to unknown causes which were believed to be connected with its environment. It is suspected that these failures were caused by the formation of water trees, or in certain cases chemical trees, as is now understood from PE and XLPE cables. Butyl rubber-insulated cable has now yielded its position to PE- or XLPE-insulated cable, partly because of economics.

Ethylene-propylene copolymer (EPM) and ethylene-propylene-dien terpolymer (EPDM) are represented by ethylene-propylene rubber (EP rubber). EPM is a saturated polymer which exhibits excellent resistance to ozone, aging, weathering, and chemicals and also possesses good electrical performance and good resistance to cold. A shortcoming is the questionable quality of its vulcanizing, resulting from its saturated molecular structure. In this respect EPDM is considered an improvement over EPM. Some American manufacturers are interested in producing EPDM-insulated cables. They claim that EP rubber lead sheath cable outperforms paper-lead cable in such important considerations as capacity, emergency operating temperature, moisture resistance, and ease of splicing and terminating.

Noncross-linked polyethylene-insulated cable is currently used in French practice, especially for EHV cable, although cross-linked polyethylene (XLPE) cables, mainly because of their temperature stability, have almost replaced them in some other countries. The maximum rated temperature is 70°C for PE cables, while it is 90°C for XLPE cables. The PE insulation is very sensitive to partial discharges which may occur in small holes in the insulation.[4] Such microscopic holes cannot be avoided entirely, especially with thick extruded insulations which are necessary for extra high voltage cables. The resistance to partial discharges can be substantially improved by adding voltage stabilizers constituted from organic substances. Such voltage stabilizers may be effective, especially for high-voltage cable which uses PE of high density (HMPE; 0.96 g/cm³). The disadvantage of the greater stiffness of high density polyethylene may be offset by improved breakdown strength and temperature stability.[3]

Polyethylene is a good insulating material for extruded cable, but cross-linked polyethylene is better as far as temperature stability is concerned. Other characteristics remain almost the same. The polymer molecules can be cross-linked by chemical reactions or by γ-ray or high-energy electron beam irradiation. In cable fabrication, chemical cross-linking is used almost exclusively. Certain peroxide compounds are added to PE granulate. The vulcanizing process is carried out within 1 min at around 170°C in a steam tube immediately after extrusion. Usually referred to as a steam curing process, this is the most popular cross-linking process for XLPE cable. It is normally carried out by HCV, CCV, or VCV equipment. Insulation used to be extruded alone, but now two layers (conductor shield and insulation) and even three layers (conductor shield, insulation, and insulation shield) are now extruded simultaneously. The latter is called the three-layer simultaneous extrusion process or the triple extrusion process; it is carried out in tandem or common head operation. Carbon-filled semiconducting materials for conductor and insulation shields have been extensively investigated to obtain a smooth surface and tight bonding.

Extruded XLPE cables have defects in their insulation which are possible initiating points for breakdown. The defects comprise contaminants, protrusions, macro- or micro-voids, and some other microscopic inhomogeneities. Much more care has been taken in the material manufacturing process, the material transfer process, and the material handling process to exclude contaminants as much as possible. There has been much interest in completely closed systems. A special metal detector is sometimes used to reject any PE pellets containing metals. A fine-meshed screen is also used, prior to extrusion, for the same purpose.

To reduce protrusions, especially from the conductor shield, the design of the triple extrusion heads and the manufacturing conditions have been improved either by trial and error, or by the aid of certain measuring instruments. It is to be expected that voids and other microscopic inhomogeneities, which might be detrimental, will be reduced by optimum cross-linking and cooling processes.

Other advanced extrusion processes have been intensively investigated, mainly to increase manufacturing line speed. Several methods have been introduced which can be called "dry-type curing processes" in contrast to the steam (wet-type) curing processes. They have, incidentally, produced XLPE with less microvoid content, which is a desirable property for insulation in higher voltage XLPE cables.

Certain peroxide compounds are added to PE granulate to initiate cross-linking and the residue of peroxide such as acetophenone may stabilize the insulation performance of XLPE. In this respect they are similar to voltage stabilizers or VS agents. Some organic compounds are sometimes added to XLPE insulation as VS agents. The effectiveness of such voltage stabilizers in cable form is still uncertain, but they have already proved to be effective in needle experiments. XLPE insulation may be filled with inorganic compounds for the same purpose. Such materials are said to be mineral-filled or simply filled XLPE; they have lower thermal resistivity than unfilled XLPE.

Most distribution power cables in the voltage range 15 to 46 kV installed in the U.S. at the present time (in the 1970s) are insulated with chemically cross-linked polyethylene.[1] They are now called XLPE cables. This type of cable is being widely used in place of oil-impregnated paper-insulated lead-covered cable. Since 1970, extensive circuits rated 69 kV, have been installed, and since 1972 some 115 and 138 kV circuits have been commissioned. The operating record of this cable over the last 10 years has been outstanding. Figure 1.1.1. shows a world trend of the maximum rated voltage for PE and XLPE insulated cables available since the 1950s. Figure 1.1.2. shows the quantity of polyethylene that has been used for electrical wires and cables since 1965.[2] The rapid increase in use since 1967 is assigned to XLPE cables.[2]

XLPE-insulated cables are easy to install, can be plowed in, and are easy to splice and terminate. They require less skill to install and are less sensitive to atmospheric conditions than the oil-impregnated paper-insulated cables. XLPE insulated cables do not require oil pressure and the associated apparatus to produce it. This, coupled with the other features mentioned above, results in lower installation cost than for self-contained or pipe-type oil-filled cable systems. These are the main reasons why XLPE-insulated cables have received wide acceptance by power utilities and extensive use for transmission circuits at higher operating voltages such as 138 kV. Even 345 kV operation is expected.

Increase in operating voltage is an important development objective. Another is to increase current capacity. This has created much interest in European countries. Polyethylene-insu-

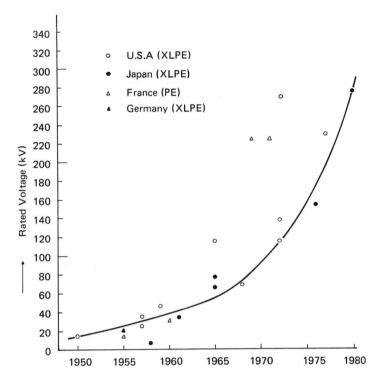

FIGURE 1.1.1.   Available maximum rated voltage for PE and XLPE insulated cables.

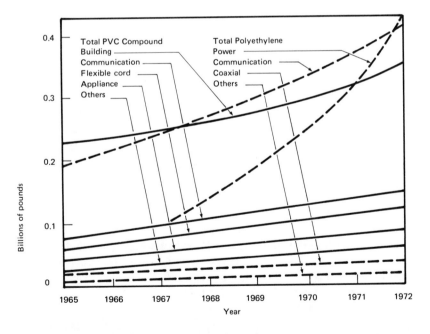

FIGURE 1.1.2.   PE and PVC used for electrical wires and cables in the U.S. (From Rosato, D. V., *Wire Wire Products*, March 1970, p. 61. With permission.)

lated cables of 225 kV to 600 MVA (1200 mm² Al conductor) and XLPE-insulated cables of 110 kV to 300 MVA (600 mm² Cu conductor) with forced external water cooling are now under development. XLPE cables of 22 kV to 400 MVA with forced internal water cooling were developed in 1974 in Japan.

Table 1.1.1.
SELECTED CHARACTERISTICS OF HIGH-VOLTAGE INSULATING MATERIALS

	Dielectric constant (ϵ)	Loss factor $\epsilon\tan\delta$ (%)	Thermal resistivity ρ (therm-ohm)	Operating temp. T_0 (°C)	Softening temp. T_{max} (°C)
Material			**Performance**		
HMPE and VSP	2.3	0.10	350	80	90
XLPE unfilled	2.3	0.10	350	90	135
XLPE filled	2.7	1.56	350	90	135
EPR	3.3	2.25	610	90	135
Oil paper	3.5	1.00	500	80	—

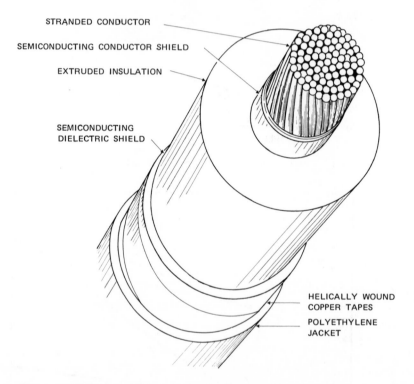

STRANDED CONDUCTOR

SEMICONDUCTING CONDUCTOR SHIELD

EXTRUDED INSULATION

SEMICONDUCTING DIELECTRIC SHIELD

HELICALLY WOUND COPPER TAPES

POLYETHYLENE JACKET

FIGURE 1.1.3.   Typical construction of an extruded dielectric cable.

Development of splicing and terminating technologies is important too, from the standpoint of easy operation, reliability, and cost. This is described in another chapter. A special kind of extruded cable, such as polyethylene-insulated sodium conductor cable, has been investigated and developed even for 138-kV cable. Table 1.1.1. shows a comparison of the most important insulating materials.[5]

1.1.2. Structures and Performances of Extruded XLPE Cables

The conductor is aluminum or copper with concentric stranding. Extruded semiconducting conductor and insulation shields are used to suppress corona discharge. The metallic shielding comprises copper tapes, helically applied, and the overall jacket is extruded polyvinyl chloride (PVC) or polyethylene. Some utilities use a lead jacket to increase short circuit capability.

Figure 1.1.3. shows the basic construction of a high-voltage cable with extruded plastic

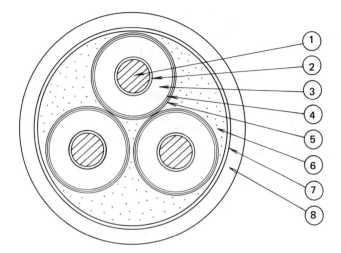

1   Conductor
2   Conductor Shield
3   Insulation
4   Insulation Shield
5   Shielding Copper Tape
6   Compound
7   Supporting Tape
8   Sheath

FIGURE 1.1.4.   6.6 kV Triple-cored XLPE insulated PVC-sheathed cable (common sheath). (Courtesy of Hitachi Cable Ltd.)

insulation. It consists of the inner semiconducting layer, the main insulating layer, and the outer semiconducting layer. One might be interested in some examples of medium- and high-voltage XLPE cables presently available.[6]

A. Three-Cored Cables Rated 6.6 to 22 kV

Three single-shielded cables in round shape are stranded together with jute compound and are then sheathed with polyvinyl chloride. An example of a common sheath type of cable is shown in Figure 1.1.4.

Cables rated above 6.6 kV are provided with a copper sheath around each of the member cables to form a uniform electric field in the cable insulation. In the design of cables rated 3.3 kV, the copper sheath can be omitted, for the electric stress is low enough. Such cables are usually provided with a common copper tape for safety. No copper sheath is used for 600 V power cables, except where it is required to prevent the electrostatic coupling to communication or control circuits. Figure 1.1.5. shows an example of a cable of triplex type which consists of three single-cored cables with vinyl sheaths. To prevent the cable structure from expanding or bird-caging due to the electromagnetic force induced in time of short circuit, the stranding pitch should be as short as possible. Characteristics of this triplexed XLPE cable are

1. Low thermal resistance and 10 to 15% larger ampacity than the common sheath-type cable
2. Easy to bend, allows small bending radius
3. Easy to splice and terminate, especially suitable for making joints with prefabricated type joint parts

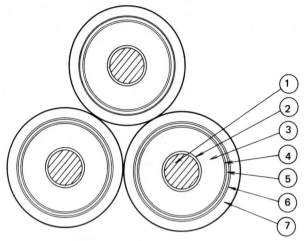

1	Conductor
2	Conductor Shield
3	Insulation
4	Insulation Shield
5	Shielding Copper Tape
6	Supporting Tape
7	Sheath

FIGURE 1.1.5.   Triplex type of XLPE insulated PVC-sheathed cable (independent sheath). (Courtesy of Hitachi Cable Ltd.)

4.   Easier to pull into a pipe than conventional cable (much easier than pulling in three single-core cables)

This relatively new cable is becoming more and more popular. On the other hand, it is 4% larger in overall radius than conventional three-cored type cables, and it turns out to be rather complicated to prevent water from pipe mouths.

B. 66- to 77-kV Cables

This type of cable is characterized by its sandwich structure which is composed of inner and outer semiconducting compound layers (resistivity 10^3 to 10^6 Ω-cm) with main insulation in between. All are extruded simultaneously so as to provide intimate contact between the components.

C. The Semiconducting Layer

The semiconducting layer serves

— To provide a uniform electric field around the concentric-cylindrical cable insulation by reducing the potential gradient over the surface of the stranded conductors and inside the metal shielding
— To prevent corona discharge at the surfaces of the stranded conductors and the insulation by maintaining close contacts between the inner and outer surfaces of the insulation
— To impede insulation from departing a conductor due to thermal expansion and contraction
— To protect a cable against damage caused by the heating of its conductor and shielding copper tape due to short-circuited current

Corona inception voltage has been increased dramatically since the triple extrusion method was introduced. At the same time, semiconducting cloth tapes can be applied instead of the semiconducting compound to cables rated below 22 kV, because they are economical and easy to handle, especially when they are used as an insulation shield in splices and terminations.

Insulation shields used to be fabricated from adhesive tapes or bonded extruded semi-conductive layers. The bonded shield proved to be superior for high voltage use, but to be rather difficult to joint. A strippable or free-stripping shield layer has been developed to make jointing easier. A cable with this free-stripping shield can now be manufactured by the triple extrusion (tandem or common) method. The exfoliation strength and stability can be controlled only by the choice of blending chemicals. This should be differentiated from free-stripping shield extruded not simultaneously with main insulation, which is comparatively unstable in the exfoliation strength.

Vinyl, polyethylene, and chloroprene are protective sheath materials. Polyethylene sheath is superior in oil and chemicals resistance, while vinyl or chloroprene sheath is superior in flame-retardant property.

D. Characteristics of XLPE Cables

— Electrical performance is excellent. Dielectric breakdown strength and volume resistivity are high, and dielectric loss tangent and dielectric constant are both low.
— Thermal resistivity is low. XLPE-insulated cables can operate continuously at a maximum temperature of 90°C, because they have excellent heat aging characteristics. This leads to larger power transmission capacity for XLPE cables than for noncross-linked PE cables.
— XLPE insulated cables are light in weight, because no metallic covering or jacket is needed. They are therefore easy to handle and install.
— They are of the dry type. Trouble associated with oil such as might be anticipated in oil-impregnated paper cables (oil leak accidents and insulation degradation which might take place in case of oil loss or flow-down) are absent and there is no problem with the cracking of lead sheaths due to vibration and thermal expansion and contraction.
— They are chemically stable.
— They are easy to splice and terminate.

As stated above, XLPE-insulated cables have excellent characteristics, but on the other hand they have some shortcomings as described below.

• Their insulation by nature is thicker and consequently larger in radius than present paper-insulated cables.
• They are inferior to OF cables in repetitive impulse performance.
• They are inferior in temperature characteristics to oil-impregnated paper-insulated cables.
• Effects of water permeation into insulation remains still unsolved.
• Cross-linked polyethylene is not as corona-resistant as oil-impregnated paper. In this respect even butyl and ethylene-propylene rubber are superior.

1.1.3. Cross-Linking Methods

As previously described, the insulation structure of an XLPE cable is composed principally of the inner semiconducting layer, the main insulation, and the outer semiconducting layer. Its manufacturing process can be divided into three parts: a mixing process, an extrusion process, and a cross-linking process.[7] In the mixing process, the main polyethylene insulation is mixed with a cross-linking agent (DCP for instance) of 2 to 3 PHR (parts per hundred

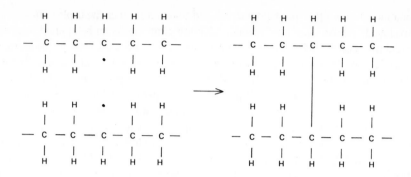

FIGURE 1.1.6.   Cross-linking reaction in polyethylene.

resin) and with a small amount of anti-aging agent. Extrusion and cross-linking are a simultaneous process. In the extrusion process, the mixed materials are extruded at the same temperature (about 140°C) as in the previous mixing stage. Thereafter, while being cross-linked, they are at the same time being pressurized in order to suppress bubbling from gas generated in the cross-linking reaction. This is a brief description of the cross-linking method used for cable manufacture.

A. Methods for Cross-Linking Polyethylene

Polymers are in most cases cross-linked by either a radiation-induced reaction or by a chemically initiated reaction. In the former method, γ-rays or high-energy electron beams are usually utilized. In the latter method, cross-linking agents such as organic peroxide promote the process chemically.

When irradiated, linear polymers are cross-linked or degraded into lower molecular weight polymers or even monomers. Polyethylene is a typical example of a cross-linking type; other examples are polyvinyl chloride, polypropylene, polystyrene, and polyacryl acid. Examples of the degradation type are polyisobutylene, polymethacrylate acid, and polyvinyliden chloride. In some cases, the type of reaction is determined by environment. For example, polyvinyl chloride cross-links in vacuum under irradiation at elevated temperatures, but degrades in air under γ-ray irradiation.

The degree of cross-linking is represented by a G value. It is the number of cross-links which are bridged for each 100 eV of radiation energy absorbed. According to Charlesby's theory, the formation of gel starts when one cross-link is generated per molecule for the average molecular weight, and therefore the G value can be experimentally obtained from the dose at which the gel formation begins. Or it may be obtained from the swelling ratio or the elasticity. G values measured so far are between two and three for polyethylene and are considered to be comparatively independent of the kind of polyethylene. They may vary due to the difference in irradiation temperature or in irradiation environment.

Chemical reaction in the cross-linking of polyethylene is considered to take place as shown in Figure 1.1.6. Polyethylene has only a small number of unsaturated bonds in its molecule. They are 15 in number per 1000 C in high-pressure polyethylene, and 0.6/1000 C in low-pressure polyethylene, while they are 200/1000 C in natural rubber. This is the reason why no sulfur accelerator can be used as cross-linking agent. On the other hand, the peroxide is utilized mainly as an agent to cross-link polymers with no double bonds. Agents such as di-α-cumyl peroxide (DCP) and associated peroxide are widely used. A chemical reaction for polyethylene via peroxide can be described by Figure 1.1.7., where a peroxide such as DCP decomposes above a certain temperature into two radicals which react with polyethylene thereby causing cross-linking. There are various kinds of cross-linking agents available as shown in Table 1.1.2.

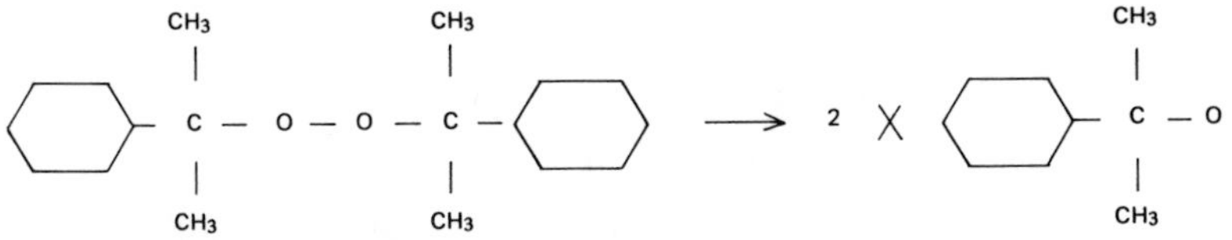

FIGURE 1.1.7. Decomposition of DCP.

Table 1.1.2.
VARIOUS KINDS OF CROSS-LINKING AGENTS

Name of agent	Chemical structure	Decomposition temp. (halflife 1 min)
Benzoyl peroxide	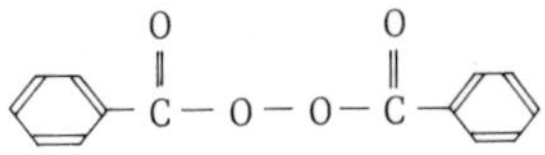	133°C
Di-t-butyl peroxide	$(CH_3)_3 C - O - O - C(CH_3)_3$	193
Di-cumyl peroxide (DCP)	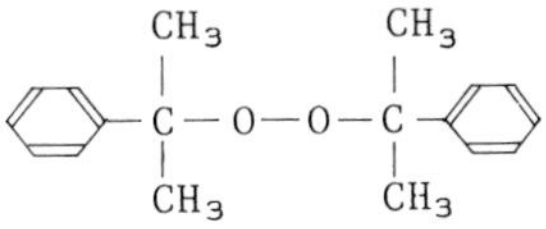	171
2,5-Dimethyl-2,5-di(t-butyl peroxy) hexane (DMDBH)	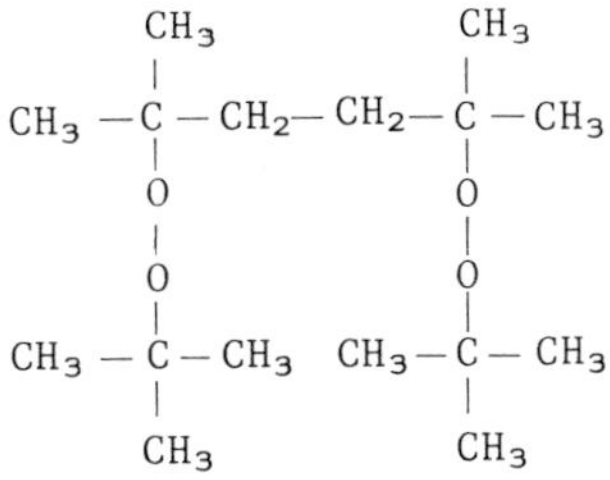	179
2,5-Dimethyl-2,5-di(t-butyl peroxy) hexane (DMDBHY)	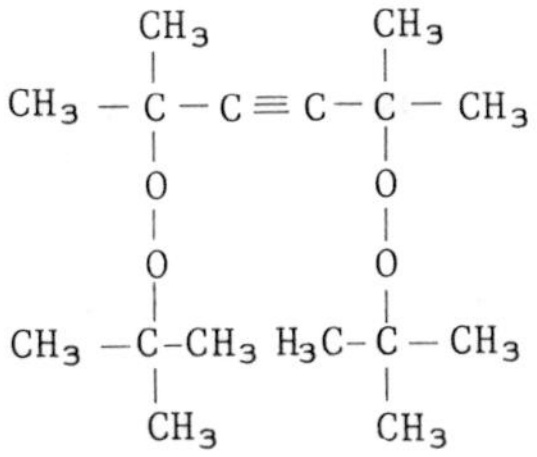	193
1,3-Bis (t-butyl peroxy isopropyl) benzene	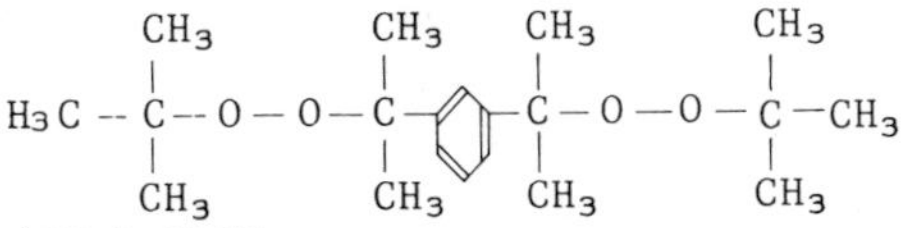	182
t-Butyl-hydro peroxide	$(CH_3)_3COOH$	>200
Cumene hydro peroxide	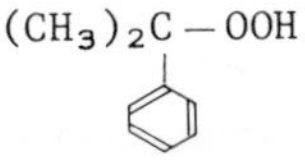 $(CH_3)_2C - OOH$	>200
Poly sulfon azide	$R - [SO_2N_3]_x$	PP use
Azide formate	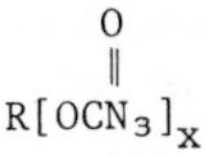 $R[OCN_3]_x$	

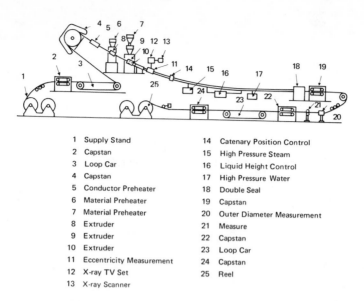

1	Supply Stand	14 Catenary Position Control
2	Capstan	15 High Pressure Steam
3	Loop Car	16 Liquid Height Control
4	Capstan	17 High Pressure Water
5	Conductor Preheater	18 Double Seal
6	Material Preheater	19 Capstan
7	Material Preheater	20 Outer Diameter Measurement
8	Extruder	21 Measure
9	Extruder	22 Capstan
10	Extruder	23 Loop Car
11	Eccentricity Measurement	24 Capstan
12	X-ray TV Set	25 Reel
13	X-ray Scanner	

FIGURE 1.1.8.   A layout of tandem CCV. (From Ohtani, K., *Jpn. Plast.*, (in Japanese), 25(5), 40, 1974. With permission.)

a. CCV and VCV Methods

Rubber vulcanization, for a long time, has been a typical example of cross-linking between polymer main chains. Rubber insulation excelled in the wire and cable industry until a new technology was introduced to vulcanize or cross-link polyethylene. For this reason a cross-linking machine is conventionally called a vulcanizer. There are now three types available to cross-link polyethylene using peroxide such as DCP: catenary continuous vulcanization (CCV), vertical continuous vulcanization (VCV), and horizontal continuous vulcanization (HCV).

A screw-type extruder is normally used to extrude a polyethylene/peroxide mixture into a vulcanizing tube, directly connected to the outlet of the extruder. It is, in general, a simple-axis screw; what is called an SK type, composed of two steps of different screw radius, has come into use recently. This is based on the TSP theory — a cable is extruded at a temperature (125 to 127 °C) below the decomposition temperature for DCP and is cross-linked in the vulcanizing tube at 200 to 400 °C. A final product is obtained after a cooling process, followed by the complete cross-linking, at which time it is wound on a drum.

The CCV method enables a precable to generate such a catenary in a vulcanizing tube as determined by its tension. This method is widely used for the outer tube diameters of 3 to 70 mmφ; a variety of equipment have been devised for this purpose. Figure 1.1.8. illustrates a layout of a typical CCV. A conductor is veered out from a supply capstan into the extruder of the first stage after it has been preheated to a certain temperature by a conductor preheater. Usually the first-stage extruder has a single screw of 65 to 90 mmφ diameter and is utilized to extrude semiconducting compound. The conductor, covered by the semiconducting compound in the first stage, is led to the second stage and subsequently to the third. The main extruder is at the second stage and functions to extrude a polyethylene cover of insulation over the conductor with its semiconducting layer. The second and third extruders form their common head (the double cross head) to conduct simultaneous extrusions of the main insulation and the outer semiconducting layer. The third extruder supplies semiconducting compound. Much background information is available to design the double cross head, which is typified for the most part by three designs.

A vulcanizing tube in catenary is generally 200 to 250 mmφ in inner diameter. It is

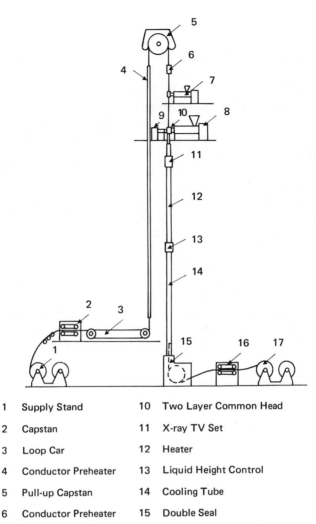

1	Supply Stand	10	Two Layer Common Head
2	Capstan	11	X-ray TV Set
3	Loop Car	12	Heater
4	Conductor Preheater	13	Liquid Height Control
5	Pull-up Capstan	14	Cooling Tube
6	Conductor Preheater	15	Double Seal
7	Extruder	16	Capstan
8	Extruder	17	Reel
9	Extruder		

FIGURE 1.1.9.   A layout of tandem VCV. (From Ohtani, K., *Jpn. Plast.*, (in Japanese), 25(5), 41, 1974. With permission.)

usually 70 to 100 m long, but many systems put into operation recently have a catenary as long as 150 to 200 m. The rate of the cross-linking reaction is determined mainly by heating temperature and time. Therefore, when line speed and temperature are held constant, this rate is determined in practice by the length of the vulcanizing tube. Logically, the longer tube has been preferred as far as this is economically acceptable. Modern CCVs have a cable manufacturing system with a catenary tube longer than 150 m. Such CCVs are capable of manufacturing cable up to 50 mmϕ in overall diameter. The eccentricity of the conductor can be serious when the cable is larger than 50 mmϕ in outer diameter.

The VCV method is useful for the production of cables of outer diameter larger than 50 mmϕ. The VCV system illustrated in Figure 1.1.9. clearly shows a vulcanizing device of the vertical type. Such an arrangement precludes eccentricity phenomenon due to the gravity. No control device is needed to keep the soft surface of a cable off the inner surface of a tube because the cable is pulled down by the gravity. These are the two main merits compared with the CCV system. However, the VCV system requires a high building to house it, which may result in high initial investment. Many of VCV housings recently built in Japan, the U.S., and Europe are as high as 60 to 80 m.

The VCV process is almost the same as that for CCV. A conductor is pulled to the top of a VCV housing by a pulling capstan, is preheated in a preheating tube and an electric preheating device, is introduced to a first-stage extruder in which it is covered with semi-conducting compound, suffers the simultaneous extrusion of an insulation layer and an outer semiconducting layer, and is finally wound on a drum after a cooling process. A typical vulcanizing tube is as large as 200 to 270 mmϕ in inner diameter. As stated above, CCV and VCV are now widely used to manufacture XLPE cables. The former works at a line speed of 10 to 200 m/min for the production of 3 to 70 mmϕ XLPE cables, and the latter at 1 to 30 m/min for 30 to 150 mmϕ XLPE cables.

There have been several attempts to utilize ultrasonic waves and microwaves to cross-link polyethylene in cable form, but the high-pressure vapor method has been the most popular in cable industry. More recently, gas is sometimes used in the curing process in place of steam, especially for manufacturing what is called void-free XLPE insulation. This method has been first adopted in Japan.

Other attempts have been made to use ultrasonic waves, microwaves, and high-energy electron beams simultaneously with high-pressure steam. Cross-linking in boiling water is now under investigation (Siloplas E); it shows a slow rate of cross-linking, but is easy to operate.

In the high-pressure steam cure method, steam pressurized to 20 to 30 kg/cm^2 is introduced into a vulcanizing tube where the cross-linking process is performed. Steam is a convenient medium for conveying heat and pressure in the cross-linking of polymers. To increase the reaction rate, a higher temperature is preferred. In order to increase the steam temperature, the pressure must inevitably increase also; that is, the pressure and temperature are inter-dependent. This leads to some economic and technical problems. A compromise is often struck with a pressure of 20 kg/cm^2. The maximum pressure available for this purpose is 29 kg/cm^2.

The steam cure method has some problems. One is that a comparatively long time is required to dry XLPE insulation so as to eliminate water molecules which have been trapped in it; another is that small voids often remain after the water is removed. These voids are thought to be responsible for water trees or electrochemical trees and to have adverse effects on dielectric breakdown. This suggests that gas rather than high pressure steam might be used and indeed some gas is applied as a heat transport and pressure medium. Two ways have been devised to do this: one uses the gas only as a pressure medium, heat being applied to the insulation by radiation. The other uses gas for both purposes.

Figure 1.1.10. shows a cable manufacturing equipment which utilizes radiant heating. It is not difficult to obtain 400 °C at the cable surface and 600 °C at the heat source under reasonable pressure conditions. Electric heaters are distributed in several zones so as to obtain the desired temperature distribution. A computer program assures the optimum temperature distribution pattern according to the dimension of a cable to be produced. Inert gases such as N_2 and even CO_2 are suitable pressurizers to minimize the bubble formation at the time of cross-linking, since they do not react chemically with the cable material. In one operation, the selected gas is pressurized to 10 to 20 kg/cm^2 and introduced to a vulcanizing tube. This method can improve the rate of cable production because of the higher temperature, and at the same time reduce the void content in XLPE insulation.

Figure 1.1.11. shows a cable manufacturing process using heated gas. A gas such as N_2 or SF_6 is preheated to 300 to 400 °C in a heating jacket situated outside the vulcanizing tube, allowing the high-temperature and high-pressure gas to be introduced into a cross-

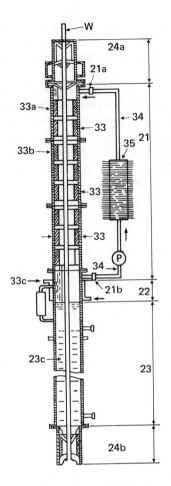

21   Radiant Heating Zone
   21a: Inert Gas Inlet Port
   21b: Inert Gas Outlet Port
22   Pre-cooling Zone
23   Cooling Zone
   23c: Cooling Water
24a   Supply Side Seal
24b   End Seal
33   Heater (Divided into Three Units a, b, c)
34   Circulation Passage
35   Condensor

FIGURE 1.1.10.   An XLPE cable manufacturing apparatus by the radiant heating method.
(After U.S. Patent 3,588,954; Figure 8.)

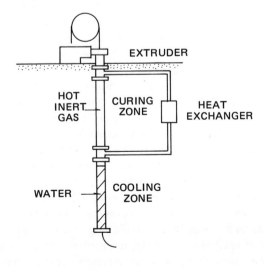

FIGURE 1.1.11.   Vertical continuous vulcanizing
equipment using hot inert gas flow.

linking zone. Shortcomings as far as heat transfer concerned are overcome by forcing the flow of the heated gas.

Cross-linking equipment for CCV and VCV consists of a mold processor and a cross-link processor as already described. One of the most important requirements for the former is that no materials used decompose or deteriorate at a specified high temperature during the long period when they are being mixed; that is the period when the polyethylene used remains uncross-linked. In order to obtain this condition, the design and operation should be optimized as far as possible, taking into account the screw, the temperature controller of the screw and the cylinder barrel, and the cross head, especially with respect to its flow and surface material dsign. It is also important for the latter that the temperature distribution in the heating and cooling zones be so designed as to optimize the heat transfer.

b. Long Land Die Methods

Polyethylene is a poor thermal conductor and therefore has some limitation in heat transfer by nature when processed by CCV and VCV methods. It requires a considerably longer time to uniformly heat polyethylene with peroxide to cross-linking temperature through its thick insulation. The manufacturing line speed is limited by thermal conduction, even when temperature is increased to the maximum at which no deterioration will take place. New tests have been made on the basis of the following requirements:

1.   To obtain as uniform a radial temperature distribution as possible
2.   To make space and height needed for an equipment as small as possible, hopefully resulting in the lower initial investment
3.   To reduce void content as far as possible

A break from convention is evident in two processes. One is called the Anaconda® process, the other the Engel process.

In the Anaconda® process, which was developed by R. F. Hinderer with cooperation of J. A. Mullen,[8] a lubricant is injected into and guided around the wall surface of a metal tube, called a long land die. Figure 1.1.12. illustrates a cross section of this process. Polyethylene is extruded together with peroxide out of a molding die #16 of a crosshead #10 of the present type, into a porous tube #25 of an extension of the die. It is then guided to flow in a metal tube #43, whose surface is coated with a lubricant by the porous tube, to be finally heated to finish cross-linking inside the metal tube. Grooves are helically cut on the outer surface of the porous tube so that the lubricant may be supplied from a high-pressure injection device through the flow paths #36 and #37. The extruder usually operates at a pressure of 70 to 90 kg/cm² with a pressure drop of 4 to 7 kg/cm² along the metal tube as long as 12 m. The metal tube is heated to 270 to 350 °C to produce a cable with an insulation thickness of 13 mm and outer diameter of 32 mmφ, at a speed of about 1 m/min.

The MDCV process[9] has been developed in Japan by adding a special technique to the Anaconda® process. The MDCV process is characterized by the use of an auxiliary agent during molding to facilitate the smooth movement of a material inside the die. The material should not gel in contact with peroxide, it should not boil even under cross-linking conditions, and the rate of its absorption into a material should be less than 100 mg/cm² at 150 °C in 45 hr. Figures 1.1.13. and 1.1.14. show two examples of the MDCV processes. The first is an example of equipment to supply the auxiliary agent into a long mold die. The second device is an example of applying a shielding layer to an extruded and molded material to prevent organic peroxide from migrating to the auxiliary agent existing on the inner surface of the sleeve.

Characteristic of the MDCV production line is its employment of a horizontal line and a complete dry curing system. An MDCV line is designed to enable the conductor to move

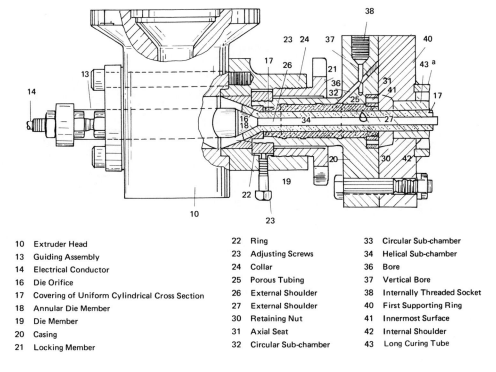

FIGURE 1.1.12.   Anaconda® process. (After U.S. Patent 3,045,142, Figure 1a, Sheet 1.)

10	Extruder Head	22	Ring	33	Circular Sub-chamber
13	Guiding Assembly	23	Adjusting Screws	34	Helical Sub-chamber
14	Electrical Conductor	24	Collar	36	Bore
16	Die Orifice	25	Porous Tubing	37	Vertical Bore
17	Covering of Uniform Cylindrical Cross Section	26	External Shoulder	38	Internally Threaded Socket
18	Annular Die Member	27	External Shoulder	40	First Supporting Ring
19	Die Member	30	Retaining Nut	41	Innermost Surface
20	Casing	31	Axial Seat	42	Internal Shoulder
21	Locking Member	32	Circular Sub-chamber	43	Long Curing Tube

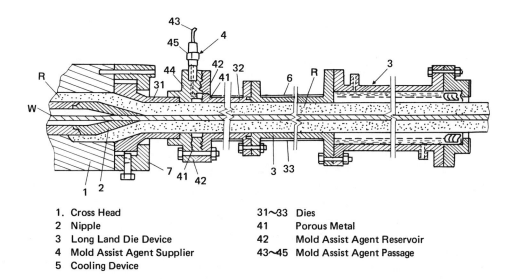

1.	Cross Head	31~33	Dies
2	Nipple	41	Porous Metal
3	Long Land Die Device	42	Mold Assist Agent Reservoir
4	Mold Assist Agent Supplier	43~45	Mold Assist Agent Passage
5	Cooling Device		

FIGURE 1.1.13.   MDCV process.

straight forward without undergoing excessive bending. This accords with the need for manufacturing XLPE cable with large-sized conductor. Since the heating or cross-linking and cooling section are separated, no water penetration or diffusion into the insulation is anticipated. The XLPE insulation is therefore almost dry and void-free. The diameter of a cable is accurate and the concentricity of the cable is at least as good as that obtained with conventional steam-cured cables. The heating and cooling sections can be controlled independently and therefore the cross-linking temperature can be easily held close to the critical

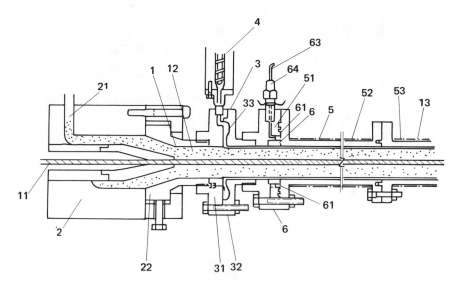

1. Not-Yet-Crosslined-PE Conductor
2. Cross Head
3. Die Extruding Shield against Peroxide
13. Peroxide Shield Layer
5. Long Land Sleeve

FIGURE 1.1.14.   MDCV process.

temperature at which polyethylene suffers from thermal degradation. This can lead to an increase in manufacturing line speed and a smaller size of cross-linking equipment.

Some problems remain with the homogeneous injection of a lubricant under pressure, the eccentricity of the conductor, the separation and recovery of lubricant, the exchange work of a long land die, and the generation of microvoids, especially in the outer layer of the insulation.

The Engel process is another development of considerable significance; it originated in Europe. This is a method of "forging" granular polyethylene (0.7 to 1.2 mmφ) by a piston movement in which it is internally heated and melted to cross-link. This technique is used for manufacturing hot water pipes of XLPE, which are typically 20 mmφ in inner diameter with a 2-mm wall thickness. The material used is high-density polyethylene (HDPE) with 0.4% cross-linking agent such as peroxide. Figure 1.1.15. shows a standard machine of this type capable of producing the pipes just described at a line speed of 10 m/hr. Unfortunately, this method has not yet been utilized for cable manufacturing.

A trial has been made to produce XLPE cables using this concept. The RX (ram cross-linking) process has been published.[10] A screw extrusion molding machine or a screw injection molding machine is combined with multiple plunger extruders. Control by servomechanism maintains the continuous extrusion mold under high pressure, resulting in homogeneous XLPE insulation. The RX process is shown in Figure 1.1.16. Polyethylene is mixed with cross-linking agent in the screw extrusion mold equipment at a temperature and pressure at which no cross-linking takes place, and then introduced into a reaction chamber by the plunger extruders. It is pressurized to 3000 kg/cm² with the temperature above 180 °C, then rammed through a special pressure-shearing device. The material thereby becomes so active that the cross-linking is completed within a short time. It has been demonstrated that even material under a cross-linking process can deform and flow which should enable the manufacture of cables by this method. Figure 1.1.17. shows an example of the head section of an RX device.

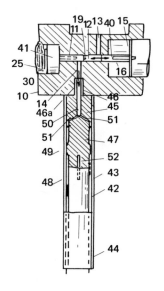
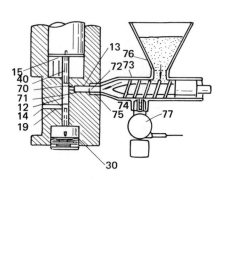

10. Housing	47. Slender cylindrical shaft
11. Splender hole	48. Tubular pass
12. Pressure chamber	49. Ringed ridge
13. Material pass	50. Axial pass
14. Material pass	51. Fire ducts
15. The first pressure chamber	70. Inlet orifice
17. Fluid motor device	71. Expanded middle part
19. The second piston	72. Expanded part with screw
25. Expanded element	73. Tube
30. Spring	74. Screw element
45. Blocking plug	75. Screw element
46. Expanded cylindrical part	76. Hopper
46a. Spigot with spring	77. Measuring pump

FIGURE 1.1.15.   Engel process. (After Japan Patent Sho-45-3568.)

The HEB® process* is another experimental process. Figure 1.1.18. illustrates a layout of the process which is characterized by its holding head. An extrusion molding machine is temperature-controlled with high accuracy by an oil-heating device. A bayonet-type extrusion head provides the next stage which connects with a final heating zone 1500 mm in length, followed by 2 zones each with their heating tubes of 2500 mm in length. Nitrogen gas is sealed in the heating tubes. The molding head is also heated by the oil-heating device. The material enters the molding head at 130 °C and leaves at 180 °C. The screw of the extrusion molding machine is specially designed on the basis of many experiments.

The MDCV, RX, and HEB processes just referred to represent new and advanced processes for cross-linking. The recent development of these processes has been certainly motivated by the fact that the modern CCV and VCV processes need high investment to meet the demand for XLPE cables. Each may very well cost almost $1 million per 1 manufacture line (1982 dollars).

These systems can be summarized as follows. Figure 1.1.19. illustrates the concept of an advanced cross-linking apparatus. Material with some cross-linking agent is melted and extruded from a screw extrusion molding device or a plunger-type molding device, then introduced to a molding head below the decomposition temperature for the cross-linking

---

* HEB® is the trademark of Harburger Works of Fried Krupp Co., Ltd., Humburg, W. Germany.

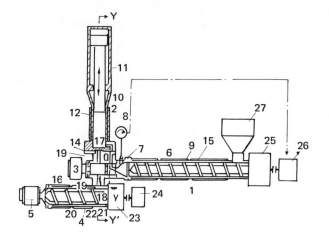

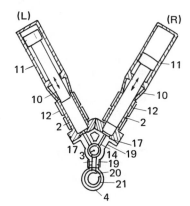

1.	Screw Extruder
2.	Reaction Chamber
3.	Change Valve
4.	Head Die
5.	Plunger
6.	Oil Pressure Cylinder
7.	Casing
20.	Resin Reservoir
21.	Resin Pressure Detector
22.	Resin Passage
23.	Change Valve Controller
24.	Resin Passage

1.	Screw Extruder
2.	Reaction Chamber
3.	Change Valve
4.	Head Die
5.	Plunger
6.	Oil Pressure Cylinder
7.	Casing
20.	Resin Reservoir
21.	Resin Pressure Detector
22.	Resin Pasasge
23.	Chnage Valve Controller
24.	Resin Passage

FIGURE 1.1.16.   RX process. (After Japan Patent Open Sho-48-8346.)

agent. It is subjected to both shear and compression due to the constricting action of an orifice at the inlet and to heating from the jacket; it is then heated uniformly to the cross-linking initiation temperature. The material flows into a die under high pressure. In the long land die, the material continues to cross-link by both internally generated and externally given thermal energy and at the same time flows in plug form. This flow is aided by a lubricant such as silicone oil and melted metal salts, which can be supplied through a porous metal. Cross-linking is completed in an atmosphere such as N_2 gas and some heated oil after the molding process.

It is expected that this method will be applied for cable manufacturing, too. The system will be compact because molding and cross-linking are conducted in the same head. It will possibly produce a high-quality XLPE cable with less voids and less foreign substances, since the reactions are conducted in a closed metal tube under high pressure.

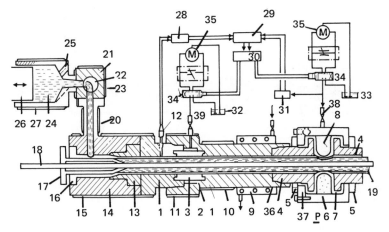

FIGURE 1.1.17.   A molding head of the RX process.

1. Dies		21. Valve	
2. Porous Metal Tube		22. Piston	
3. Heat Medium		23. Heater	
4. Die Adapter		24. Melted Resin	
5. Flange		25. Cylinder	
6. Main Body		26. Plunger	
7. Throttle Valve		27. Heater	
8. Pressure Medium		28. Pressure Changer	
9. Cooling Device		29. Feedback Bus	
10. Heater		30. Adjuster	
11. Heater		31. Pressure Sensor	
12. Resin Pressure Gauge		32. Heat Medium Tank	
13. Head Ring		33. Pressure Medium Tank	
14. Cross Head		34. Servo Valve	
15. Heater		35. Motor	
16. Sleeve		36. O-Ring	
17. Nipple		37. Flange	
18. Conductor		38. Valve	
19. Cross-linked Polymer		39. Valve	
20. Heater		P Pressure Adjusting Device	

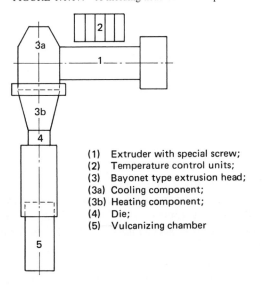

(1)   Extruder with special screw;
(2)   Temperature control units;
(3)   Bayonet type extrusion head;
(3a) Cooling component;
(3b) Heating component;
(4)   Die;
(5)   Vulcanizing chamber

FIGURE 1.1.18.   HEB process. (After Europlastics,
July 1972.)

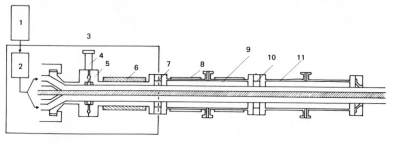

1. Molding Machine
2. Orifice, Pressure Control
3. Mold Head
4. Lubricate Injection Device
5. Porous Metal
6. Heater
7. Seal
8. Heater
9. N₂ Gas or Heating Oil
10. Seal
11. Cooling N₂ Gsa or Oil

FIGURE 1.1.19. An advanced molding and cross-linking apparatus. (From Ohtani, K., *Jpn. Plast.*, 25(5), 50, 1974. With permission.)

c. Radiation Cross-Linking

Radiation cross-linking technology has been applied to electric wires, thermo-contraction films, and foam mold products and more recently some significant progress has been made in the application of this method to the manufacture of power cables.

A high-energy electron beam is the most popular in industrial use among many radiation sources. It is obtained from an electron accelerator. Electron beams can yield ionization enough to initiate chemical reaction, but unfortunately they have low penetration which is considered a shortcoming. For this reason, they have been used only to irradiate thin films. However, some trials have been made, on the basis of the recent progressive development, to apply the method to thick power cables. This technique is now available for low-voltage (600 V) cables with polyethylene insulation. Features of this technique are

1. Cross-linking at room temperature
2. Short reaction time resulting in an increase in manufacturing line speed
3. No cross-linking agent required

Many thin extruded wires can be irradiated at the same time, but they should be rotated in order to receive a uniform irradiation from a high-energy electron beam source. Some experiments have been made to investigate the application of this method to the manufactured 3- to 22-kV class XLPE cables. There are some technical and economic problems to be solved. Among these are

1. Optimum heating to release residual charge in the insulation
2. Optimum pressurization method to suppress foaming of polyethylene when subjected to large-dose, short-time irradiation
3. Design of a deflecting magnet to enable uniform irradiation to be carried out without rotation, from all directions (360°)
4. Reducing the cost of electron accelerators (a 2.5 MeV × 20 mA machine may cost nearly $1 million)

Figure 1.1.20. shows an example of the irradiation equipment with its deflection magnet for thick cables. The magnet is so designed as to deflect electrons from an accelerator and

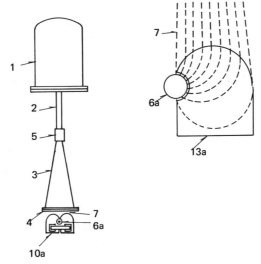

1   Electron Accelerator
2   Evacuated Extension Tube
3   Spreading Portion
4   Electron Aperture
5   Beam Scanning Device
6   Object to be Irradiated
7   Electron Current
10 Magnet Pole Face
13 Vacuum Pump

FIGURE 1.20.   Electron irradiation to a cylindrical material. (After Japan Patent Sho-35-435.)

irradiate the cable insulation over its entire circumference. The polyethylene cable is molded by an extrusion molding machine and then subjected to electron irradiation at the appropriate temperature for cross-linking. Inevitably charge remains in the insulation after the cross-linking stage; it is removed by heating the insulation to a suitable temperature, after which it is cooled down. It is necessary to carry out cross-linking and heating under pressure to suppress foaming in the insulation.

It may be possible in the future to cross-link polyethylene as thick as 6 to 10 mm by a large accelerator, rated at perhaps 3.0 MeV × 30 mA. Multiple irradiation systems may be one solution for this purpose as shown in Figure 1.1.21. A toroidal irradiation system is now under consideration to develop irradiated XLPE cables rated 138 kV and 230 kV.

1.1.4. Imperfections in XLPE Cable Insulation and Its Associated Problems

An XLPE solid insulated cable, although it has had much in-service experience, is a defective cable[11-17] to be improved. As there are over 100,000 mi of buried XLPE power cables at voltages above 15 kV in the U.S., many studies have been made of their dielectric properties in both wet and dry environments.

There are abundant theories on the dielectric breakdown mechanisms of such cables, but it is clear that the breakdown voltages attained are substantially less than one would anticipate by extrapolation from laboratory results on thin samples. The AC short-time breakdown stress of XLPE insulation is extremely high, being in excess of 20,000 V/mil (800 kV/mm), but actual XLPE cables give far lower values for breakdown. This is thought to be due to imperfections in the cable.

The extruded cables considered here differ from the well-known cables with lapped insulation, where the dielectric is applied in thin layers from the inside outwards. A defect

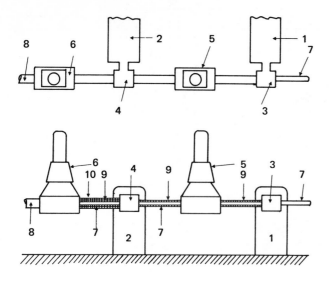

1. First Extruder
2. Second Extruder
3. First Molding Head
4. Second Molding Head
5. First Irradiation Device
6. Second Irradiation Device
7. Conductor
8. Insulated Conductor (Cable)
9. First Insulation Covering
10. Second Insulation Covering

FIGURE 1.1.21.   A multiple-stage irradiation apparatus.

in one of the layers has little effect on the total quality of the dielectric. In an extruded cable, the dielectric is applied lengthwise, a full insulation wall is manufactured at one time. Such solid cables are more subject to deterioration and breakdown, which can start from even the smallest defects than are lapped cables. A fatal imperfection may occur at one spot only, whereas the rest of the insulation is sound. Lengthwise application of insulation thus leads to isolated defects and causes the characteristics typical of extruded dielectrics. When tested at high voltage, breakdown values are found to scatter appreciably both regarding voltage and the sites of breakdown. These defects can be called ''macro-defects'', they can be eliminated to a great extent by improving present cable manufacturing and material handling processes. There are other types of defects which are demonstrably ''micro-defects''. Such defects have received attention recently for the following two main reasons:

1.   Macro-defects are now minimized by the introduction of new techniques such as completely closed clean systems of material flow to eliminate the inclusion of foreign substances, a three-layer simultaneous extrusion process to reduce asperities from the semiconducting layers to the insulation, an optimum extrusion operation control to suppress the formation of macro-voids in the insulation, and so on.
2.   There is a trend for extruded dielectric cables to supersede oil-impregnated paper-insulated cables at higher voltages. In order to achieve high-voltage XLPE cables, attention must be paid to what are called ''micro-defects''.

It is difficult to define macro-defects and micro-defects rigorously. Perhaps the dividing line lies at a few hundred microns. There are other classifications of defects in XLPE cables, according to their physical characteristics. For example, there appear to be four types of defects as follows:

1.   Voids
2.   Protrusions from the semiconducting layers
3.   Contaminants (metallic and nonmetallic)
4.   Inhomogeneities in crystallite scale

Figure 1.1.22. illustrates some defects such as voids, protrusions, and contaminants.[12] It is

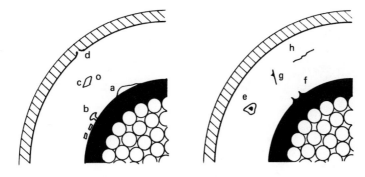

a) loose semi-conductive screen,

b) bubbles caused by gas-evolution in the conductive screen,

c) cavities due to shrinkage or gas-formation in insulation,

d) defects in the core-screen,

e) inclusion of foreign particles that separate gases, often due to moisture in the particles,

f) projections or points on the semi-conductive screen,

g) splinters and

h) fibers.

FIGURE 1.1.22.   Defects in extruded cable dielectrics. (From Kreuger, F. H., *Endurance Tests with Polyethylene Insulated Cables*, CIGRE Paper 21-02, CIGRE, Paris, 1968, 2. With permission.)

considered that the most critical imperfections in decreasing order of importance are (1) protrusions at the conductor shield, (2) contaminants comprising conductive particles and insulating particles having significantly higher conductivity than the insulation, and (3) voids which discharge during the application of the AC voltage.

A. Voids

The voids referred to above may be a consequence of (a) a loose semiconductive screen, (b) bubbles caused by gas evolution in the conductive screen, (c) cavities due to shrinkage or gas formation in insulation, (d) defects in the core screen, (e) inclusion of foreign particles that separate gases, often due to moisture in the particles, and finally, (f) what are called "micro-voids" which form inherently during the cross-linking process, especially by the steam-curing method. A void is a potential space for an internal discharge to deteriorate the insulation. A void equal to or larger than that which allows an internal discharge to occur is considered detrimental. For practical convenience, the maximum allowable void size can be obtained, with some safety factor, from the corona inception voltage across existing voids. Values of 80 μm and 50 μm have been measured for critical voids in 66- to 77-kV cables and 138- to 154-kV cables, respectively. Voids larger than these values can be called "macro-voids".

Micro-voids are characterized by the fact that they appear uniformly over the entire length of the insulation, which is really different from the characteristics of "macro-voids", stated above, which are included as one-spot smears by accident. Microvoids have a radial distribution which may be determined by the condition of the cross-linking operation. A microphotograph of some micro-voids and their radical distribution are shown in Figure 1.1.23. and Figure 1.1.24.[18] Figure 1.1.24. also shows a comparison of void content between steam-cured XLPE insulation and long land die-cured XLPE insulation.

Macro-voids often can be found with the aid of a discharge detector. Scanning equipment has been devised to detect and locate these voids.[19-26] No distinctive correlation has yet been established between the size of voids and breakdown voltages or lifetime of extruded cables. There are classical data, frequently cited, which demonstrate the coincidence between the life of polyethylene insulation in the presence of cavities and the life of actual polyethylene

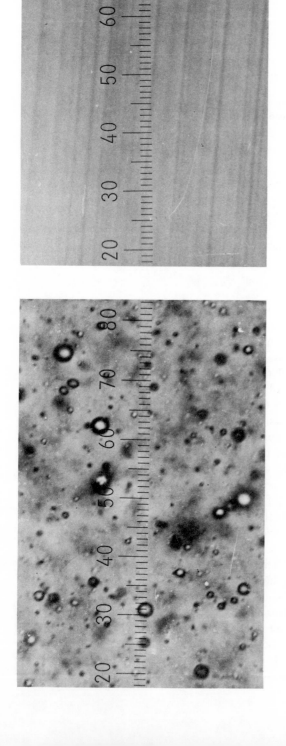

FIGURE 1.1.23. Photomicrographs of XLPE cable insulation. (A) Conventional process. (B) MDCV process. Photos taken under differential interference microscope with the smallest division 2.7 μm wide. (Courtesy of Dai-nichi-Nippon Wire & Cable Co.)

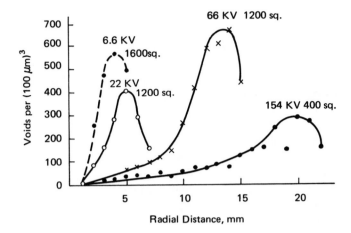

FIGURE 1.1.24.   Radial void distribution in conventional XLPE cable insulation. (From Yamaguchi, H., Okada, M., and Fuwa, M., *Voil Distribution in Crosslinked Polyethylene Insulated Cables Made by Conventional and New Processes*, IEEE/Pes Meeting C72 504-9, IEEE, Piscataway, N.J., 1972, 5. With permission.)

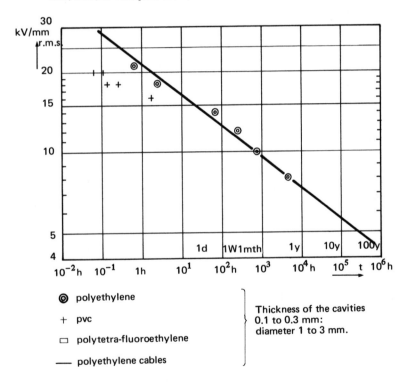

FIGURE 1.1.25.   Voltage-life of polyethylene and other dielectrics in the presence of cavities. After Kreuger, F. H., *Endurance Tests with Polyethylene Insulated Cables*, CIGRE Paper 21-02, CIGRE, Paris, 1968, 3. With permission.)

cable, as shown in Figure 1.1.25. The laboratory experiment was carried out for a test arrangement in which a number of plastic foils were stacked, some of them being punched to form artificial cavities in the dielectric. The life curve, i.e., voltage-time (V-t) characteristic, seems to conform to a straight line on double logarithmic scale. The slope of the line is given by n. This number depends on the type of material, n being 9 for polyethylene.

The number n is not, however, a material constant, but may depend on the size of the cavities, applied voltage, cavity environment, and so on. This phenomenon is treated in detail in Chapter 2 of Volume I. At any rate, an important result here is that the voltage life of polyethylene samples coincides with that of particular samples of actual polyethylene cables. The coincidence of the results does not mean that actual cables have as large cavities as do the laboratory test samples.

No presently available XLPE cables show the same curve as obtained in Figure 2.3.25. (Chapter 2, Volume I). The number n is significantly larger than 9. The reason for this is unclear, perhaps cable insulation has been improved to a certain degree. The cross section of a cable made with natural insulation, freshly cured in steam, displays an opaque concentric halo a short distance in from the outer shield. Under an optical microscope, the halo appears to comprise many small voids, or micro-voids, ranging from about 2 to perhaps 50 μm diameter with concentrations up to $10^5/mm^3$. Other locations in the cross section — close to the conductor, for example — show fewer and smaller voids. Each of the observed cavities is probably partly, or wholly, filled with water, air, methane, or other peroxide decomposition products. These voids are generally much smaller than those defined in AEIC power cable specifications.

Micro-voids are probably formed by the same process as that which forms mist in the atmosphere. They constitute sensitive regions of the polyethylene bulk, which can be expected to be more vulnerable to the hazard of micro-discharges than the crystalline or amorphous regions which surround them. Effects of such micro-voids have been ascertained by two kinds of experiments. These show that impregnation by gas or liquid results in an increase in the breakdown voltage of extruded cables, and that a correlation exists between void diameters and breakdown voltages.

The fact that SF_6 gas impregnation shows a substantial increase in breakdown voltage was first reported in 1970.[27] It is now concluded that SF_6 gas diffuses, dissolves, and permeates into any macro-voids and/or micro-voids to suppress such gaseous discharges in the voids as might cause dielectric breakdown of the cable replacing existing gas by higher dielectric-strength gas or higher dielectric-constant oil. A modern manufacturing process can exclude macro-voids with adequate care. Micro-voids, inherent in the steam-curing methods, can be reduced to around 100 μm in size. The next question is whether or not voids of around 10 μm are detrimental. The answer seems to depend on the choice of cable insulation thickness or the maximum or average design electric field strength. The maximum allowable void size is chosen to be less than (with some safety factor) the value at which no internal discharge takes place. The calculated value for this is 80 μm for 66- to 77-kV cable and 50 μm for 154-kV cable.[28,29] The void size produced can be reduced to far less than 10 μm by a new process called the dry-type cross-linking process. It is interesting to note that low-density polyethylene in general use for cable insulation has free volume in which molecular chains in amorphous regions undergo micro-Brownian movement. In addition it has inherent micro-voids in the amount of about 0.02 cm^3/g.[30]

The effects of voids on AC breakdown strength of XLPE cables was substantiated by experiments carried out on XLPE cable insulation with various sizes of voids. Figure 1.1.26. adduces evidence to show that reduction in void size increases AC breakdown strength.[31] Similar experiments were also carried out to demonstrate the effect of stress in XLPE cables, as shown in Figure 1.1.27.[31] This reference holds that AC breakdown is followed by the inception of gaseous discharges in micro-voids, which may obey Paschen's law.

B. Protrusions

Asperities from the inner semiconducting layer (conductor shield) and even from the outer semiconducting layer (insulation shield) of PE and XLPE cables can produce, when energized, an electric field high enough to initiate a dendritic breakdown path called a tree,

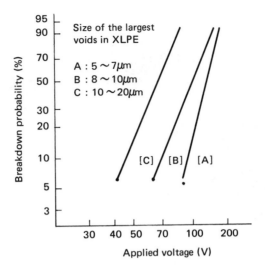

FIGURE 1.1.26. AC strength characteristics of 6.6-kV XLPE cables with 3.2-mm-thick insulation 5 kV/30 min. Step after 35 kv, 1 hr application.

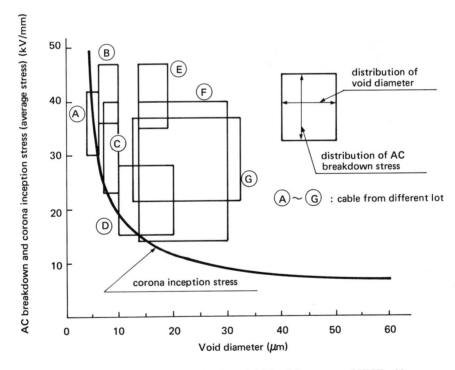

FIGURE 1.1.27. Relation between void size and AC breakdown stress of XLPE cables.

which may possibly lead to final breakdown of the cable insulation. This is an example of the treeing phenomenon described Volume I, Chapter 2. There are three types of semiconducting materials for these layers:

1. Semiconducting fabric or cloth tapes
2. Semiconducting polyethylene or cross-linked polyethylene
3. Semiconducting EVA copolymer

FIGURE 1.1.28.   Photograph of trees generated from asperities around semiconducting layers of an XLPE cable. (Courtesy of Pacific Gas & Electric Co.)

All of them usually contain carbon black (fine carbon powder) in order to obtain their semiconducting characteristic. Cloth tape is inferior to carbon-filled polymers because the former possesses by its nature a fuzzy, lapped edge which can be an easy starting point for treeing. Furthermore, it adheres poorly to the insulation and cannot be freed from voids which may precipitate internal discharges (corona). Such internal discharges can result in what is generally called "corona deterioration" of the insulation through tree initiation. Figure 1.1.28. shows trees growing from both semiconducting layers of an XLPE cable. Improvement was achieved by utilizing semiconducting tapes of semicured butyl or nylon which minimize butt spaces and fuzz.

In order to eliminate these defects in XLPE cable insulation systems more completely, the extruded, semiconducting, polymeric shielding layer referred to above was devised to replace the tapes, especially for high-voltage cables rated 66 to 77 kV and above. This practice has now become more and more popular, even for the lower voltage classes of extruded cables. It is expected to give smooth, protrusion-free surfaces. It may be the technological breakthrough which opens up the era of high voltage and even extra-high voltage XLPE cables. It is appropriately referred to as the second generation of solid dielectric cables.

The surface smoothness of the semiconducting layer certainly depends upon its manufacturing process, which will be described in a later section. Table 1.1.3. demonstrates that the surface roughness of the semiconducting layer, especially for conductor shielding, creates

Table 1.1.3.
BREAKDOWN VOLTAGE OF CROSS-LINKED POLYETHYLENE
CABLES WITH ROUGH AND SMOOTH SURFACE OF INNER
SEMICONDUCTIVE LAYER

Surface condition	77kV (100 mm² insulation thickness 17 mm)		33 kV (100 mm² insulation thickness 11 mm)	
	AC voltage (kV)	Impulse voltage (kV)	AC voltage (kV)	Impulse voltage (kV)
Irregular (rough)	110	420	100	380
Regular (smooth)	360	1100	210	700

From Fujisawa, Y., Yasui, T., Kawasaki, Y., and Matsumura, H., *IEEE Trans. Power Appar. Syst.*, 87(11), 1900, 1968. With permission.

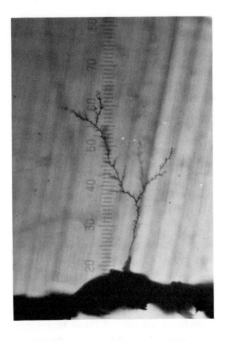

FIGURE 1.1.29.   Impulse tree growing from an asperity of semiconducting compound layer of an XLPE cable. (Courtesy of Hitachi Cable Co.)

conditions conducive to tree initiation, which can in turn lead to the final breakdown of the insulation by either AC or impulse voltage.[32] Figure 1.1.29. shows an example of tree generation from a semiconducting, carbon-filled, polyethylene layer under the action of a high-impulse voltage.

Filled carbon blacks can form asperities on the surface of a semiconducting layer, because they inevitably contain foreign substances or contaminants such as hydrocarbons, fine sands, and clays. As a matter of practicality, it is difficult to exclude contaminants smaller than 150 μm in size in compound form. These may be detrimental especially for EHV extruded cable. The possibility of particles of contaminant coming out on the compound surface depends upon the degree of draw-down in the compound thickness during the extrusion

Table 1.1.4.
LIFETIME (IN HOURS) AT 13.5 kV/mm WITH THE UNDERMENTIONED INCLUSIONS

	Glass needles	Wooden needles	Steel needles	Reference cable (no inclusions)
Polyethylene A	10	25	>1200	>1200
Polyethylene B	0.02	1.5	2	>1000
Polyethylene C	0.1	3	20	—
Variant B1	—	—	5	0.5
Variant B2	—	—	0.01	1[a]

[a] This reference length was four times as long as the other ones. If the curve is corrected the lifetime is 5 hr.

From Kreuger, F. H., *Endurance Tests with Polyethylene Insulated Cables*, CIGRE Paper 21-02, CIGRE, Paris, 1968, 6. With permission.

process. To reduce this degree would require a greater fluidity of the compound which depends on the carbon black content. It is desirable to reduce this content, but to do so might be impossible with ordinary carbon blacks, because to do so could result in an increase of electrical resistivity of the compound beyond the specified upper limit. There is now a newly developed carbon black available, which shows three to four times larger conductivity in compound form than ordinary carbon black as described in Volume I, Chapter 2, Section 2.2. Semiconducting compound with the smaller amount of carbon black is expected to be available. This should endow it with a flexibility similar to that of extruded insulation material.

C. Foreign Substances (Contaminants)

It was suspected that contaminants in cable insulation would reduce short-time breakdown voltage and deteriorate insulation around them, thereby shortening the service life of the cable. But it was recognized that it was extremely difficult to find out how such contaminants initiate breakdown. An example of the laboratory study for this purpose is shown in Table 1.1.4.[12] Contaminants such as glass, wood, and steel were embedded in polyethylene. The size of the contaminants was not so small as to exaggerate their effect: 0.7- to 1-mm long and about 0.1 to 0.2 mm in diameter were typical. The radius of the points of such needle-like contaminants was not entirely controllable, but was less than 5 μm. The figures in the fourth column of the table refer to samples without inclusions. These reference samples were made of cable that was extruded before the contaminants were added, in order to check whether the contaminants would actually cause breakdown.

Polyethylene may be contaminated during the

a.   Material manufacturing process
b.   Transfer process
c.   Mixing process
d.   Extrusion process

Cable manufacturers are now interested in contaminants larger than 10 mil or 250 μm. The conclusion at the moment is that no contaminants larger than this critical size would affect cable characteristics in the present insulation designs.

Data on contamination in polyethylene resin immediately after manufacture are available.[33] Contaminants included in polyethylene are generally classified for the purpose of production

Table 1.1.5.
ANALYSIS OF CONTAMINANTS
IN POLYETHYLENE

**Grouping from the appearance of
contaminants**

Group	A	B	C
Ratio %	92	7	1

Elementary Analysis

Group			
	A	B	C
Burnt resin	20/23	23/29	3/10
Contaminants in water or air	3/23	3/29	0
Ferrous substances	0	3/29	6/10
Cuporous substances	0	0	1/10

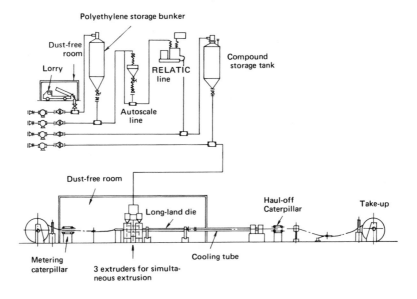

FIGURE 1.1.30.   A contamination-free system.[18a]

control according to their appearance. It is a matter of urgency that some method be found to reduce contaminants in this material. The visual classification has limited success as indicated in Table 1.1.5. Contaminants are divided into three groups: amber (A), black (B), and metal (M); they are rechecked by elementary analysis. Almost all the contaminants judged to be in group "A" were amber. The B group contaminants were burnt resin with little exception. Some of the materials ascribed to metals were burnt resin. Almost all of the contaminants investigated here were smaller than 250 μm, a few were 280 to 360 μm in size and they were thought to have come from water or air. One concludes it is possible to obtain quasi-pure polyethylene from raw material supplies.

As a next stage, a new total clean system for contamination-free manufacture has been developed. An example is shown in Figure 1.1.30. A metal container can be used for

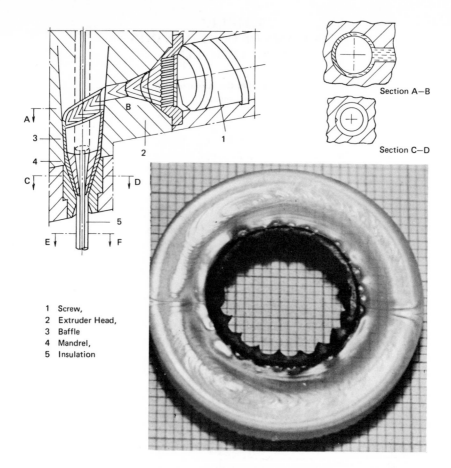

FIGURE 1.1.31.   Flow of material in the transition zone from extruder to extruder head; (1) screw, (2) extruder head, (3) baffle, (4) mandrel, (5) insulation. (Courtesy of AEG-Telefunken.)

delivery from a raw material manufacturer. But even in this closed system, it is advisable for quality control to check the XLPE compound for contamination. It should be noted that use of a metal detector and a fine-screen mesh in an extruder can exclude contaminants to a certain degree.

D. Inhomogeneities

A flow pattern can be observed in the cross section of PE or XLPE cable insulation,[16] representing the flow history of material extruded from the extruder head as illustrated in Figure 1.1.31. Two streams, formed to wrap the stranded conductor, coalesce, finally giving a seam structure. These flow and seam structures produce some inhomogeneities which one might suspect would affect insulation characteristics of this type of cable. The formation of these structures can be described as follows.

A polymer melt has structural viscosity, i.e., its macromolecules are oriented according to the shear velocity. Contrary to Newtonian liquids, the velocity profile of which is parabolic in tubes, non-Newtonian polymer melt displays a velocity distribution that is more uniform in the central section of the flow, while the velocity gradient close to the boundary surface is higher. As a consequence, salient orientations and stretchings of the macromolecules occur in these zones. The orientations are initially retained in the cable and are frozen during the cooling process by reduction of the chain mobility and crystallization.

CABLE BREAKDOWN VS TIME
COMPARISON OF STANDARD AND IMPREGNATED CABLE

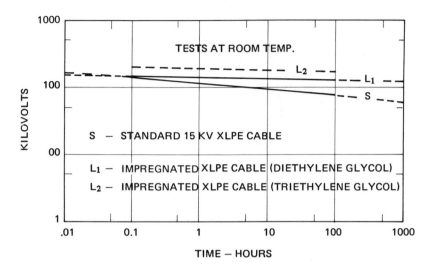

FIGURE 1.1.32.   Voltage life of impregnated XLPE cable. (After McKean, A. L., *IEEE Trans. Power Appar. Syst.*, 95(1), 255, 1976. With permission.)

The effects of these inhomogeneities in the electrical properties of the insulation are unclear. Treeing experiments on test specimens from various zones of 2 PE cables for 20 kV did not exhibit any significant difference. Nor has any evidence been obtained indicating that the seam is a weak region. Nevertheless, some correlation between breakdown path and flow structure was obtained from experiments of breakdown between needles inserted into the insulation as artificial defects and the inner semiconducting layer. Other data appear to show breakdown paths to be selective among flow streams.

Liquid impregnation is also effective in increasing breakdown voltage. Standard and impregnated cables were compared under a variety of test conditions. Figure 1.1.32. shows that both diethylene glycol and triethylene glycol impregnants substantially increase the dielectric strength of XLPE cables.[15] Perhaps more importantly, they flatten the slope of the volt time curve, thus improving the life characteristic. Figure 1.1.33. shows that the breakdown voltage at any one time step is increased and the scatter decreased. In these experiments, 2-min step tests, 1-hr step tests, and 24-hr step tests were carried out. Figure 1.1.34. compares the impulse breakdown performance of a standard XLPE cable and an XLPE cable impregnated with diethylene glycol. The average breakdown stress of standard XLPE cables was 2000 V/mil (80 kV/mm) (350 kV), whereas the breakdown stress of the impregnated XLPE cable was 3100 V/mil (124 kV/mm) (540 kV), representing a 50% improvement. These tests clearly indicate that, even under transient conditions, impregnation of the insulation with a liquid produces a substantial increase in the breakdown strength. Unfortunately, no discernible corona discharges have been observed so far for good quality XLPE cables subjected even to extremely high AC stresses. This measurement indicates that even if corona discharges occur in the microporous regions, their energy levels are below the sensitivity of modern detectors. Theoretical estimates have been made of the corona that would be measured from discharges in small (1 to 20 μm) micro-voids operating at stresses above 200 V/mil (8 kV/mm). Utilizing Paschen's curves, it is clear that discharges can occur at very high stress, but the corona energy would correspond to less than 0.01 C, substantially below the sensitivity of modern corona detectors.

The second evidence suggesting some effect of micro-voids comes from breakdown tests

PROBABILITY OF CABLE BREAKDOWN AFTER 1 HOUR
SINGLE STEP VOLTAGE AT ROOM TEMPERATURE

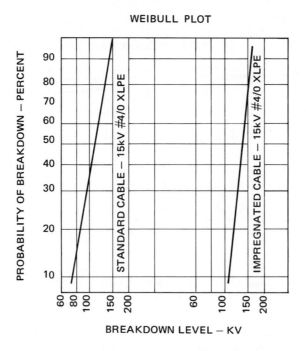

FIGURE 1.1.33.   Probability of cable breakdown. (After McKean, A. L., *IEEE Trans. Power Appar. Syst.*, 95(1), 256, 1976. With permission.)

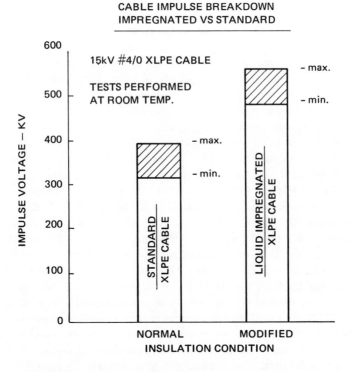

FIGURE 1.1.34.   Impulse strength characteristics. (After McKean, A. L., *IEEE Trans. Power Appar. Syst.*, 95(1), 256, 1976. With permission.)

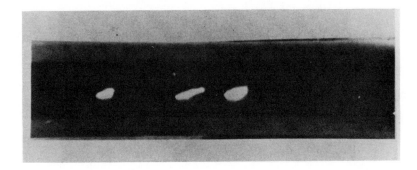

FIGURE 1.1.35. Skips in conductor shield of a 15-kV XLPE cable detected by "Skip Detector" during factory production. (After Bahder, G., Eager, G. S., Silver, D. A., and Lukae, R. G., *IEEE Trans. Power Appar. Syst.*, 95(5), 1153, 1976. With permission.)

on XLPE cables with different micro-void size distribution. Three groups of 6.6 kV XLPE cables were tested, having micro-voids of 5 to 7 μm, 8 to 10 μm, and 10 to 20 μm in diameter, respectively. Figure 1.1.27. shows the result of breakdown probability vs. applied voltage for the 3 groups of cables which had been subjected to 5-kV, 30-min step tests. It is clear from this figure that the increase in the size of representative micro-voids shifts AC breakdown voltages downwards to lower values. Seven kinds of XLPE cable with different insulation thickness (3.2 to 11.0 mm) and with different void size distribution were also examined. Some correlation may be found between AC breakdown mean stress and void diameter as shown in Figure 1.1.26.

E. Shielding

Skips in the conductor shield can occur during manufacture due to improper extrusion conditions. These can permit internal discharges, leading to failure of the cable. Skips are normally not detected by partial discharge measurements because the insulation fits tightly to the conductor as manufactured. In normal service under load cycle conditions, a cavity will form at the skip and a discharge will occur which may ultimately lead to cable failure. It is claimed that the skip can be detected by a "skip detector", as shown in Figure 1.1.35.[14]

1.1.5. Effects of the Environment on PE and XLPE Cable Insulation

Impurities can diffuse into PE or XLPE cable insulation since such cables usually possess no lead sheath but rather a PVC or PE jacket and metal tape shielding. They are therefore open to their environment. Phenomena so far attributed to environmental effects on PE and XLPE cables are

1. Water treeing or electrochemical treeing
2. Sulfide treeing — chemical treeing

The former phenomenon takes place when the insulation is subjected to both water and electric field, while the latter is a consequence of sulfide attack on insulation. Both form tree-like patterns which generally reduce insulation breakdown voltage.

A. Water Tree Pattern

Extensive investigations into failure of medium-voltage PE and XLPE cables due to unknown causes have been made in both the U.S. and Japan. The U.S. studies included comprehensive investigations by a utility company.

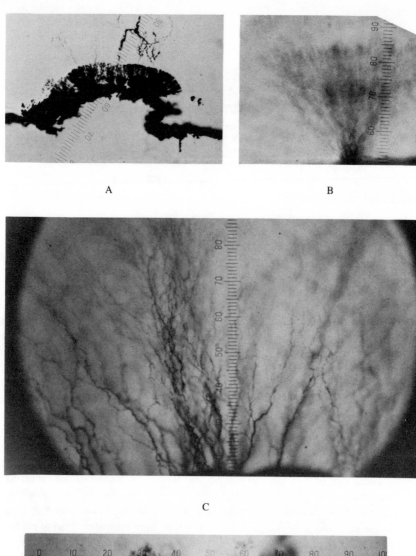

FIGURE 1.1.36.   Various patterns of trees developed in polyethylene cables: (A) broccoli, (B) strings, (C) delta, (D) bow tie, (E) plume, (F) dendrite, (G) spike, and (H) fans. (Courtesy of Pacific Gas & Electric Company.)

   Microscopic observation on PE and XLPE insulation samples from high-voltage cables recovered from service (1 with nearly 20 years of service) revealed that some samples exhibited specks which could be called trees. Typical tree-like patterns found in some samples are shown in Figure 1.1.36. According to the reference,[34] they are classified as

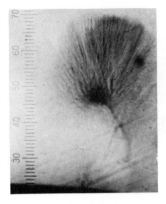

Figure 1.1.36.E

Figure 1.1.36.F

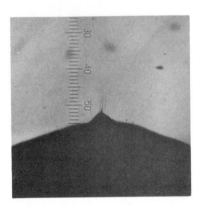

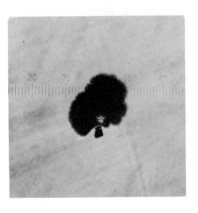

Figure 1.1.36.G

Figure 1.1.36.H

a. Broccoli, appearing only at tape-type fabric conductor shields, typical lengths being up to 250 μm (10 mil)

b. Strings (streamers), appearing mainly at tape and occasionally at extruded conductor shields, with tree lengths varying up to a total projection through the insulation

c. Delta, observed at conductor and at insulation shields, with tree lengths here also varying up to a total projection through the insulation. In particular, they have been observed eminating from the insulation surface on cables containing a loose-fitting extruded insulation shield

d. Bow tie, appearing at contaminants within the insulation, the majority being shorter than 250 μm (10 mil) but a few up to 3 mm (120 mil), and at micro-voids; this type of tree is probably now predominant with modern, all extruded cable (including shielding system)

e. Plumes, appearing at contaminants within the insulation in the neighborhood of the conductor shield, having tree lengths up to 500 μm (20 mil)

f. Dendrites, discovered rarely, being located in the vicinity of failures, initiating at either the conductor or at the insulation shield and growing to considerable length; these are probably electrical trees

g. Spikes, appearing at extruded conductor shields, with lengths up to 120 μm (5 mil)

h. Fans, growing from contaminants within the insulation, being less than 120 μm (5 mil) long; perhaps they comprise some material such as antioxidant

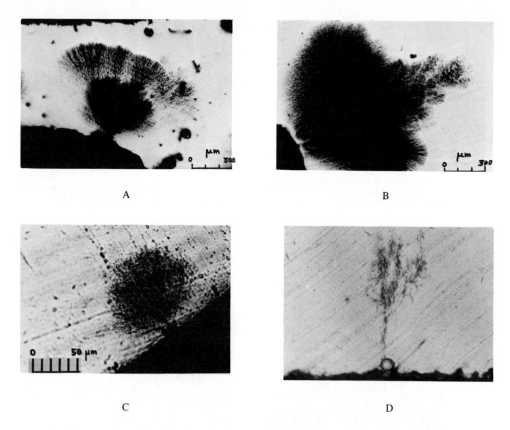

FIGURE 1.1.37.   Water trees in XLPE cables. (A) 6.6-kV XLPE cable, 4(E/√3), 95°C water, acceleration test 14,000 hr; 5/8 in. = 300 μm. (B) 6.6-kV XLPE cable, 4(E/√3), 95°C water, acceleration test 14,000 hr; 5/8 in. = 300 μm. (C) 6.6-kV XLPE cable, 3(E/√3), 40°C water, hydrostatic pressure 6 kg/cm² G, acceleration test 12,000 hr; 6/8 in. = 50 μm. (D) Water tree starting from a void near inner semiconducting layer (XLPE cable).

The direction of growth of tree patterns (a) to (f) was essentially parallel to the electric field lines while trees (g) and (h) indicated a random direction of their longest dimension. The spikes and fans described above are produced during the cable manufacturing process and consist of foreign materials in the insulation. Figures 1.1.37. and 1.1.38. show water trees and bow-tie trees grown in XLPE cables, respectively; they are selfexplanatory.

A case study of a number of PE and XLPE cables removed from service was helpful in clarifying how some particular types of trees are formed. This is shown in Figure 1.1.39.[35-37] With the exception of a very few cases, trees developed in service are not of an electrical type, originated by partial discharges, rather their formation must be linked to some kind of penetration of foreign material such as water into the insulation under voltage stress. They are called "electrochemical trees" in the U.S. and some other countries, whereas in Japan the name "water tree" is used because water is always involved. The term "water tree" seems more logical than the term "electrochemical tree", for this reason it has become widely used recently.

Water trees should be carefully differentiated from electrical trees, which are simply referred to as trees. Water trees do not exhibit corona discharges; in contrast, electrical trees are accompanied by intensive partial discharges. The propagation time of water trees is measured in years, whereas an electrical tree very quickly propagates through the insulation, once it is generated. The advent of a water tree does not lead to rapid cable failure. From an appearance point of view, the two types are usually different from each other, but occasionally they are difficult to distinguish. A practical method for differentiating between

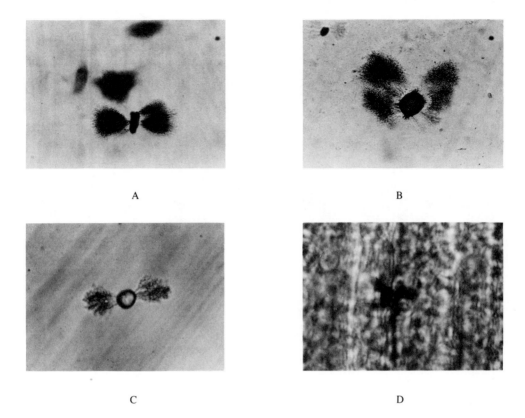

A

B

C

D

FIGURE 1.1.38.   Bow tie trees in XLPE cables. (A) Bow tie tree from contaminant 6.6 kV XLPE cable, $4(E/\sqrt{3})$, 95°C water, acceleration test 14,000 hr. (B) Bow tie tree from contaminant 6.6 kV XLPE cable, $4(E/\sqrt{3})$, 95°C water, acceleration test 14,000 hr. (C) Bow tie tree from void 6.6 kV XLPE cable, $4(E/\sqrt{3})$, 95°C water, acceleration test 14,000 hr. (D) Bow tie tree from amber 22 kV XLPE cable, $3(E/\sqrt{3})$ 75°C water, acceleration test 10,000 hr.

the two types of trees is to examine them after insulation is dried. Water trees then become invisible while electrical trees are clearly visible.

B. Origin of Water Trees

Protrusions, contaminants, and voids, or even micro-voids, favor the formation of water trees. This emphasizes the regulation of contaminants and voids in cable insulation for its quality control, as is described in another chapter.

Reduction in breakdown voltage due to the formation of water trees is a serious concern, as indicated in Figure 1.1.40.[36] This figure shows a V-t characteristic for XLPE and PE cables with inner and outer semiconducting layers of cloth tapes. The data group with open marks was obtained as short-time AC breakdown stress for some XLPE cables which had been subjected to the acceleration test, and a cable removed from service. This V-t characteristic is somewhat different from the characteristic normally obtained by the application of various constant voltages. This latter represents the breakdown strength of a cable aged under certain conditions for a certain time interval. Lifetime for power cables is generally estimated from its dependence on the $-n$th power of the applied voltage. The value of the n is around 15 to 25 for XLPE cables, while, as already described, it is 9 or less for polyethylene subjected to internal discharges.

If water trees are involved, lifetime cannot be expressed in terms of a simple $-n$th power of the voltage. Once water trees are created, it must be recognized that a drastic decrease in breakdown voltage is to be anticipated. This is quite different from, and really worse than, the V-t characteristic under dry conditions.

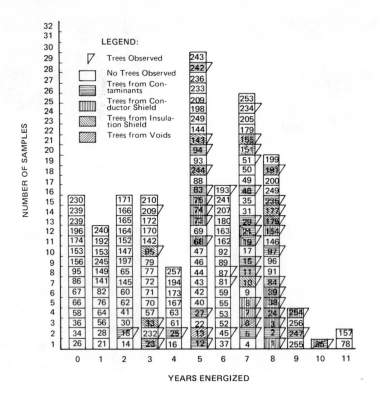

FIGURE 1.1.39.   Origin of trees. (From Kreuger, F. H., *Câble Isolé au Polyéthylène à Imprégnation Gazeus,* CIGRE Paper 21-02, CIGRE, Paris, 1970, 825. With permission.)

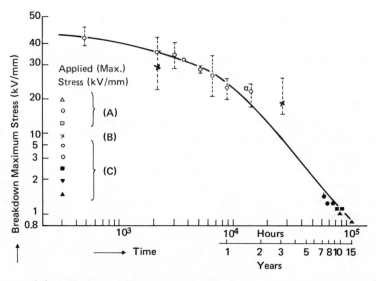

(A)   Cables subjected to an acceleration test
(B)   Cables removed from service
(C)   Cables to have failed due to water trees during service.

FIGURE 1.1.40.   A V-t curve to estimate a remaining lifetime of XLPE and PE cables with water trees. (From Tanaka, T., Fukuda, T., and Suzuki, S., *IEEE Trans. Power Appar. Syst.,* (95)6, 1898, 1976. With permission.)

The water trees shown in Figure 1.1.37. were obtained for cables with cloth semiconducting layers which had some irregularity and hygroscopic property. Recent progress in cable technology makes it possible to reduce irregularities or asperities through the simultaneous extrusion of semiconducting layers and polyethylene insulation. This makes the time for nucleation of water trees longer.

C. Countermeasures

Various countermeasures are available to protect cable against water tree formation;[39] a lead sheath which makes a perfect seal against water is an extreme case. Such sheaths have been used for EHV extruded cables and sometimes for HV submarine XLPE cables. However, this approach may not be economical for either high- or low-voltage cables, and it surely makes splicing more difficult. Thus, in spite of its excellence in precluding water, it leaves much to be desired when manufacturing extruded cables. Perhaps a better idea would be to develop and use a water-tight plastic sheath. Recently, an improved metallic shield has been introduced with the principal objective of increasing short circuit capability. This shield consists of a longitudinally folded and corrugated copper tape applied with an overlap under an extruded polyethylene overall jacket. This new metallic shield, in combination with the extruded polyethylene overall jacket, virtually precludes moisture ingress from the environment and thereby renders the cable highly resistant to water tree formation in its insulation structure.

Reduction of voids, protrusions, and contaminants is another way to protect cable insulation from water tree formation. This has been a traditional approach to achieve improved designs at higher voltages and to ensure greater reliability in service through consistency of quality effective against water trees. It has also proved voids make room for water and become a potential water supplier for water tree formation. It has been suggested that water-filled voids can even generate bow-tie water trees, though this is controversial. As mentioned already, cloth tape conductor-shields can provide a source for water trees by being hydrophilic and by projecting fuzz into the insulation from its semiconducting surface. From this point of view, a nylon tape is better than a cloth tape; its surface is smoother and it is less hydrophilic.

Semiconductive polyethylene or cross-linked polyethylene has superceded tapes because of its superior performance. Asperities can be minimized by the proper extrusion of such a semiconducting layer. The polyethylene should be as pure as possible and it is especially important to exclude hydrophilic and metallic contaminants with rough surfaces.

Care should be taken to prevent water permeation through the exposed ends of extruded cable during storage and installation. Similar precautions against the ingress of water should be taken during splicing.

D. Sulfide Trees

Tree-like paths have been found in polyethylene-insulated control cables installed in a chemical plant and in an XLPE-insulated submarine cable laid in marine algae.[40] Such manifestations are referred to as sulfide trees. A cross section of a cable with sulfide trees is shown in Figure 1.1.41. They grow from a conductor or a conductor shield even when a cable is not energized. The cable is likely to fail when they approach the outer surface of the cable insulation.

Sulfide treeing is the only chemical treeing process which has been found so far. The proposed mechanism for this phenomenon is as follows. Sulfides, such as hydrogen sulfide (H_2S), permeate through the jacket and insulation of a cable and finally reach its metallic conductor. The copper conductor reacts with the sulfides to produce cuprous sulfide (Cu_2S) which crystallizes and grows back into the insulation to form dendritic deposits or trees. If the dendrites of cuprous sulfide, which is a conductive material, continue growing through

Distribution of Sulfide Trees

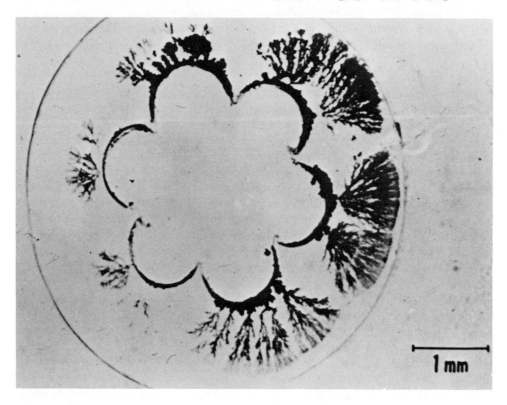

FIGURE 1.1.41.   Distribution of sulfide trees in a core. (Courtesy of Furukawa Electric Industries Ltd.)

the insulation, failure will ultimately take place. It is to be noted that sulfide trees usually originate from the convex part of the stranded conductor (or at the concave part of the insulation) and very rarely at the concave part of the stranded conductor (or at the convex part of the insulation).

Since sulfide trees are caused by a chemical reaction between permeating sulfides and copper, some preventive methods can be devised on the basis of the considerations described in Figure 1.1.42.[41] One excellent solution is to replace the copper conductor by an aluminum conductor. Plating of noncopper metals and coating the conductor with organic materials such as enamel have been investigated with a certain degree of success.

Another idea is to block sulfide permeation. A metallic sheath is very good for this purpose, but it is expensive and makes cable handling more difficult. An alternate approach is to use a sulfide-capturing material. Such materials have been devised to protect cables against sulfide attack.[40] A layer of this material is located between the cable insulation and the jacket. It operates on the principle of transforming active sulfides into stable compounds in the capture layer.

1.1.6. Improvement of "Defect" Cables — Recent Technological Development

It is not the intent of this chapter to discredit polyethylene and cross-linked polyethylene insulation, by using the term "defect" cables. On the contrary, the author believes that PE or XLPE insulations are among the best cable insulations presently available. Moreover, it is most likely that they will be subjected to improvements and innovation.

The life expectancy of XLPE-insulated power cable is determined by the level of imper-

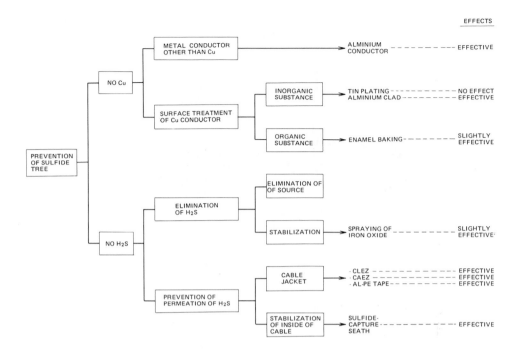

FIGURE 1.1.42.   Preventive measures against sulfide trees. (From Tabata, T., Nagai, H., Fukuda, T., and Iwata, Z., *IEEE Trans. Power Appar. Syst.*, (91)4, 1359, 1972. With permission.)

fections present in the insulation structure of the cable. As stated in previous sections, these critical imperfections lead to breakdown by ionization or water trees comprising in decreasing importance: (1) protrusions from conductor shields, (2) contaminants, and (3) voids. It is fortunate that the prevalence of these imperfections is in the reverse order. Quantitative correlation between these imperfections and insulation characteristics of the cable is still in question, but the goal is clear.

A. Reduction in Protrusions

As long as fabric tapes are utilized for conductor shields, nylon tapes are to be preferred. With careful manufacture, they are acceptable for low- and medium-voltage extruded cables. However, the use of semiconducting compounds has become standard practice most recently, especially for HV and EHV cable designs. Many manufacturers adopt the semiconducting compound even for low-voltage (3.3 or 6.6 kV) extruded cables in view of the product reliability.

Simultaneous extrusion techniques have been developed using semiconducting polyethylene copolymer and cross-linkable polyethylene for a conductor shield and/or an insulation shield. Two layers, i.e., main insulation and conductor shield compound, are extruded simultaneously, and the remaining layer, i.e., the insulation shield, is made of semiconductive fabric tapes. The ultimate design should comprise an extruded insulation shield as well; it is usual to manufacture this in a tandem operation at the present time. This process, called three-layer simultaneous extrusion, is expected to provide almost perfect performance as regards smoothness and bonding between the layers. The arrangement of a three-layer tandem extrusion by a vertical continuous vulcanizer is schematically illustrated in Figure 1.1.43.[32] Figure 1.1.44. shows an example of test data which indicate the improvement of construction and electrical properties for 66- to 77-kV XLPE cables.[32] The tandem operation may be called a ''pseudo-simultaneous triple extrusion process''. Some manufacturers utilize a ''perfectly simultaneous triple extrusion process'' which is provided by a three-layer

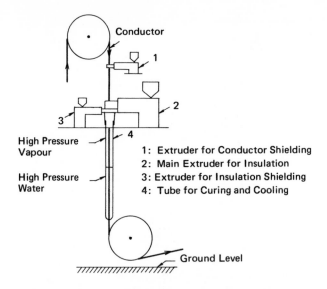

FIGURE 1.1.43.   Arrangement of tandem three-layer extrusion system (common three-layer extrusion system is also available). (From Fujisawa, Y., Yasui, T., Kawasaki, Y., and Matsumura, H., *IEEE Trans. Power Appar. Syst.,,* (87)11, 1900, 1968. With permission.)

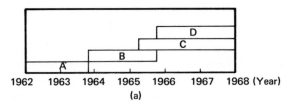

FIGURE 1.1.44.   Improvement of construction and electrical properties of 66- to 77-kV XLPE cables. E — extruded shielding, T — tape shielding. (From Fujisawa, Y., Yasui, T., Kawasaki, Y., and Matsumura, H., *IEEE Trans. Power Appar. Syst.,* (87)11, 1901, 1968. With permission.)

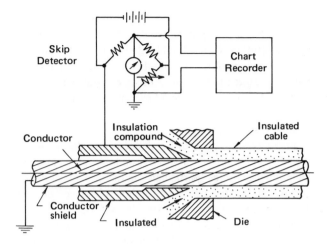

FIGURE 1.1.45.   Skip and protrusion detector-principle of operation. Schematic shows application of the skip detector to monitor the conductor shield resistance in the tip of the insulation extruder during cable manufacturing process. (After Bahder, G., Eager, G. S., Silver, D. A., and Lukae, R. G., *IEEE Trans. Power Appar. Syst.*, (95)5, 1559, 1976. With permission.)

common extruder head system. This improved method of manufacture is becoming increasingly important for higher voltage cables due to the difficulty in achieving the desired bond between the insulation and the insulation shield when the latter is applied in a separate operation. The improvement brought by the newer technique is attributed in part to the smoother insulation surface resulting from improved extrusion techniques. Although the bond between the insulation shield and insulation is more uniform and more tenacious when the insulation structure is manufactured in one operation, the insulation shield should nevertheless strip cleanly from the insulation surface for jointing operations. This can be accomplished by moderate application of heat.

A nondestructive test device[14,42] has been developed which can detect the existence of irregularities in the form of skips and protrusions of all significant sizes and shapes in the conductor shield of extruded dielectric cable during the tandem- and triple-extrusion processes. To detect such irregularities (which can occur both at the extruder for the conductor shield and in the main extruder for applying the insulation and insulation shield, in the case of triple-extruded cables), the detector is located in the head of the main extruder. This equipment is capable of detecting and facilitating the determination of the exact location of irregularities in the conductor shield which, if not eliminated, can adversely affect the quality and service performance of XLPE-insulated cable. Use of the skip and protrusion detector shown in Figure 1.1.45. will allow serious conductor shield irregularities to be detected at an early stage of manufacture. Moreover, since this device permits accurate location of imperfections and is nondestructive to the cable, the imperfections, after being removed, can be examined and their cause identified. Corrective action can then be taken to prevent recurrence of this condition with consequent improvement in cable quality and no increase in rejected material when testing at higher test voltages. It is now considered essential for the manufacture of high-quality cable and improvement in performance in service that the conductor shield be monitored continuously in the tip of the insulation extruder to detect and eliminate significant perturbations to the conductor shield.[14]

It is really required of the cable industry that it manufacture XLPE cable with its shield completely bonded over the whole length of the cable. Since this is rather difficult, an alternative idea has been introduced. This is based on the beneficial effect of oil impregnation on the stability of such cable parameters as breakdown voltage, resistance to corona, and resistance to treeing. It is embodied in a liquid-filled tape XLPE cable which has excellent voltage stability as a consequence of reinforcing the efficiency of the semiconductive shield. Silicone oil is selected as impregnant because of its compatibility with the insulation and semiconductive material.[43] Such complicated XLPE cables are reserved for special purposes.

B. Reduction in Contaminants

Polyethylene is contaminated on three different occasions; during material manufacturing processes, during transfer and handling processes, and during cable manufacture. The permissible size of contaminant particles (particularly in the U.S.) is considered to be 250 μm. In order to minimize the size and number of foreign substances enclosed in the finished cable insulation, all contamination should be strictly excluded to whatever extent possible in the aforementioned processes. This is the only solution. One must start with the purest possible supply of granular polyethylene. Cable manufacturers should negotiate with their material suppliers. A totally enclosed system should be established for material transfer and handling. The raw material should be kept in as clean a room as possible so as to be supplied to the extruder in the same clean state as originally received. Cross-linking agents and fillers should be contamination-free. The maximum value of 250 μm is a temporary figure established for the control limit by cable manufacturers. A metal detector may be useful in the final stage, prior to extrusion, to reject all pellets that contain ferrous or nonferrous metals.

C. Reduction in Voids

Macro-voids need not be considered since they can be excluded through careful treatment with modern cable manufacturing equipment. There has been controversy as to whether micro-voids as small as 50 μm can affect short- and long-term electric properties of cable insulation and if so, to what extent. Some evidence is described in a previous section concerning the correlation between void size distribution and breakdown voltage. Bow-tie trees have been found, but their detrimental effect is still in question. They are associated with water in almost all cases, and therefore they should be called "water bow-tie trees" or "electrochemical bow-tie trees".

Technical development is still in progress to obtain more reliable cross-linked polyethylene for higher voltage use by minimizing voids. There are two potential causes of micro-voids in insulation walls; these are improper extrusion parameter and improper cross-linking cycles. Proprietary screws and cross heads have been designed and installed on an extruding machine to improve the extrusion parameters.[17] It is hoped this equipment will insure proper compacting of all molecules under controlled pressure conditions. Such higher pressure conditioning minimizes micro-voids.

Proper cross-linking of the polyethylene requires optimization of the steam exposure, the hot water gradient cooling, and the cold water exposure of the insulation. The steam cycle is selected by a computer program which assures the proper time-temperature chemical reaction. The correct gradient during the cooling period is essential in controlling the crystalline structure development within the cross-linked polyethylene wall. The final water cooling leg is necessary to bring the temperature of the insulation wall and conductor down to ambient before the insulated cable exits the pressurized vessel.

Thus, there remains a potential for improvement in steam-cured XLPE insulation walls; micro-voids may be the inevitable outcome from the conventional steam-curing method used for cross-linking. Water in the vapor state is involved in the formation of the microvoids, so an effort should be made to halt water or vapor permeation into insulation in the curing process.

There are several ways to achieve this. It may be useful to apply high-energy electron beams for cross-linking. This method has been applied successfully for cross-linking low-voltage insulated wires with thin-wall insulation, but not for high-voltage cables with thick-wall insulation. A challenging research program has been undertaken in the U.S. to develop 138 kV and 230 kV irradiated cross-linked polyethylene cables. So far this effort has been directed toward the design of a toroidal irradiation system.

Several advanced methods for chemical cross-linking have been developed, some of which have established quite a reputation, especially for high-voltage cables. Methods under development can be classified as follows:

1.   Gas-curing method
2.   Modified steam-curing method
3.   Long-land die method
4.   Ultra-high-pressure cross-linking method

Inert gases are a logical replacement for high-pressure steam to transfer heat and pressure. Two ways are available for doing this. One is to utilize pressurized, heated, inert gases such as SF_6 and N_2. This is similar to the steam-curing method except that pressure and temperature can be selected independently. The alternate approach is to use the gas as a pressurizing medium only, heat being supplied by other means such as an infrared heater. For this reason, it is referred to as the radiant-curing process to differentiate it from the former gas-curing process. Both gas-curing systems result in a substantial reduction in the size and number of voids, with consequent improvement in AC and impulse breakdown voltages and V-t characteristics. It is possible to use melted metal (sodium or potassium) salts instead of steam, but it may be difficult to apply this technique to XLPE cables, although it is already in practical use for manufacturing low-voltage, rubber-insulated wires at room temperature. Oil, such as silicone oil, can also be used as a heating and pressurizing medium.

A proposal has been made to mitigate the effect of moisture on a finished insulation including material cross-linked in high-pressure steam. This idea is to impede the diffusion of moisture into the insulation during the cross-linking stage. It can be called a modified steam-curing method. A water-trapping agent is added to the compound used for the outer semiconducting layers. It has some advantage over the other methods in that existing equipment can be utilized without modification.

A technique has been developed to irradiate with high-frequency (400 kHz) ultrasonic waves, via a special medium, which causes heating by the viscoelastic friction of molecular vibration; this results in the cross-linking of the polyethylene. It is highly advantageous to heat material uniformly in the radial direction. This is particularly useful when cross-linking thick insulation walls. It is unfortunate that the absorbed energy is insufficient to complete the cross-linking by this method alone; however, it can be used jointly with the steam-curing method or even with the modified steam-curing method. All the new methods described above can be looked upon as modified VCV methods.

The long land die method first appeared as the MDCV method, as already described. The original goal of this development was a modification of the CCV or HCV method, which had problems in the control of concentricity and outer diameter and with water involvement. The polymer melt is never in contact with the gaseous or liquid heating medium, only with the metal of the long land die. Consequently no void formation is, in principle, to be expected, although micro-void formation is probable near the surface of cable insulation due to the confinement of volatile products produced during the heating process.

New cable manufacturing processes such as the gas-curing process and the MDCV process can be characterized as ''dry type'' cross-linking processes, in contradistinction to the steam-curing process which is ''wet''. Cross-linked polyethylene cable insulation manufactured by any of the dry-type methods is typified by

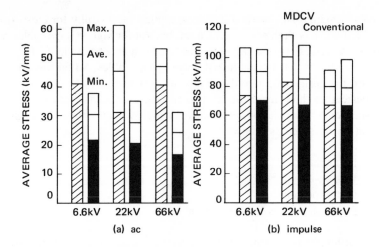

FIGURE 1.1.46.   Comparison of short-time breakdown stress between conventional and improved (MDCV) XLPE cables.

1.   Smaller and fewer voids
2.   Smaller content of water
3.   Better electric characteristics

compared with insulation made by the wet-type method. Nonsteam-curing methods allow only 0.01% by weight of water content, while steam-curing methods usually give 0.2 to 0.3% by weight.

About 10^5 to 10^6 voids per mm^3, ranging in size from less than 1 μm to 30 μm, are usually observed in the conventional steam-cured XLPE insulation. In contrast, in insulation manufactured by a nonsteam-curing process, for instance the MDCV process, voids larger than 1 μm but less than 1.3 μm are rarely observed except in the outermost layer of insulation. Figure 1.1.23. shows a marked difference in microscopic observations of void distribution profiles between the two insulations verifying that void content is virtually zero in dry-type cured XLPE insulation. This is reflected in significantly improved short-time AC breakdown characteristics, as shown in Figure 1.1.46. and long-term characteristics as shown in Table 1.1.6.[18]

Finally, an ultra-high-pressure cross-linking process is now under development. This features the simultaneous extrusion of molding and cross-linking under ultra-high pressure. Thermal energy is supplied in part or fully by the frictional heating of material caused by the shock of a piston movement or by shear stress. Cross-linking time can be reduced drastically. One might call this a kind of plastic forging process.

Newly developed technology has been described, for the most part, and perhaps unduly, from the standpoint of void reduction. One should also take note of the other technical and economic aspects, specifically, such technical points as making radial temperature distribution as uniform as possible to increase manufacturing line speed with high fidelity, and/or to produce higher voltage cable with "thick" wall insulation. This may also improve the economics of production. Cost reduction is surely an economic objective for much new equipment, since a modern VCV installation requires an enormous amount of investment; perhaps, as mentioned, close to $1 million.

D. Effects of Additives and Fillers

The addition of "medicines" into the "defect" cable to improve the reliability of cable insulation has long received much attention. These additives are in almost all cases organic

Table 1.1.6.
WATER IMMERSION TEST RESULT UNDER HEAT CYCLE

Cable	Conductor section (mm²)	Insulation thickness (mm)	Applied voltage (kV)	Applied time (hr)	PVC sheath	tan δ[a] (%)	Water tree (per cm) Maximum length (μm)	Water tree (per cm) Tree number
MDCV	100	4.0	28	8,021	With	0.020	Not observed	
					Without	0.019	Not observed	
Conventional	100	4.5	12.5	7,983	With	0.035	90	5
					Without	0.025	120	10

[a]   At room temperature at 12.5 kV.

From Yamauchi, H., Okada, M., and Fuwa, M., *Voil Distribution in Crosslinked Polyethylene Insulated Cables Made by Conventional and New Processes*, IEEE PES Meeting C72 504-9, IEEE, Piscataway, New Jersey, 1972, 342. With permission.

Table 1.1.7.
EFFECT OF A VS AGENT ON LONG-TERM CHARACTERISTICS OF PE AND XLPE INSULATION

Applied voltage	Composition	Records on tree occurrence
10 kV	Control	Detected on 1 specimen at 245 hr and 2 at 325 hr
	PE + VS trinity	No tree detected on 4 specimens by 325 hr
	XLPE + VS trinity	No tree detected on 4 specimens by 325 hr
14 kV	Control	Detected on 1 specimen at 70 hr and 1 at 230 hr
	PE + VS trinity	No tree detected on 4 specimens by 325 hr
	XLPE + VS trinity	No tree detected on 4 specimens by 325 hr

From Itoh, Y., Hayashi, K., and Kawasaki, Y., *Effects of Additives and Contaminants on Electrical Properties of Polyethylene,* IEEJ-EIM Study Meeting, IM-74-43, (in Japanese), Inst. of Electrical Engineers of Japan, Tokyo, 1974, 6. With permission.

compounds.[44,45] Massive impregnation, rather than addition of compounds different from the main insulation material, is of interest, too.[46] These are named inorganic fillers or simply fillers.

The additives just referred to are usually called VS agents or voltage stabilizers. Many voltage stabilizers comprise compounds with electron-rich structures such as heteroatoms, halogens, and conjugated π-electron systems. It is a popular speculation, though not substantiated experimentally, that these types of additives are effective against both water trees and electrical trees, and what is more important, are chemically and thermally stable for a considerable period of time. In fact, voltage stabilizers have not as yet earned their reputation, nor seen much practical use, although XLPE power cables with some additives are now available.

Impregnation of inorganic fillers may improve insulation characteristics of cable insulation through resistance to corona, AC V-t characteristics, and impulse V-n characteristics. How mineral fillers achieve this is not fully understood, but on the basis of the technical development, it is interpreted in terms of space charge effects and perhaps their inherent physical and chemical properties, also.

Tests have been recently conducted on additives comprising more than two ingredients, each of which is expected to play its respective role. Consider "VS-trinity" for example. This comprises 3 kinds of "medicines"; ferrocene, siloxane oligomer, and 8-hydroxyquinoline. Ferrocene is to trap the electrons in the dielectric, thereby preventing them from gaining energy from the electric field. The siloxane oligomer is expected to precipitate around defects such as voids and contaminants and so reduce the electric field. The ingredient 8-hydroxyquinoline will hopefully trap metallic ions, especially cuporous ions, by chelatifying them. Table 1.1.7. shows some evidence for thermal stability of the proposed additive in cable insulation. The physical properties of the cable insulation are not influenced by the additive. It is claimed that this additive can provide trapping centers for high-energy electrons accelerated by applied electric fields, and consequently reduce the probability of dielectric breakdown.

There is another type of additive available which is expected to congregate around voids and foreign substances, thereby reducing the local high electric field in their vicinity. Many former additives are chemically unstable, leaving their efficacy in doubt. The latter additives are being developed in order to find ingredients which have less affinity with the parent material, and have a higher dielectric constant than the parent material. Some prospective additives appear to have problems as far as thermal stability and life are concerned, although they may have already proved effective in laboratory treeing experiments.

Although only a few manufacturers produce mineral-filled XLPE power cables, they have some advantage over unfilled cables as regards to thermal conductivity. It is a matter of selecting suitable filling minerals such as calcinated clays, titanium oxides, talc, and calcium carbonate. This type of cable is now under development for DC use because mineral fillers are expected to reduce the influence of space charge formed in the cable insulation, especially near the conductor and the shield.

1.1.7. EHV Extruded Cables
A. France

The first challenge in the EHV cable field from an extruded cable came from a French cable manufacturer.[47] These early products were PE rather than XLPE cables, which have been the principal topic in previous sections. As a matter of course, a determined effort was made to eliminate any imperfections in the polyethylene insulation to the extent that the technology permitted at that time. Successful service performance of polyethylene-insulated cables, albeit in lower temperature operation than XLPE cables, encouraged the manufacture of the first 225 kV in 1966. The first 225 kV feeder with extruded polyethylene insulation was connected to the French network in June 1969. This feeder provides the connection between an overhead line and the bus-bars at the Chevilly substation in the Paris region. The feeder is 90 m long, corresponding to 270 m of cable. The conductors are of copper and their cross section is 805 mm^2. The normal transmitted power is 300 MVA. A cross section of the cable is shown in Figure 1.1.47. According to Reference 47, the composition of the extruded-polyethylene-insulated 225-kV cables is as follows.

The conductor is composed of copper or aluminum strands, whose cross section may attain 1000 mm^2. Where the cross section exceeds 600 mm^2, the conductor is manufactured in segmented form. A semiconducting polyethylene sheath is extruded over the conductor. The purpose of this electrical shield is not simply to eliminate strand effect, but mainly to prevent ionization of the air at the surface of the conductor. For the same reason, a semiconducting sheath is extruded over the polyethylene insulating wall. The characteristics of the semiconducting material must satisfy a number of conditions. Perhaps surprisingly, its mechanical and chemical properties are more important than its electrical properties. It will suffice for its resistivity to be no greater than $10^6 \Omega/cm^2$. After extrusion under normal conditions, elongation should exceed 150%. This elongation is controlled at room temperature within the normal elongation rate for plastic material. To optimize the electrical properties of the cable, extrusions of the semiconducting sheaths and the polyethylene insulation are carried out in the same operation and therefore at the same temperature, i.e., slightly over 200°C. This temperature should not result in any deterioration of the compound and, in view of hot spots in the extruder, it should be able to resist a temperature of 250°C. One of the most important factors is the absence of freed gas at the maximum extrusion temperature. The semiconducting compound must adhere tightly to the polyethylene insulation so that the whole semiconducting sheath, insulation wall, and semiconducting sheath presents no mechanical discontinuity. This condition assumes that any one of the semiconducting compound components may penetrate into the polyethylene insulation. Triple extrusion results in a high-quality cable, and has proven to be essential for EHV polyethylene cables. A 225-kV, polyethylene-insulated cable has been produced by an extruded cable equipment with a 50-m high tower. The system is similar to the tandem operation for XLPE cables described earlier. Precautionary measures are taken to avoid penetration of dust

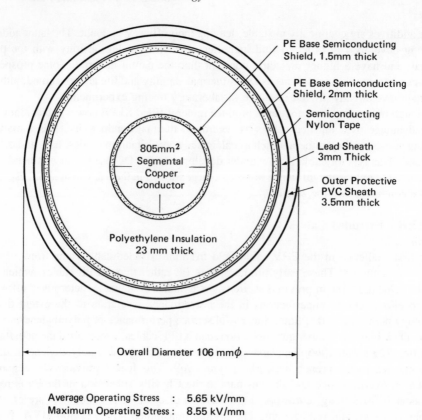

FIGURE 1.1.47. Cross section of a single-core extruded polyethylene cable with copper conductor for a 225-kV service voltage.

particles between the single and double heads. The cable is first cooled in a water-filled pipe, then heated to various temperatures according to the height, and finally passed through a horizontal trough.

The outer semiconducting sheath is covered with a semiconducting nylon tape, a metal shield of either copper tapes or a lead sheath. The cross section of this shield is dependent on the characteristics of the short circuit current which may flow through it. The outer protective sheath is of polyvinyl chloride.

In September 1971, the total length of 225 kV polyethylene-insulated cables in service or being installed in France amounted to 5 km.

B. Japan

Japan, like other countries, has utilized a considerable amount of XLPE cable since 1960, as illustrated in Figure 1.1.48.[29] Long-term acceptance tests of 15-kV XLPE cable were carried out from 1971 through 1976 by certain electric utilities and the 6 major cable manufacturers in Japan. On the basis of their successful field test results, one of the utilities decided to use XLPE cables for its major 154-kV underground systems starting in 1977. Needless to say, cable accessories were developed and tested at the same time.

In this project, like other projects for HV XLPE cables, the following three items were pursued:[29]

1.   Identification of a method to determine the design stress of the insulation according to the Weibull distribution of breakdown stress instead of the method which uses the

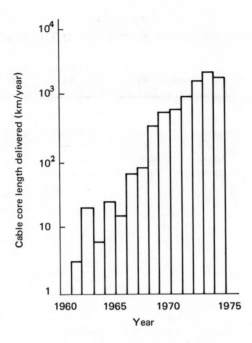

FIGURE 1.1.48.   Deliveries of XLPE cables with ratings from 22 kV up to 154 kV by 6 major cable manufacturers in Japan.

mean value and standard deviation of the normal distribution. A reasonable high-
voltage testing method is proposed for XLPE cables on the basis of agreeable statistical
treatment.

2.   Voids or contaminants in the insulation are probable origins of bow-tie trees, and
therefore some conditions for their generation and their critical influence on cable
performances should be clarified.

3.   In order to improve the performance of XLPE cables, especially void and water content,
newly developed "dry-curing process" — long land dies curing, radiant curing, gas
curing, ultrasonic curing, etc. — are used. The third approach appears to be typical
of the Japanese cable manufacturers.

The outline of 3 tests of 154-kV XLPE cables and accessories which were carried out in
this project are shown in Table 1.1.8.[29] A photograph of one of the 154-kV XLPE cables
developed in this project is shown in Figure 1.1.49. It is manufactured with the MDCV
process. Development of XLPE-insulated cables rated 275 kV and even higher is now
underway, as shown in Figure 1.1.50.

C. U.S.

Development of 138-kV cable has long been considered a practical and desirable objective
in the U.S. in view of the excellent success of XLPE up to 69 kV.[48] Immediate possible
applications envisioned include station get-away, under-pass and aerial ties, as well as main
transmission runs. At 138 kV, however, the relatively heavy insulating walls required
increase the developmental challenge and demand increasingly sophisticated technology in
design and manufacture. The practical aspects of design at the 138-kV level emphasized the
importance of employing higher operating stresses than at 69 kV in order to limit the insulation
thickness to a reasonable value (less than 1 in. or 25 mm). This requirement truly challenges
the manufacturer, since it requires advanced extrusion technology to process satisfactorily

Table 1.1.8.
OUTLINE OF 154-kV XLPE CABLES FIELD TESTS

	Field test #I		Field test #II	Field test #III	
Sample					
Cable					
Conductor area	1000mm²	1200 mm²	1000 mm²	400 mm²	400 mm²
Insulation thickness	18 mm	18 mm	20 mm	21 mm	18 mm
Cable length	600 m	500 m	140 m	100 m	100 m
(approximate)	(in water)	(in air)	(in air and in water)		
Joint					
Tape-wrapped type	3 Sets		1 Set	—	
Prefabricated type	4 Sets		2 Sets	—	
Molded type	6 Sets		—	6 Sets	
Outdoor termination					
Prefabricated type	3 Sets		2 Sets	—	
Epoxy bell-mouth type (OF cable type)	—		—	6 Sets	
SF₆ gas insulated termination					
Prefabricated type	2 Sets		—	—	
Epoxy bell-mouth type (OF cable type)	1 Set		—	—	
Test condition					
Applied voltage (50 or 60 Hz)	116 kV		150 kV	145.5 kV	
Load current	1100—1200 A (16 hr on, 8 hr off)		1000 A (12 hr on, 12 hr off)	700 A (12 hr on, 12 hr off)	
Voltage applied time	7950 hr		8710 hr	3172 hr	
Test period					
Start	March 1973		Sept. 1972	Jan. 1971	
Finish	Under testing (1976)		Feb. 1974	Dec. 1972	
Test location	The Tokyo Electric Power Co., Inc., Higashi Tokyo Substation		The Kansai Electric Power Co., Ltd. Technical Research Center	The Chubu Electric Power Co., Inc. Central Technical Research Laboratory	

a heavy-wall insulation; especially when it is applied over a large compact copper conductor (1000 kcmil), with optimum electrical and physical properties of both insulation and semiconductive shields. Examples of 138-kV cables are shown in Figure 1.1.51. and Figure 1.1.52. In this design, a lead sheath is preferred as the metallic shield rather than helically applied lapped metallic shielding tape. It is claimed that it functions as a reliable shield under fault current conditions and does not impose uneven restraint against expansion of the cable during load cycling, particularly under emergency overload conditions. The lead sheath has an advantage in providing an impervious barrier against ingress of water and earth chemicals into the insulation structure of the cable. It is recognized, however, that such a comprehensive construction may present a more difficult thermal dissipation problem when load cycling. Figure 1.1.53. shows another design of a 138-kV XLPE cable which uses mineral-filled XLPE insulation.

The sophisticated steam-curing process for cross-linking seems to be the American approach to manufacture and design of EHV XLPE cable. As described in a previous section,

FIGURE 1.1.49.   154-kV Cross-linked polyethylene insulated power cable manufactured with MDCV process. (Courtesy of Dainishi-Nippon Cable Co., Ltd.)

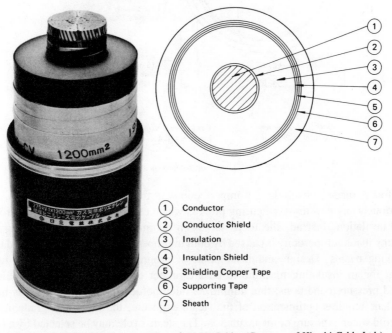

1. Conductor
2. Conductor Shield
3. Insulation
4. Insulation Shield
5. Shielding Copper Tape
6. Supporting Tape
7. Sheath

FIGURE 1.1.50.   275 kV 1 × 1200 mm² XLPE Cable. (Courtesy of Hitachi Cable Ltd.)

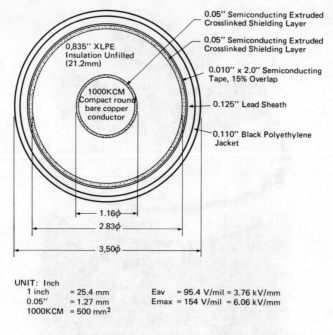

UNIT: Inch
1 inch	= 25.4 mm	Eav = 95.4 V/mil = 3.76 kV/mm
0.05″	= 1.27 mm	Emax = 154 V/mil = 6.06 kV/mm
1000KCM	= 500 mm²	

FIGURE 1.1.51.   Cross section of a 138-kV cable.

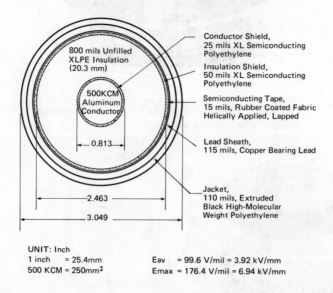

UNIT: Inch
1 inch	= 25.4mm	Eav = 99.6 V/mil = 3.92 kV/mm
500 KCM	= 250mm²	Emax = 176.4 V/mil = 6.94 kV/mm

FIGURE 1.1.52.   Cross section of a 138-kV cable.

every effort is made to exclude any imperfections in the cable insulation. No one has yet taken seriously any new dry-type curing process which claims essentially zero microporocity for wall insulation. Instead, the focus has been on improving the steam-curing process, recognizing that microporocity is caused by both improper extrusion parameters and improper cross-linking cycles. Design and installation of proprietary screws and cross heads are important for an insulating machine to ensure proper compacting of all molecules under controlled pressure conditions; this minimizes microporocity, also. Proper cross-linking of polyethylene requires optimization of the steam exposure, the hot water gradient cooling, and the cold water exposure of the insulation. The steam cycle may be selected by a computer

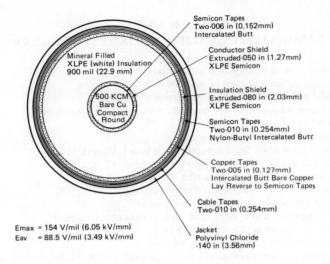

Semicon Tapes
Two·006 in (0.152mm)
Intercalated Butt

Conductor Shield
Extruded·050 in (1.27mm)
XLPE Semicon

Insulation Shield
Extruded·080 in (2.03mm)
XLPE Semicon

Semicon Tapes
Two·010 in (0.254mm)
Nylon-Butyl Intercalated Butt

Copper Tapes
Two·005 in (0.127mm)
Intercalated Butt Bare Copper
Lay Reverse to Semicon Tapes

Cable Tapes
Two·010 in (0.254mm)

Jacket
Polyvinyl Chloride
·140 in (3.56mm)

Mineral Filled
XLPE (white) Insulation
900 mil (22.9 mm)

500 KCM
Bare Cu
Compact
Round

Emax = 154 V/mil (6.05 kV/mm)
Eav   = 88.5 V/mil (3.49 kV/mm)

FIGURE 1.1.53.   Cross section of a 138-kV cable.

program which assures the proper time-temperature chemical reaction time. A correct gradient cooling period is essential in controlling the crystalline structure development within the cross-linked polyethylene wall. The final water cooling leg is necessary to bring the insulation wall and conductor temperature down to ambient before the insulated cable leaves the pressurized vessel.[17]

Development of the EHV XLPE cable is checked by a laboratory qualification test program and further verified by an acceptance test program at Waltz Mill. A rigorous qualification and screening test program has been developed by the ERC Steering Committee with guidance provided by U.S. cable manufacturers. Essential details of the program are outlined in Table 1.1.9.

Some 138-kV XLPE cables have been installed and are now in service. Improvement in manufacturing technology and design are required, but the next real challenge is for 230- and 345-kV cables. Projects for this purpose are now underway. The principal objectives are to reduce insulation thickness and to increase the reliability of the cables.

By way of summary, characteristics of some EHV XLPE and PE cables from various sources around the world are presented in Table 1.1.10.

1.2. OIL-FILLED AND PRESSURIZED OIL-FILLED CABLES

1.2.1. Increase in Transmission Power Capacity

Requirements for the development of oil-impregnated paper-insulated cables, represented by self-contained low-pressure, (medium-pressure and high-pressure) oil-filled cable systems and by pipe-type high-pressure oil-filled cable systems, have been paralleled by the ever-growing demand to transmit larger blocks of electric power to metropolitan areas. Generally speaking, underground cables are inferior to overhead lines as regards transmission capacity, being lower in both voltage and current ratings. Yet, as one power transmission system, they are often required to coordinate with each other.

Thus, bulk power transmission is one of the most important tasks recently facing underground cable technology today. There are two basic ways to increase the power transmission capacity of cables; these are shown in Figure 1.2.1. They are

1. To increase transmission voltage
2. To increase current capacity

Table 1.1.9.
LABORATORY QUALIFICATION TEST PROGRAM

I.   Impulse strength on cable and joint at rated temperature
  a. Three reverse bends on cable at minimum radius
  b. Impulse withstand at BIL, 10 impulses positive and negative
  c. Impulse breakdown, 25% steps from BIL, 6 impulses negative each step
II. DC withstand on cable at 400 kV (5 Eo)
III. Load cycle test program on cable and joint
  a. Measure initial power factor and corona to at least 173% rated voltage
  b. Load cycle at 85 C and 105 C conductor, according to ERC schedule
  c. Measure power factor and corona after 30 load cycles
  d. Measure power factor and corona after 60 load cycles
IV. AC breakdown of cable and joint — after load cycling
  a. Step voltage test, start at 120% rated, 24-hr steps at rated temperature
  b. Steps to be 20% of rated voltage to 160 kV.
  c. Then steps of 10% rated voltage to breakdown.
V.   Physical, aging, and electrical tests according to pertinent IPCEA specifications including eccentricity,
     solvent extraction, and AC/DC resistance determinations in 12-in. steel pipe
VI. Factory tests on reel length cable
  a. Conductor DC resistance at 25°C
  b. Cable power factor at room temperature, measurements to be made at 50% and 173% rated voltage
  c. Strand shield and insulation shield resistance at room, normal, and emergency conductor temperatures
  d. AC withstand, 173% rated voltage for 5 min
  e. DC withstand, 5.2 times rated voltage for 15 min
  f. Corona extinction level at room temperature
  g. Corona discharge magnitude up to at least 173% rated voltage

From Bahder, G., McKean, A. L., and Carrol, J. C., *IEEE Trans. Power Appar. Syst.*, 91(4), 1428, 1972. With permission.

Extruded cables have already replaced some OF-type cables in certain low-, high-, and even extra-high voltage applications as a consequence of recent significant technological developments. It is therefore reasonable to expect that high-current capacity, EHV, and UHV power transmission for bulk power transmission will become the domain of the OF and POF cable.

During its long history spanning nearly 90 years, OF type cable has been the subject of considerable technological innovation and development. It is generally agreed that the fundamental advances were completed during the 1st 50 years. Noteworthy milestones include:[49]

1. The use of impregnated paper
2. The invention of a flexible cable with stranded conductor, lapped-paper insulation impregnated with oil, and directly extruded lead sheath
3. The use of a screening material in intimate contact with the outer surface of the core insulation — carbon paper and metallized paper
4. The application of the principle of pressurization to eliminate the electrical weakness of voids in the insulation

Clearly these developments were directed principally at one of the requirements in Figure 1.2.1., namely, gaining higher dielectric strength and high stability of the cable insulation system.

Technological rather than fundamental advances have characterized cable development during the following 40 years. Three areas in particular can be identified:[50]

1. Increase in dielectric strength
2. Progress with thermal problems
3. Forced cooling and low-loss materials

Table 1.1.10.
EXAMPLES OF EHV EXTRUDED CABLES

	United States of America 1	2	3	4	5	6	France	Sweden	W. Germany	Italy	Holand	Holand	Japan 1	2	3
Conductor — Nominal Voltage (KV)	138	138	138	138	138	138	225	145	110	150	150	154	154	154	154
Material	Copper	Aluminum	Copper	Aluminum	Aluminum	Copper	Copper	Aluminum	Copper	Copper	Copper	Copper	Copper	Copper	Copper
Cross Section (mm²)	500	250	250	200	500	250	800	500 1,000			300	1,000	1,200	1,000	400
Shape	Compact round	Concentric Stranded	Compact round	Concentric Stranded	Concentric Stranded	Compact round	Compact Segmented round					Compact Segmented round	Compact round	Compact round	
Insulation — Material	XLPE	XLPE	XLPE (Filled)	XLPE	XLPE	EP-Rubber	PE	XLPE	PE	EP-Rubber	PE (with VS Agent)	XLPE (Some Cables with VS Agent)	XLPE (with VS Agent)	XLPE	XLPE (with VS Agent)
Net Thickness (mm) (Conductor Shield Thickness)	21.2 (1.3)	20.3 (0.7)	22.9 (1.3)	22.6	20.3 (1.5)	20.3 (0.9)	24.0	20.0 19.0	18.0	22.0	25.0	18.0 (2.0)	20.5 (1.5)	20.0 (2.2)	17.0 (2.2)
Insulation Shield — Material	XLPE	XLPE	XLPE (Filled)	XLPE	XLPE	EP-Rubber	PE	XLPE	PE	EP-Rubber	Varnish + Semi-conducting Tape	XLPE	XLPE	XLPE	XLPE
Thickness (mm)	1.3	1.3	2.0		2.0							0.4~3.0	1.0	1.0	1.0
Metal Shield	Led Sheath 3.2mm	Led Sheath 2.9mm	Copper Tape 0.13mm ×2	Copper Tape	Copper Tape	Flat Square Copper, 34 Longitudinal Lay		Copper Wires Longitudinal Lay	Carbon Paper + Copper Wires	Copper Wires + Braid	Copper Tape + Copper Wires	None Under Water	Copper Tape 0.1mm ×1.2	Copper Tape 0.1mm ×1	Copper Tape 0.1mm ×1
Sheath — Material	PE	PE	Vinyl	Vinyl	Vinyl	Vinyl	Vinyl	Vinyl			Vinyl	None	Vinyl	Vinyl	Vinyl
Thickness (mm)	2.8	2.8	3.6			2.8							4.5	4.5	4.0
Overall Diameter (mm)	89.0	77.5			95.2	76.3		84.0 95.0			92.0~93.0	92.0~97.0	96.0	81.5	74.0
Approximate Weight (kg/km)	18,700				8,300	9,300		7,500 9,900			13,300~14,400	16,000~18,800	16,100	8,600	7,700
Maximum Operating Stress (kv/mm)	5.3	6.9	6.5	6.4	5.9	6.9	8.1	6.9	6.5	6.0	7.0	6.9	6.3	7.1	7.9
Average Operating Stress (kv/mm)	3.8	3.9	3.7	3.7	3.9	3.9	5.4	4.2	3.5	3.9	3.5	4.9	4.3	4.4	5.2

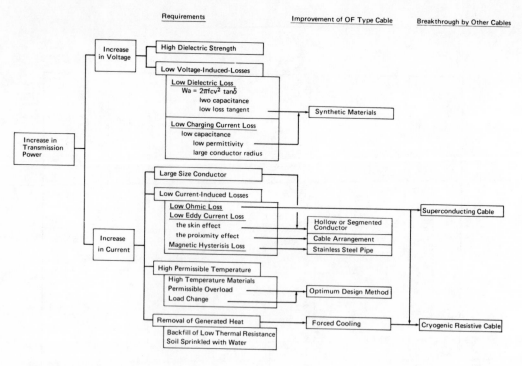

FIGURE 1.2.1.   Various factors associated with the development of bulk power transmission underground cables using oil-impregnated paper insulation.

In the early stages, improvement in the dielectric breakdown characteristics of oil-impregnated paper was a principal concern. It was sparked by the need for higher-voltage cables. This effort was based on the notion that higher-density paper, less than 50 μm thick, when lapped adjacent to a conductor, enabled the maximum operating stress to be increased from 11 kV/mm to 15 kV/mm, with a consequent increase in transmission voltage. This advantage was offset to a certain degree by increases in capacitance and dielectric dissipation factor. This limited cable installation length from the standpoint of economic transmission.

The urgent need for long-distance underground power transmission at high voltages focused attention on item (2) above. In this regard it was recognized that the thermal behavior of impregnated paper was a limiting factor. Lower dielectric loss and the lower permittivity turned out to be the main factors in making such transmission systems a reality, as shown in Figure 1.2.1. Every possibility for reductions in both these properties was extensively investigated. The result was the successful production of low-loss oil-impregnated paper insulation with sufficient dielectric strength. This was accomplished by:

1. Lowering tanδ through washing the paper with deionized water, i.e., by removing ionic impurities from the paper
2. Lowering ϵ and tanδ by reducing the density of paper
3. Maintaining high impulse strength by reducing the air-permeability of paper (reduction in density would lead to a reduction in dielectric strength)
4. Lowering tanδ of impregnant mineral oil or replacing conventional mineral oil by synthetic oil such as polybutene for POF cable and dodecylbenzene for OF cable, and thereby improving the dielectric stability and life-confidence of the system

These items were realized not only through improvement of paper manufacturing processes, but also by development of the low-density paper lapping technique, by lead sheath

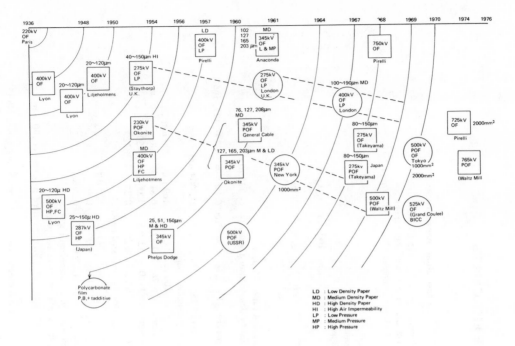

FIGURE 1.2.2.   Evolution of EHV cables.[50]

covering in vacuum, and by mass impregnation. The refining process was improved with respect to conventional mineral oil; at the same time, it was endowed with low viscosity.

Success at this stage was marked in 1961 by the first development of 275-kV OF cables in the U.K. and 345-kV POF cables in the U.S. Subsequently, one witnessed the development of 400-kV OF cables in the U.K. in 1967 and 275-kV OF and POF cables in Japan in 1968. Figure 1.2.2. and Table 1.2.1.[50] show some pioneering examples of EHV cables and their development characteristics which depend on cellulose paper technology.

Heat losses are now recognized as a critical problem for high-power cables in their third stage of development. Every possibility for reducing heat losses from cable systems should be investigated. A number of development items are listed in Figure 1.2.1. One sees from the figure that losses are caused either by current (magnetic field) or by voltage (electric field). The current can induce losses in the conducting materials within a cable system, while the voltage generates dielectric losses.

A. Dielectric Losses

The loss tangent of the best oil-impregnated cellulose paper is in the range of 0.15 to 0.25%, which seems to be the ultimate attainable by technical innovation. One should probably not expect any better oil-paper composite to appear in the near future. Since the losses increase as the square of the voltage, they impose a limit upon the applied voltage which is in the range of 400 to 750 kV; this is indicated in the curve of Figure 1.2.3.[1]

The relatively high capacitance of a cable compared to an overhead line is another limiting factor. The dielectric constant of low-density paper impregnated with oil is 3.4. The cable behaves like a capacitor. It must be charged and discharged cyclically, whenever it is energized. This produces no useful power, only losses. Since capacitance increases with the length of the cable, there is a critical length, which is usually between 20 and 100 km, where the charging current reaches the thermal limit of the cable. This critical length is, of course, dependent on the design and the voltage; in particular, the higher the voltage, the shorter the critical length. Cable capacitance can be compensated by shunt reactors, but these are expensive components.

Table 1.2.1.

CHARACTERISTICS OF REPRESENTATIVE EHV CABLES

Year	Voltage (kV)	Manufacturer/cable type	Pressure (kg/cm²)	Insulation thickness (mm)	Service stress max. (kV/mm)	Insulation characteristics
1936	220	Lyon OF	2.2	24	9.3	AC 38 kV/mm, impulse 0 ~ 115 kV/mm
1947	220	Lyon OF	2.2	18	11.3	Development and utilization of thin paper
1950	380	Lyon OF	15	24	15.7	AC 56 ~ 64 kV/mm, impulse 170 ~ 190 kV/mm
1950	380	Liljeholmens	10	28	14.4	
1955	420	Lyon OF	15	22	15.7	Forced cooling
1957	500	Lyon OF	15	28	14.7	Forced cooling
1957	400	Liljeholmens OF	15	28	14.3	Medium density paper

	50 μm	100 μm	130 μm
ε	4.2	3.8	3.2
Tanδ (%)	0.33	0.28	0.19

Year	Voltage (kV)	Manufacturer/cable type	Pressure (kg/cm²)	Insulation thickness (mm)	Service stress max. (kV/mm)	Insulation characteristics
1958	400	Pirelli OF	2.0	26.5	13	High density thin paper (25 μm)
1960	287	OF	15	21 ~ 22	10.8 ~ 12.8	Paper washed by deionizated water (40 ~ 125 μm)
1961	275	OF	Max. 5.4 / 5.4	16.1	13.0	
1961	345	Okonite POF	14	26.3	11.7	Deionized paper 0.71 g/cc (ε = 3.4) 100°C tanδ = 0.22%, 127, 165, 203 μm
1961	345	General Cable, POF	14	25.8	11.9	Deionized paper, 0.75 ~ 0.80 g/cc, 76, 127, 203 μm
1961	345	Anaconda® OF	5.3	27	12.3	Medium density deionized paper

Thickness	100 μm	125 μm	160 μm	200 μm
Density (g/cm³)	0.78	0.82	0.81	0.81
Impermeability	1920	3200	3200	2450
100°C tanδ	0.22	0.29	0.24	0.22%

Year	Voltage (kV)	Manufacturer/cable type	Pressure (kg/cm²)	Insulation thickness (mm)	Service stress max. (kV/mm)	Insulation characteristics
	345	Phelps Dodge, OF	14	26.3	11.4	Medium and high density deionized paper

Thickness	25 μm	50 μm	150 μm
Density (g/cm³)	1.10	0.90	0.80
100°C tanδ	0.34	0.31	0.27%

Year	Voltage (kV)	Manufacturer/cable type	Pressure (kg/cm²)	Insulation thickness (mm)	Service stress max. (kV/mm)	Insulation characteristics
1964	345	POF	14	26.3	11.7	Medium and low density deionized paper
	500	POF	15	31		Mineral oil for both impregnant and pipe oil
1967	400	OF		22.9	1.5 in.² 15	100 μm high airtightness plus MD deionized paper up to 190 μm
					1.75 in.² 13.5	
1968	275	POF	15	19.5	11.4	Low density, deionized paper, Sun #6 for pipe oil, polybutene HK15E for impregnant

Year	Voltage	Manufacturer / Type				Details
1968	275	POF	15	19.5	11.4	Low density, deionized paper, Sun #6 for pipe oil, polybutene HK15E for impregnant
1968	275 / 500	OF / Okonite POF	6.0 / 200 psi / 14	19.0 / 34	11.3 / 15.8	
1969	500	Anaconda® POF	14	32	17	Polybutene impregnated low density paper conductor shield — metallized Mylar® / metallized paper insulation shield — metallized Mylar®
1969	500	Phelps Dodge POF	14	34	15.8	Impregnant oil — Tridodecyl benzene plus 10 vol% polybutene / Pipe oil — TDB plus 5 vol% polybutene (Amoco® D-50) / Shield — carbon paper; Low density paper — 120 μm, 200 μm / Impregnant oil — polybutene / Pipe oil — polybutene 500~600 SSU / Conductor shield — carbon paper / Insulation shield — metallized paper
1969	500	General Cable POF	14	34	15.8	Conductor shield — carbon paper / Insulation shield — MP / Impregnant oil — polybutene / Pipe oil — polybutene
1974	525	BICC, OF installed in the U.S.	Max. 15.5	30.5	16.3	Conductor 2000 kcmil (or 1000 mm²) / 645 MVA / Smooth sided aluminum sheath
1975	725	Pirreli OF		31.5	20.7	Conductor 2000 mm² / Helically corrugated aluminum sheath

Sub-table (1968, 275 POF):

Thickness	70 μm	100 μm	125 μm	150 μm
Density	0.88	0.71	0.67	0.66
100°C tan δ		0.21	0.23%	

Sub-table (1969, 500 Anaconda POF):

Thickness	100 μm	125 μm	160 μm	200 μm
Density	0.81	0.80	0.74	0.70
Impermeability (Gurley · sec)	3200	2800	1400	1300

Sub-table (1969, 500 General Cable POF):

Thickness	100 μm	150 μm	200 μm
Density		0.75~0.80	
Impermeability		900	

100°C tan δ = 0.01%, 100°C 150 SSU

100°C tan δ = 0.01%, 100°C 90 SSU

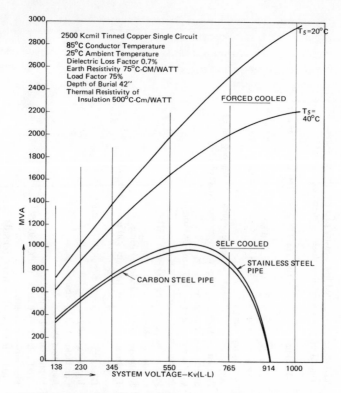

FIGURE 1.2.3.   MVA ratings for single circuit 2500-kcmil copper HPOF pipe cable.[1]

The transmission capacity of a cable is obtained from the steady-state permissible current. Equation 2.1.9. in Section 2.1. of Volume I is about all that is needed to calculate it. The AC resistance of the conductor may be approximate before installation since it includes the losses caused by skin effect, proximity effect, charging current, and if any, the magnetic hysterisis loss. According to AEIC specifications, the conductor temperature should not exceed 80°C for 40- to 345-kV impregnated paper-insulated cables; this is indicated in Table 2.1.5. in Section 2.1. of Volume I.

B. Current-Induced Losses

One way of meeting the requirement for increased current capacity in bulk transfer circuits has been by increasing the conductor cross-sectional area. The maximum conductor size presently manufactured is 2500 to 3000 mm² (5000 to 6000 kcmil). With such large conductors, losses due to the skin effect become critical, and an extensive technological effort has been made to reduce them. The conductor should be segmented.

For OF cables, two designs of hollow conductors are common.[51] One, known as the Conci (or Voussoir) type, is formed by layers of segmental strips stranded in alternate direction and with the selfsupporting inner layer providing the oil duct. The other design is an adaptation of the well-known Millikan construction. It usually has six segments insulated from each other, laid around a central duct, often containing a supporting spiral, and bound together with bronze tapes. One obvious way to minimize the skin effect is to increase the duct diameter, an approach which is particularly effective with the Conci type. Thus a 1600-mm² copper conductor of this kind is found to have a skin effect at 85°C of 21% with 12-mm duct and of 6% with a 40-mm duct. With Millikan conductors, because the elemental wires of each sector are transposed, the increase in duct diameter has a very small influence. The skin effect depends essentially on the contact resistance between the elemental wires and hence can vary considerably with the manufacturing technology used. A representative value for a good quality 2500-mm² copper conductor at 85°C with a 25-mm duct, is about 14%, but values as high as 30% are unfortunately likely in the present stage of development.[4]

Eddy currents also induce losses in the screens and sheaths of the conductors. In a three-phase, single-conductor cable system, sheath losses are significant and can be larger than dielectric, ohmic, skin effect, and proximity effect losses. They can be reduced by bonding, i.e., single point bonding or cross bonding as described in Section 2.1.6. of Volume I.

POF cables usually possess higher losses than comparable OF cables, owing to the proximity effect of the three close conductors and the eddy currents and magnetic hysterisis in the steel pipe. Because of the increased losses, the largest size of copper conductor that can be used is 2500 kcmil (1250 mm²).[1] The pipe loss and the AC resistance of the conductors may be reduced by using stainless steel pipe. When these losses are reduced, it is possible to use larger than 2500 kcmil copper conductors; sizes up to 3500 kcmil (1750 mm²) become feasible. The use of a stainless steel pipe in place of carbon steel increases the transmission capacity as indicated in Figure 1.2.3. Even glass-reinforced epoxy-resin pipe can be used where economics permit, although further technical-feasibility study is really needed. Conductor losses due to skin effect can be reduced by using enamel on the individual strands. If aluminum conductors are used, sizes up to 3500 kcmil are permissible because the high contact resistance between strands due to the inherent oxide coating reduces the skin effect.

C. Limit of Transmission Voltage Set by Thermal Runaway

Impregnated paper-insulated cables can exhibit thermal runaway as described in Section 2.3.3. of Volume I. EHV or UHV cables especially, with their thick insulation, may be subject to thermal breakdown originating in the positive temperature dependence of tanδ. This was confirmed by experiments which showed that a 500-kV OF cable failed at 930 kV under open-air installation, but at 680 kV when it was wrapped with glass wool. Critical conditions can be derived from the Equation 2.3.8. of Volume I, the temperature dependence of tanδ, and some consideration of thermal conductivity. One calculation shows[52] that thermal runaway is initiated in a 275-kV POF cable with conductor temperature 80°C when tanδ reaches 0.35%. The loss tangent is usually 0.15 to 0.20% for a new cable but probably increases with time due to thermal degradation of oil-impregnated paper. Such thermal breakdown takes days to weeks, and is therefore called a long-term thermal breakdown. Figure 1.2.4. compares the calculated dependence of the long-term thermal breakdown voltage on the external thermal resistance for a 500-kV POF cable with 2 experimental results.[52] It is also claimed, on the other hand, that under certain conditions the permissible current determined by thermal instability is greater than that determined by thermal degradation. Careful consideration should be given to this new limiting factor, especially when developing UHV cables with conventional cellulose paper.

D. Heat Dissipation

All the losses described so far are ultimately transformed into heat. Therefore, besides the reducing losses, one should aim at increasing the heat dissipation as another means for enhancing bulk power transmission. The ampacity is thus determined by the permissible loss per unit length, which depends on the capability of the cable to dissipate the heat caused by the losses to the environment.

Average values for thermal resistivity are in the range of 70 to 150°C cm/W, but may rise to 300°C cm/W or even higher when the soil dries out at temperatures above 40°C. The thermal resistivity should be kept constant during cable operation, which requires that the surface temperature of the cable not significantly exceed 40°C. This corresponds to a conductor temperature below 85°C, which however, unfortunately means a reduction in the ampacity to less than half its maximum value. It is also necessary to provide the cable with a stable longitudinal thermal environment; otherwise, high-temperature regions called hot

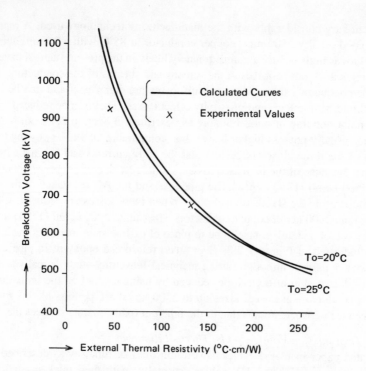

FIGURE 1.2.4. Relations between a long time AC breakdown voltage and external thermal resistivity in 500-kV POF cables. (From Fukagawa, H., *Research on Transmission Capacity Enlargement of EHV Cables*, (in Japanese), Review of CRIEPI No. 28, Central Research Institute of Electric Power Industry, Tokyo, 1975, 1. With permission.)

spots are created which limit the ampacity of the cable system. For this purpose, the cable in heat removal are to be expected from advanced technologies which incorporate forced cooling with oil, water, or even liquid nitrogen.

1.2.2. Designs of 500-kV Class Cables
A. Self-Contained Cables

During the past 10 to 15 years, the AC maximum service voltage has been increased to 550 kV and tests are in progress at 750 kV. The sustained power rating per 3-phase circuit has reached a figure of 1300 MVA when naturally cooled and about double this figure when force cooled.[51]

Figure 1.2.5. shows a cross section of 525-kV cables installed at Grand Coulee Dam on the Columbia River in the state of Washington. It is approximately 6400 ft (1950 m) long.[53,54] It is a self-contained aluminum-sheathed oil-filled cable with cross-bonded sheaths to eliminate sheath circulating currents. A concentric stranded conductor of 2000 kcmil (1000 mm²) was used in the belief that it has a more circular cross section and a better surface than a compact segmental conductor of the same cross-sectional area, thereby producing a cable with a superior electric strength. The conductor comprises a layer of self-supporting plain copper segments forming the central oil duct, surrounded by four layers of flat copper wires. The insulation comprises papers with thickness from 75 μm next to the conductor to 250 μm at the outer screen. Other important characteristics of the papers are similarly graded through the insulation wall; the inner papers next to the conductor have a high impermeability to air and a high density to meet impulse voltage requirements, while those at the outside have a low density to minimize dielectric loss. To ensure low losses at high temperature,

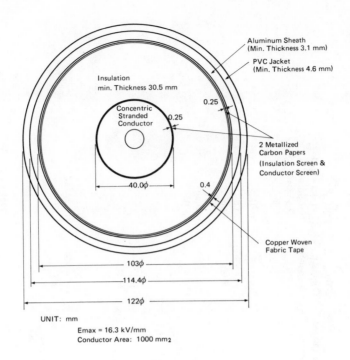

FIGURE 1.2.5.   Cross section of a 525-kV self-contained oil-filled cable.

papers made with deionized water are used. The impregnant is a low-viscosity mineral oil. The power factor of the cable system is 0.23% at 360 kV and ambient temperature. An aluminum sheath is used to accommodate the maximum hydraulic pressure, 1550 kN/m^2 (15.8 kg/cm^2), and transient pressure, 2250 kN/m^2 (23.0 kg/cm^2), that could occur in the cable system. A smooth-sided sheath is preferred here to a corrugated sheath because of its proven reliability.

After several years of cooperative research and development between the Tokyo Electric Co. and leading cable manufacturers in Japan, a number of getaway 500-kV OF cable systems have been installed in several places.[55-59] Table 1.2.2. shows some specifications and characteristics of 5 designs of 500-kV selfcontained oil-filled cables installed at 3 different locations.

More details are presented on design B, but the other designs are similar. A cable circuit with a transmission capacity 1100 MVA/cct (ampacity 1270 A/cct), comprises 500 kV 1 × 2000 mm^2 aluminum-sheathed and PVC-jacketed oil-filled cables, sealing ends with 5.4 m porcelain bushings, oil-immersed type sealing ends, and an oil-feeding system. The conductor consists of six-segmented, compact, round, tin-plated, copper wires bound by stainless steel tapes and carbon papers. The insulation, made of deionized papers, is graded from 100 μm to 200 μm and is impregnated with hard-type alkylbenzene synthetic oil. Double-layered paper, which is semiconducting on one side and insulating on the other side, is applied both between the conductor shield and the insulation and between the insulation shield and the insulation. The insulation is 33 mm thick and is therefore subjected to 14.2 kV/mm (355 V/mil). The insulation shield is made of a combination winding structure of metallized paper and carbon paper. The aluminum sheath must be 155 mm in outer diameter, but it should be stated that there has been little experience with such a large corrugated tube. For this reason, the pitch and height of the corrugation were determined from the results of repeated bending tests with reference to the permissible bending radii as shown in Table 1.2.3. Jackets are made of semi-hard PVC protective layers to a thickness of 5.5 mm which is 1 mm larger than 4.5 mm for 275-kV OF cables. This was chosen by taking account of the outer diameter

Table 1.2.2.
LIST OF SPECIFICATIONS FOR FIVE DESIGNS OF 500-kV SELF-CONTAINED OF CABLES

Items	A[55a]	B[56b]	C[57c]	D[58b]	E[59b]
Conductor					
Nominal cross section (mm²)	1600	2000	1000	2000	1200
Shape	Compact round, 6 segments	Hollow, 6 segments	Compact round, 6 segments	Compact, 6 segments	Compact round, 6 segments
Outer diameter (mm)	54.0	59.1	42.7	60.4	47.9
Binder thickness (mm)		~0.7			0.5
Oil duct					
Inner diameter (mm)	18.0	18.0	14.0	18.0	18.0
Thickness (mm)	0.8	0.8	0.8	0.8	0.8
Outer diameter (mm)	19.6	19.6	15.6	19.6	19.6
Insulation					
Total thickness (mm)	33.0	Grading 33.0	33.0	33.0	Grading 33.0
Inner carbon paper (mm)			0.2	0.2	
Outer carbon paper (mm)			0.3	0.3	
Screens (for conductor and insulation)					
Carbon paper (mm) }	0.85	~0.25	—	0.15	0.3
Metallized paper (mm) }				0.15	
Copper woven fabric					
Tape thickness (mm)	—	—	—	0.5	0.5
Sheath	Al, corrugated	Al, corrugated	Pb	Al, corrugated	Al, corrugated
Thickness (mm)	3.1	3.2	4.3	3.2	2.9
Height (mm)	8.0	~8	Reinforced by stainless and cloth tapes	6.4	
Pitch (mm)		~49			Corrugate radius 33.8
Outer diameter (mm)	144	~155	119		
PVC Jacket					
Thickness, min (mm)		5.5	6.0	5.5	5.5

Cable diameter (mm)	157	~165	145	~180	~149
Cable weight (kg/km)	—	~41,000	51,000	~42,000	~32,900
Conductor resistivity 20°C max. (Ω/km)	—	—	—	0.00913	0.0151
Capacitance max. (μF/km)	—	—	—	0.247	0.24
Insulation resistivity 20°C min (MΩ/km)	—	—	—	110,000	30,000
Max. electr. stress (kV/mm)	—	14.2	—	—	15 ~ 16

[a] Fukushima Nuclear Power Generating Station.
[b] Sodegaura Thermal Power Generating Station.
[c] Okutataragi Pumped Power Generating Station.

Table 1.2.3.
PERMISSIBLE BENDING RADII OF CABLES

Kind of cable	Core number	Permissible bending radius
SL	—	7.5 d
Lead sheath	Single	15 D
	Three	10 D
Aluminum sheath	Single[a]	15 D
	Three	12 D

Note: d: Cable outer diameter; D: Outer diameter including metal sheath.

[a]   Below 154 kV.

(165 mmφ) and weight (41 kg/m) of the cable and the anticipated surge voltage on the sheath. Some test results are shown in Table 1.2.4.[56] Figure 1.2.6. shows a photograph of a 500-kV OF cable.

B. Pipe-Type Cables

The dominant design for bulk transmission of power in the U.S. has been the high-pressure, oil-filled, pipe system, which accounts for fully 90% of the 3000 circuit miles of cable from 69 to 345 kV. Since successfully completing research and development on 345-kV cables in 1965 at the Cornell University site, approximately 170 km of 345-kV pipe-type cable have been installed and successfully operated, although several failures have occurred in joint casings after 7 years of service.[60] Considering its limitations, which include costly terminations, splicing times that measure in weeks, and the problem of heat dissipation from three single-phase conductors within a common thermal envelope, the popularity of pipe-type cable might seem questionable. Nevertheless, its reliability has been proved, and excavations in urban areas where undergrounding is used most, are opened long enough to install the required steel pipes.[61]

Pipe-type cable consists of three single-phase conductors, each individually insulated and shielded and contained within an oil-filled steel pipe. Usually, the conductor is made of compact segmental strands for sizes larger than 600 mm² (1200 kcmil) and of compact round strands for smaller sizes. A layer of high-tensile strength bronze or stainless steel tape holds the segmented strands together. There is no need to provide any oil duct as is the case for self-contained OF cable. The maximum electric stress governs the choice of insulation thickness as it does for self-contained OF cable. EHV cables use high-grade cellulose paper washed by deionized water. No metal sheath is needed; a moisture seal is used instead. This is necessary to prevent the cable insulation from absorbing moisture while the cable is exposed to the atmosphere prior to completion of its installation The self-contained OF cable has no such problem. A single-sided, metallized, polyester tape is often applied as a moisture seal since it has excellent properties as regards moisture impermeability and oil-resistance. Also, because it is metallized, it facilitates electrical contact between the electrostatic screen and the reinforcing layer. Requirements for the moisture seal and for radial oil flow resistance are clearly contradictory. One of the two layers are often applied, each layer having two tapes. A reinforcing layer, made of stainless steel tape and polyester tape intercalated with each other is frequently applied. This also helps to seal against moisture. A skid wire is wound around each single-phase cable to reduce friction and thereby promote the easy and safe pulling-in of the cable. The pitch and number of skid wires should be determined

Table 1.2.4.
PERFORMANCES OF A 500-kV OF CABLE

Test Items		Test Results		Standard Requirements	
Frame Tests					
Conductor Resistance (20°C) (Ω/km)		0.00906 ~ 0.00913		<0.00943	
Cable Capacitance (20°C) (μF/km)		0.252 ~ 0.253		<0.26	
Insulation	Resistance (20°C) (MΩ–km)	67,000 ~ 77,000		>38,500	
	Withstand Voltage (kV·min)	420·10	Good	420·10	
Jacket	Resistance (20°C) (MΩ–km)	67 ~ 79		20	
	Withstand Voltage (kV·min)	5·1	Good	5·1	
Loss Tangent	Measuring Voltage (kV)	318	420	318	420
	tanδ (%)	0.160 ~ 0.161	0.168 ~ 0.170	>0.20	>0.22
Gas Test Values		0.0251 ~ 0.0255		0.05	
Bending Test Values		Good		<Outer Radius × 25	
Sample Tests					
Pulling Eyes Tensile Strength (ton)		Good, Fracture at 46.5 tons		24ton·2 hrs	
Insulation* Withstand Voltage	Long–Time AC (kV·hrs)	Good, Pass at 810 kV·3 hrs.		690 kV·6 hrs.	
	Switching Surges (kV×times)	Good		–1,310kV × 3 times	

Table 1.2.4. (continued)
PERFORMANCES OF A 500-kV OF CABLE

Test Items		Test Results				Standard Requirements					
Insulation Withstand Voltage	Lightning Impulse (kV•times)	Good, BD at 2,310kV × 3 times									
	"	Good, BD at 2,400kV × 1 time				30kV•1 min.					
Jacket W. Voltage	Short-time AC (kV•min)	Good, BD at 60kV•25 sec									
	Impulse Voltage (kV×times)	Good, BD at −135kV × 3 times				−60kV × 3 times					

	Measuring Voltage (kV)	318	500	550	635	318kV ∿ 635kV
	(8°C)	0.170	0.176	0.180	0.195	<0.22
Temperature Dependence of Loss Tangent (%)	40°C	0.155	0.161	0.165	0.182	"
	60°C	0.150	0.157	0.162	0.179	< 0.20
	70°C	0.149	0.156	0.161	0.178	"
	80°C	0.149	0.157	0.162	0.178	"
	85°C	0.151	0.160	0.166	0.183	"
	95°C	0.160	0.169	0.175	0.193	"

Ionization Coefficient (%)		0.010 ∿ 0.015						<1% (550kV ∿ 318kV)				

Insulating Oil	Viscocity	Temperature(°C)	0	20	40	60	80	90	0	20	40	60	80	90
		(CP)	50.8	17.1	7.4	4.0	2.6	2.2	<110	<34	<14	<7.8	<5.0	<3.6
	AC Breakdown Voltage (kV)	60.5kV (85°C), 60.9kV(95°C)							>30 KV/2.5 mm					
	Corrosion Resistivity	1A							Color Change Code <1					

Test Items		Test Results	Standard Requirements
Vinyl	Tensile Strength (kg/mm^2)	2.03 ∿ 2.22	> 1.8
	Elongation (%)	240 ∿ 260	> 200
	Hardness (Shore-D)	64 ∿ 65	> 40
Cable Oil Pressure Test		Good, 51kg/cm^2·1 min. Facture.	11kg/cm^2·30 min.

* Cable with its terminations.

FIGURE 1.2.6.   View of a 500kV OF cable. (Courtesy of the Furukawa Electric Co., Ltd.)

according to the outer diameter of the cable of interest; the number naturally increases with increase in cable size. Preferably 30 to 40° should be chosen for the angle between the skid wire and the cable axis. In actuality the pitch is 100 to 150 mm for double entry skid wires, around 200 mm for triple entry, and 200 to 250 mm for quartic entry.

Table 1.2.5. illustrates several representative designs of EHV POF cables;[6,62] the last two cables were designed for the Cornell cable testing project.[62] These cables were designed to carry 500 MVA or more depending on governing condition such as soil, ambient temperature, and load factor, being installed without forced cooling or any other means for temperature control.

As shown in Table 1.2.6., 4 designs (A—D) of 500-kV HPOF (high-pressure, pipe-type, oil-filled) cables were developed in connection with the acceptance test program for 500-kV class cables at Waltz Mill.[63-70] These cables can carry 800 MVA without forced cooling. We shall briefly describe the design C[66] which is shown in Figure 1.2.5. The 2000-kcmil segmental copper conductor is made up of 4 segments, each having 244 tin-coated strands. The compact conductor was selected for its advantageous space factor. Individual segments are prespiralled to facilitate assembly into a single unit. The conductor shield comprises 5 carbon black paper tapes and 1 duplex tape butted together over a 5-mil tinned bronze binder. There are 22 5-mil tapes, 64 6.5-mil tapes, and about 32 8-mil tapes, with 12 reversals throughout the wall. The insulation shield comprises 2 5-mil metallized paper tapes, intercalated face out, then 2 intercalated metallized Mylar® tapes, over which are applied 2 5-mil tinned copper tapes with opposing lay, each intercalated with a Mylar® tape. This shield construction also acts as a moisture seal. Completing the cable assembly are 2, half-round,

Table 1.2.5.
SOME DESIGNS OF EHV POF CABLES

Nominal Voltage (kV)	154	275	345	345
Conductor				
Nominal cross section	400 mm²	1400 mm²	2000 kcmil	2000 kcmil
Outer diameter	24.0 mm	46.0 mm	1.63 in.	1.63 in.
Shape	Compact round strands	4-Segmental compact round strands	4-Segmental coated copper conductor	127 Wire concentric coated with enamel, except 6-in. outer layer tin coated, conductor
Binder			Coated bronze tape	
Conductor shield	CBP, MP	CBP, MP	Metalized paper strand shielding	Metalized paper and CB strand shielding
Insulation				
Nominal thickness	13.5 mm	19.5 mm	1.035 in.	1.000 in.
Maximum stress	~ 9.83 kV/mm	~ 11.24 kV/mm	~ 298 V/mil	~ 305 V/mil
Insulation shield			Metalized paper	Metalized paper
Reinforcing layer			Coated copper tape, Mylar® tape and Mylar® intercalated with dacron tape	
Moisture seal				
Thickness	~ 0.3 mm	~ 0.3 mm	Coated copper tape intercalated with metalized Mylar® tape	Metalized Mylar tinned copper tape intercalated with plain Mylar® tape and plain Mylar® tape
Reinforced layer				
Thickness	~ 0.5 mm	~ 0.5 mm		
Skid wires			Coated brass	Polyethylene
Height	2.5 mm	2.5 mm		
Single cable				
Outer diameter	60 mm	92.5 mm		

Note: CPB = Carbon black paper; MP = metalized paper.

high-density polyethylene skid wires, 100 mil × 200 mil, applied double entry with 3.0-in. lay.[66]

Design D has a covering comprising a moisture seal and shielding assembly. This consists of two pairs of aluminum, foil-backed, polyester tapes, intercalated face-to-face, together with another two pairs of tin-coated, copper tape, intercalated with a polyester tape. The polyester tape of the outer layer is extra wide in order to provide complete outer surface coverage of its companion tin-coated copper tape. All of this is in addition to the insulation shielding which has 2 3.5-mil perforated, foil-backed, paper tapes, intercalated with metal faces out.

Only for the design B, was the five-segment conductor chosen.[65] This was because of concern that the common four-segment conductor might go out of round, thereby creating a disturbance in the uniform taping with the resultant formation of soft spots. In place of paper tapes, traditionally used, this design makes use of thin nylon tapes which it is hoped will reduce the size difference between insulated and bare segments, and the tape crushability.

1.2.3. Possible Improvement of the Oil-impregnated Paper System

A. Electrical Performance

Oil-impregnated paper insulation based on the present cable technology certainly provides

Table 1.2.6.
FOUR PROPOSED DESIGNS OF 550-kV HPOF CABLES

Items	Design			
	A	B	C	D
Conductor Size and type	2000 MCM (1000 mm²) compact segmental tinned copper			
Fillers	4 Segments 2 Insulated	5 Segments 3 Insulated with thin nylon tapes	4 Segments 2 Insulated paper fillers	4 Segments 2 Insulated jute fillers
Segment Binder	Tinned bronze tape (TB tape)			
Conductor shield	3 MP tapes	CBP tape strand, 30 mil	One 5-mil TB tape intercalated with one 6-mil paper tape / 5 CBP tapes and one Duplex CBP tapes	One 5-mil TB tape intercalated with one 5-mil paper tape / Two 5-mil CBP tapes and one 5-mil Duplex CBP tapes
Insulation/impregnant Thickness	1.340in. (34 mm)	1.250 in. (31.8 mm)	1.340 in. (34 mm)	1.340 in. (34 mm)
Maximum stress	400 V/mil (15.8 kV/mm)	425 V/mil (17 kV/ mm)	400 V/mil	400 V/mil
Impregnant oil	Polybutene	Polybutene with tridecylbenzene	Paraffinic oil	Polybutene
Insulation shield	Metalized paper tapes plus tinned copper tapes			
Moisture seal	Metalized Mylar® tapes			
	Coated copper tapes intercalated with metalized	Metal foil backed CBP tapes; tinned copper tapes	Two intercalated 5-mil MP tapes	Two 3.5-mil perforated foil backed paper tapes inter-

	Mylar® tapes (MM tapes)	Two Intercalated 2.5-mil MM tapes; two 5-mil tinned Copper shielding tapes, each with a 2-mil Mylar® tape; tinned copper tapes applied in opposite lay	Double layers of 2 intercalated 2.5-mil foil backed Mylar® tapes; double layers of a tinned copper tape intercalated with a Mylar® tape
			calated metal faces out
Skid wires	Stainless steel wires insulated with high density polyethylene		
	High density polyethylene	High density polyethylene	High density polyethylene
Diameter			
Conductor diameter	1.63 in.	1.61 in.	1.632 in.
Plus insulation	4.13 in.	4.32 in.	4.342 in.
Nominal cable diam.		4.68 in.	4.704 in.
Pipe diameter	12.75 in. (323.8 mm) O.D.		

Note: CBP = Carbon black paper; MP = metalized paper; MM = Metalized Mylar®.

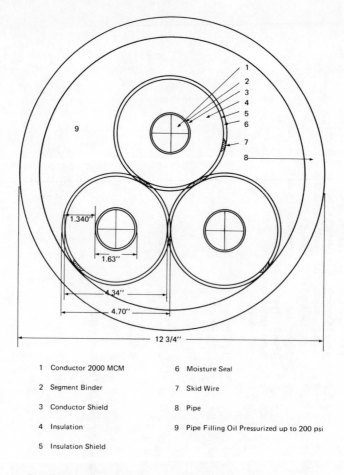

1 Conductor 2000 MCM   6 Moisture Seal

2 Segment Binder   7 Skid Wire

3 Conductor Shield   8 Pipe

4 Insulation   9 Pipe Filling Oil Pressurized up to 200 psi

5 Insulation Shield

FIGURE 1.2.7.   500-kV HPOF cable with 2000 MCM conductors.

500-kV cable with a transmission capacity of 950 MVA in the selfcooled mode. This arises partly because it withstands an electric stress of 15 kV/mm or more at the conductor surface and has a dissipation factor of less than 0.2%.

For a 750/765 kV cable, assuming a BIL of 2000 kV and dimensions appropriate to a maximum stress of 100 kV/mm, the maximum AC stress at the operating voltage is 20.7 kV/mm.[72] As far as dielectric strength is concerned, it is well established that high-grade cellulose paper withstands the electric stress. Such a high AC working stress causes a significant increase in dielectric loss, which turns out to be the limiting factor as far as transmission power capacity is concerned. It is therefore imperative to strive for as low a loss angle as possible for the insulation. Searching for new materials with low loss tangent is one research field, but as of now it is reasonably claimed that there is no better compromise than pure cellulose paper. It is certainly worthwhile making an effort to obtain high-quality cellulose paper with a loss tangent approaching 0.1% for this UHV cable, notwithstanding the fact that forced cooling methods have been exploited to remove the heat generated by power losses in both the conductor and the insulation.

Considerable improvements have been obtained by adjustments in values of air-imperme-ability and density in multi-ply papers. One such paper consists of a relatively thick low-density layer to give a lower dielectric constant, sandwiched between thin very-high im-permeability layers to give the high dielectric strength required. The reduction in dielectric loss is in the range of 15 to 25%. A value of 0.13 to 0.15% at 80°C was obtained for 132-

kV cable. Unfortunately, as expected, the thermal resistivity of the insulation was about 5% higher owing to the lower density of the paper. This was coupled with some reduction of mechanical strength.[51]

In the event that there is no further improvement in dissipation factor of cellulosic paper, increasing the voltage to 765 kV would yield no increase in MVA capability of HPOF cable. This is a consequence of the rapidly increasing dielectric loss with temperature, or in other words, from thermal instability of the insulation. Thermal instability is more likely to occur in cable splices because thicker insulation increases thermal resistance.[70]

At present it is generally accepted that the conductor temperature of oil-filled cables should not exceed 85°C under normal operating conditions. This value, though somewhat arbitrary, has been chosen to limit thermal deterioration of the cellulose paper insulation to an extent which is compatible with a desired life expectancy of around 40 years. Perhaps it is time to reexamine whether or not this temperature limitation is too restrictive; increasing the permissible temperature would increase the cable transmission capacity.

With this in mind, thermal aging experiments have been carried out in order to reach some conclusions concerning:[73]

1. The effect of temperature on the life expectancy of oil-filled cable insulation under continuous conditions
2. The significance of short-term overloading in relation to the life expectancy of the cable system

The conclusions were as follows, if the criterion for cable life is taken as:

• Either the time to reach 10% of the original value of any of the mechanical or physico-chemical characteristics
• Or the time taken to saturate the impregnant with CO (the least soluble gas generated),

then at 100°C, short lives between 3 and 5 years, are indicated. Alternatively, if a 40-year life is required, then operation at 100°C should be limited to around 1 month per year. The data obtained suggest that more than 10 years of life should be expected when the cable is operated at 90°C. This is probably a conservative estimate, the actual life expectancy is most likely greater than this.

A by-product of the thermal aging of cellulose is water, which in turn accelerates the aging. The ability of alkylates to absorb water vapor under electrical stress produces a protective action for the cellulose. It is claimed that, with alkylate impregnants, an increase of operating temperature from 85°C to 90°C is amply justified and that an emergency temperature of 120°C to 125°C for a total of a few hundred hours in the life of the cable is acceptable.[51]

B. Taping

In the taping design of an EHV cable there is only one substantial trade-off — compactness. A tightly taped cable has three advantages:[65]

1. It resists damage in handling
2. It has higher dielectric strength than even a perfectly, but loosely, taped cable
3. It has a higher thermal conductivity than a loosely taped cable

On the other hand a compactly taped cable has the disadvantage of higher dielectric constant.

EHV cables must include reversals in the direction of taping — to produce a torque-free cable for one reason. Usually the tape direction is reversed at each section (every 10 to 12 tapes) in the taping machine. At a reversal there is a multiplicity of diamond-shaped oil-filled gaps with a radial dimension of two tape thicknesses. Such gaps are considered to be dielectrically weaker than impregnated paper. It is better to push the first reversal out some distance from the conductor to a region of lower stress. Applying several sections unidirectionally runs the risk of a complete tape registration, which results in a mechanically and electrically weak spot. However, this effect may be minimized by modifying the taping machine.[65] One example has the first reversal at the 34th tape; this is about 25% of the way down the stress slope.

1.2.4. Thermal Expansion and Contraction[4,6,69,74]

A cable must be designed in such a manner as to absorb or withstand the mechanical thrusts developed due to temperature change caused by daily and seasonal load cycles. Therefore the control of the mechanical forces and movements resulting from heating and cooling impacts design in thermo-mechanical terms. High-power cables which necessitate large conductor size require more sophisticated design criteria than low- and medium-power cables.

Design concepts are determined by the type of cable installation which, in thermo-mechanical terms, is either rigid or flexible over the entire length of the cable, as shown in Table 1.2.7. In the rigid system, the cable is held in such a manner that virtually no lateral movement occurs and it absorbs the thermal expansion by developing a high internal compressive force. Nevertheless, relative movement between cable cores and sheaths is possible. In the flexible system, expansion is permitted but it is controlled in such a manner as to preclude excessive strain in any of the cable components, and hence a short fatigue life, as a consequence of movement. The component usually most affected by these strains in self-contained cable is its sheath.[74]

In most instances, there are several possible systems of cable support which can be used; the choice of system will depend on power-loading and economic factors. Further, there may be positions along a cable route at which changes in lateral support occur. Such transition points require special consideration.

For a flexible system, the limiting factor is likely to be sheath strain or fatigue, while for rigidly supported cables it is generally the force exerted on accessories, or cleats, or on the cable insulation at bends. The former exhibits a dependence on the frequency and amplitude of the temperature cycles which the system will experience. The latter is affected mostly by the maximum temperature reached by the conductor and by the rate of temperature rise. It is normal practice to design the system to withstand the maximum force resulting from the switching of full load onto a new cable system. A lower compressive force will occur with subsequent application of full load because of conductor creep.[74]

In a cable system comprising rigid and flexible sections, the considerable difference in conductor thrust between the two sections will lead to core movement from the rigid to the flexible section as the conductor temperature increases. Such relative movement between conductor and sheath can damage the insulation and core screens and can cause excessive sheath strains at the point of discontinuity.

A. Calculation of Thermal Expansion and Contraction
a. Horizontal Installation

Let us consider the horizontal cable installation that is divided into two parts at a fixed point, as shown in Figure 1.2.8.(A). When heated, a conductor tends to expand due to the thermal force. If creep relaxation of the cable conductor is ignored, this force is proportional to the conductor size A, the coefficient of linear expansion α, the effective modulus of elasticity of the conductor E, and the temperature rise ΔT. Under practical conditions, heating transients are generally of such long duration that creep relaxation cannot be neglected.

Table 1.2.7
CABLE TYPES AND INSTALLATION METHODS

Cable type	Installation method	Disposition	Devices	Most likely movement and problems
			Flexible	
POF		Horizontal	Flexible joint for pipe U-shaped or flat bellows	Paper tape displacement in the neighborhood of manholes
OF	Air Unfilled troughs	Horizontal snake Horizontal	Widely spaced cleats Horizontal sagging Vertical sagging — concrete sadles Swival cleats, trefoil cleats	Snaking movement Horizontal movement
	Ducts, pipes	Horizontal	Large manholes, offset training of cables and accessories	Cable movement into manhole when heated and into duct when cooled
			Rigid	
OF	Buried	Horizontal and incline	Compacted backfill	Relative movement of conductor and sheath especially near cable ends
	Filled troughs Unfilled troughs	Horizontal and incline Incline vertical	Compacted backfill cement bound sand Closely cleated	Vertical movement to lift trough lids
	Air	Deep vertical shafts	Supports	High internal compressive force Spacing — Euler's buckling theory

Expansion is resisted by all mechanical restraints on the conductor. These include restraint K_A at the manhole on the first half of the cable, the frictional force μWx between the insulating papers and sheath, and, if any, residual stress H on the cable. It is clearly seen that in the middle of a long length of cable each element of the cable is in equilibrium, but toward the ends the frictional forces are unable to counteract the conductor expansion. As a result, the conductor nearest the ends will move in relation to the sheath if not prevented by appropriate anchoring.

The force exerted on the point 0 in the first half length of the cable is given by

$$F = EA\alpha\Delta T - (\mu Wx + K_A) - H \qquad (1.2.1.)$$

where μ = coefficient of friction and x = the distance from one end of the cable or from the manhole. The incremental expansion Δm in the small length Δx is therefore given by

$$\Delta m = \frac{1}{EA} \left\{ EA\alpha\Delta T - (\mu Wx + K_A) - H \right\} \Delta x \qquad (1.2.2.)$$

The residual stress may be zero at the first expansion, but is equal to $(\mu Wx + K_A)$ in subsequent cycles. Lengths of thermal expansion and contraction are then

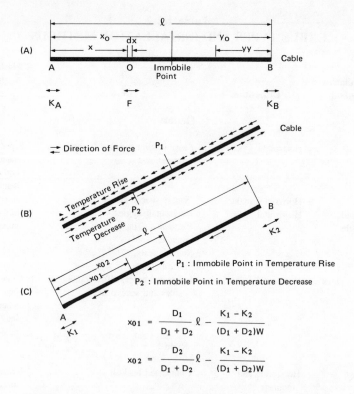

FIGURE 1.2.8. Force analysis for linear and incline installation; (A) linear installation, (B) concept of slide-down, (C) incline installation.

$$m = \frac{1}{4EA\mu W} (EA\alpha\Delta T - 2K_A)^2 \qquad (1.2.3.)$$

$$n = \frac{1}{4EA\mu W} (EA\alpha\Delta T - 2K_B)^2 \qquad (1.2.4.)$$

for the first and remaining half cables, respectively, when the temperature rise is smaller than the critical temperature rise, which will reduce the immobile region to the immobile point.

A different expression is valid when the temperature rise exceeds the critical value, this may be stated

$$m = n = \frac{1}{EA} \left\{ \frac{L}{2} (EA\alpha\Delta T - 2K) - \frac{\mu WL^2}{2} \right\} \qquad (1.2.5.)$$

when, for simplicity, $K_A = K_B = K$.

b. Incline Installation

Phenomena of expansion and contraction in inclined installations can be treated in much the same way as for a horizontal installation. Since the counter force due to friction is different in expansion from contraction, the fixed point P_1 during expansion is located higher than the fixed point P_2 during contraction, as shown in Figure 1.2.8.(B) and (C). The cable section between P_1 and P_2 tends to move downwards when the cable expands, so the cable tends to slide down as a whole [see Figure 1.2.8.(C)]. Where there is no immobile region, expansion lengths at the lower and upper ends are given, respectively, by

$$m = \frac{1}{2\,EA}\,(2\,EA\alpha\Delta T - D_1\,WL - 3K_1 - K_2)\,\frac{D_1\,WL + K_2 - K_1}{(D_1 + D_2)W}$$

$$+ \frac{1}{EA}\,(EA\alpha\Delta T - D_1\,WL - K_1 - K_2)\,\frac{D_1 - D_2}{D_1 + D_2}\,L \qquad (1.2.6.)$$

and

$$n = \frac{1}{2\,EA}\,(2\,EA\alpha\Delta T - D_1\,WL - K_1 - 3K_2)\,\frac{D_1\,WL + K_1 - K_2}{(D_1 + D_2)W} \qquad (1.2.7.)$$

where $D_1 = \mu\cos\theta - \sin\theta$ and $D_2 = \mu\cos\theta + \sin\theta$.

The sliding-down length of the cable per cycle is given by

$$S = \frac{1}{EA}\,(EA\alpha\Delta T - D_1\,WL - K_1 - K_2)\,\frac{D_1 - D_2}{D_1 + D_2}\,L \qquad (1.2.8.)$$

The critical temperature rise ΔT_c is obtained by setting $S = 0$:

$$\Delta T_c = \frac{D_1\,WL + K_1 + K_2}{EA\alpha} \qquad (1.2.9.)$$

The sliding phenomenon can be prevented by designing the cable system in such a manner as to assure its having an immobile region. For this situation, lengths of expansion and contraction are given by

$$m = \frac{1}{2EA(D_1 + D_2)W}\,(EA\alpha\Delta T - 2K_1)^2 \qquad (1.2.10.)$$

$$n = \frac{1}{2EA(D_1 + D_2)W}\,(EA\alpha\Delta T - 2K_2)^2 \qquad (1.2.11.)$$

There are several methods for preventing sliding of cable, for example

1. Clamping the cable ends to increase restrain K_1 and K_2, thereby forming an immobile region
2. Fixing clamps to the pipe mouth wall and the cables at the upper end of pipe to prevent the cable from sliding down
3. Fitting stoppers to the cable at the upper pipe end to allow the expansion of cable but to prevent the contraction from exceeding the expansion length

Figure 1.2.9.(A) and (B) show a comparison between the above methods with respect to minimum restraint force and the length of expansion and contraction vs. cable lengths. In practical application, the most suitable method will be adopted according to cable length, slope, manhole structure, and so on.

B. Offset Design

The main thermo-mechanical problem associated with cables installed in ducts and pipes is that movement of the cable into the manhole occurs during the heating cycle, followed by movement back into the duct or pipe when the cable cools. The movement can be offset in a manhole by movement at right angles to the longitudinal axis of the cable. This sometimes

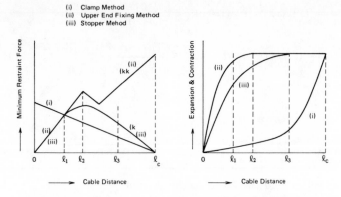

(i)   Clamp Method
(ii) Upper End Fixing Method
(iii) Stopper Mehod

FIGURE 1.2.9.   Comparison of three counter measures to prevent a cable from sliding down.

requires extremely large manholes, which presents another problem, how to reduce their size. Corrugated aluminum-sheathed cables have good fatigue performance and their use makes possible a significant reduction in manhole size compared with some lead alloy-sheathed cable systems.

The offset or manhole size can be determined by the permissible sheath strain and the permissible bending radius (see Table 1.2.7.). Figure 1.2.10. shows some offset designs such as linear, bending, and L-shaped offsets. It is important to estimate sheath strain; several simplified formulas have been proposed and used for this purpose. The recent development of computer calculation methods has facilitated the more accurate estimation of sheath strain for a given design of offset.

C. Snake Installation

The most usual type of flexible system employed for cables which are substantially horizontal is to offset the cable and design the system so that the thermal expansion of the cable between the supports is taken up by movement at right angles to the longitudinal axis of the cable. In such an arrangement, cable lies in a zigzag line like a snake. Two basic methods are used: those in which the cable is permitted to move vertically and those in which the cable is constrained to move in the horizontal plane.

The following points are characteristic of a snake installation:

1.   The thermal expansion and contraction are absorbed by the snaking movement, and offset surge is reduced as a consequence
2.   If cleated at a node or at the top of a loop, the cable is fixed by an extremely small restraining force, considerably less than in a linear installation; this makes the cleat structure simpler
3.   The strain in the metal sheath is not concentrated in the offset but is distributed over the whole length of the cable
4.   No detrimental free snaking of the cable should be observed. The restraining force required for cleats is much reduced in inclined installation; again, the result is simpler cleating

The snake installation allows what is called a nonoffset or a semi-nonoffset manhole to be used, which is much smaller than the usual offset manhole. Minimum snake pitch or inter-cleat distance is typically 4 m for 1200 mm^2 size of conductor, 5 m for 1500 mm^2 size, and 6 m for 2000 mm^2 size in 154-kV, single-core, aluminum-sheathed OF cable with vinyl protective covering.

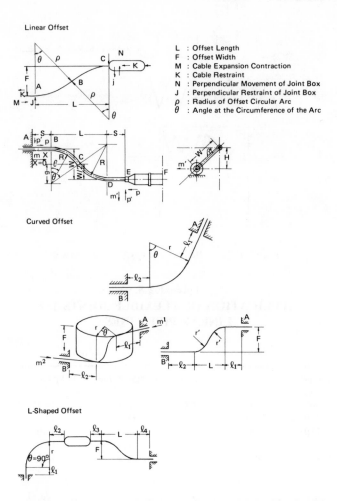

Linear Offset

L : Offset Length
F : Offset Width
M : Cable Expansion Contraction
K : Cable Restraint
N : Perpendicular Movement of Joint Box
J : Perpendicular Restraint of Joint Box
ρ : Radius of Offset Circular Arc
θ : Angle at the Circumference of the Arc

Curved Offset

L-Shaped Offset

FIGURE 1.2.10. Several kinds of offset designs. (From Iizuka, K., Ed., *Handbook of Power Cable Technology*, (in Japanese), Denki-Shoin, Tokyo, 1974. With permission.)

Snaking installation systems can be used for deep vertical shafts by supporting the cable weight at fixed or swivel cleats, fairly widely spaced, and by catering for thermal expansion by allowing the cable to deflect in alternate directions giving it an overall sine wave form.

D. Pipe-Type OF Cable

Both the pipe and the cable core expand and contract in pipe-type OF cable. Since this cable system has no metal sheath and is able to move relatively freely in the pipe, its design is considered much simpler than that of a self-contained OF cable.

Since a steel pipe has a greater modulus of elasticity, a larger cross-sectional moment of the second order, and higher rigidity than a cable core, it is subjected to large thermal stress as temperature varies. When the pipe is installed in air, either indoors or outdoors, the following precaution should be taken as far as the steel pipe is concerned:

1.   Determine the inter-fulcrum distance of the pipe in its linear portion so as to assure that it does not buckle due to thermal stress
2.   Choose the shape of the pipe in its curving portions so that thermal stress, shape variation, and oil-pressure-induced stress do not exceed their individual permissible criteria for repetitive behavior

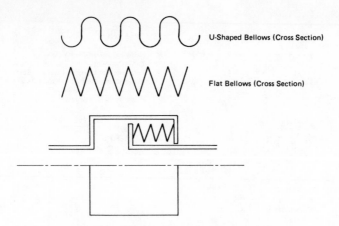

FIGURE 1.2.11.   Bellows and flexible joint structure.

Table 1.2.8.
APPLICATION OF FLEXIBLE JOINTS TO
LINEAR PIPE LINES

Joint type	Inner diameter (mm)	L_1 (m)	L_2 (m) 10 kg/cm² line	20 kg/cm² line
8B	200	0.8	13.5	10.5
10B	250	1.0	18.5	13.5
12B	300	1.2	20.0	15.0
14B	350	1.4	22.0	17.0
16B	400	1.6	24.5	18.5

Note: L_1 = Maximum length between flexible joint and guide.
L_2 = maximum length between the neighboring two guides.

3.   Insert flexible joints to absorb the expansion and contraction of the pipe, or restrain it by anchors to reduce the effect of thermal stress induced when it is difficult to apply the above two methods, or when it is impossible to offset such effects by the two methods alone

U-shaped formed bellows and flat welded bellows are typically employed as flexible joints for steel pipe as illustrated in Figure 1.2.11. Some systems in which flexible joints are applied to linear pipe systems are tabulated in Table 1.2.8.

As a result of operating experience on 345-kV POF cable systems in the U.S., it is suspected that long-time cyclic expansion and contraction due to daily loading causes mechanical stress to concentrate in manholes, thus leading to paper tape displacement, soft spots, reduced insulation in the walls, and ultimately to failure. It is further claimed that the location of the service faults is generally about 3 ft (or 1 m) beyond the end of a splice.

Field experiments were conducted at the Waltz Mill cable test site for 500-kV POF cables

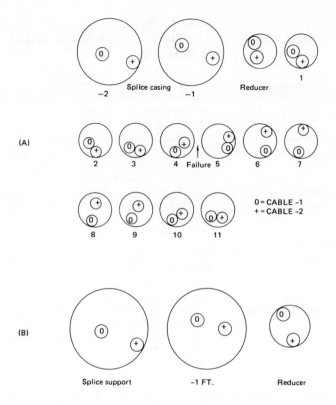

(A) Cross-section of Splice Casing and Pipe Showing Position of Cables at 1 Foot Intervals from Reducer (Ambient Temperature)

(B) Cross-section of Splice Casing with Cable Heated to Conductor Temperature of 75°C — Prior to Removal of Halt Section of Pipe

FIGURE 1.2.12.   Cross-sectional views of cable snaking in a pipe. (From McIleen, E. E., Waldron, R. C., and Garrison, V. L., *IEEE Trans. Power Appar. Syst.*, (95)1, 282, 1976. With permission.)

(2000 MCM compact segmental coated copper tapes, 1.340-in. low-density paper-polyethylene, oil-impregnated, metallized paper shielding, plus coated copper tapes, and two D-shaped insulating plastic skid wires) to investigate this thermo-mechanical behavior.[69] From these field experiments and other laboratory tests, it is concluded that severe snaking of the cables and severe bending in a splice sleeve can occur, and these can possibly lead to cable failure, although some people still question whether the snaking or bending is the cause or the result of the failure. Snaking behavior of the cables in this particular case is pictured in Figure 1.2.12. It may be desirable to provide adequate supports for the cable within the joint casings so as to prevent a concentration of bends adjacent to or at the splices. Improvement in design method is required for large-size conductor heavy-wall paper-insulated cables.

1.2.5. Synthetic Paper-Insulated Cables

As mentioned already and indicated in Figure 1.2.3., the dielectric loss of oil-impregnated cellulose paper insulation used in selfcooled POF cables limits the useful voltage to approximately 650 kV due to the excessive heat generated in the insulation.[1] Much experimental work has been carried out in recent years in an attempt to develop a low-loss material to replace cellulose paper as the dielectric in oil-filled cables. Many materials have been investigated in the form of films (plain or embossed), synthetic paper made with or without the addition of cellulose pulp, and cellulose paper/film laminates.

Table 1.2.9.
CLASSIFICATION OF SYNTHETIC TAPE INSULATION FOR UHV CABLES

Structure	Materials investigated
Polymer film only	PET, PS, PSu, PC, PPO, PP
Polymer film-cellulose paper	PE film-cellulose paper
alternate lapping	PC film-cellulose paper
Polymer film/cellulose	PE film/cellulose paper
paper laminate	PP film/cellulose paper
	FEP film/cellulose paper
Polymer-cellulose	PE-cellulose
mixed paper	PET-cellulose
	PC-cellulose
	PP-cellulose
Synthetic paper only	
Single polymer	PE paper (Tyvek®, FHCP)
	PP paper
	P$_3$O paper (Tenax®)
	Normex® paper
More than two plymers	PET-PC paper (PAP)
	Resin-impregnated paper
	(PC or PPO) — (PET nonwoven paper) (SSP)
Polymer film/synthetic paper	PP film/PP paper
or mixed paper laminate	PP film/PP-cellulose mixed paper

To be successful, materials to be developed must meet all foreseeable requirements regarding

1. Dielectric strength
2. Mechanical adequacy
3. Compatibility with impregnants

Of more than 50 materials examined, fewer than 10 have an impulse strength high enough to be considered for cables in the 750-kV and higher voltage range. Two or three materials were shown to withstand 135 kV/mm (3375 V/mil) impulse strength, these being grades of polypropylene film/cellulose paper laminates and high-density polyethylene.[51]

A. *Proposed Synthetic Papers and Their Performances*

Table 1.2.9. classifies synthetic tape insulation investigated for UHV cable application. In the first place, it is natural to try to use polymer films as insulation for UHV cables. Actually, they were investigated prior to synthetic papers. Such materials chosen for study were polyethyleneterephthalate (PET), polystyrene (PS), polycarbonate (PC), poly (2,6-dimethyl-*p*-phenylene oxide) (PPO), and polypropylene (PP). Impregnants used were mineral oils, fluid paraffin, polybutene, silicone oil, dodecylbenzene, and trichlorodiphenyl. Unfortunately the excellent dielectric properties of these films are offset by the following disadvantages:[77]

1. Impulse and AC long-time breakdown strengths are high enough for thin films, but decrease dramatically with increasing thickness. Taping requires the application of films thicker than 100 μm, but such thick films show lower dielectric strengths than expected.
2. PC films are stable against fluid paraffin, but susceptible to swelling and/or crazing in mineral oils including aromatic compounds. PPO and polysulfone films are resistant to oil but liable to craze.

3.   Oil impregnation is rather difficult, because it takes place only through film interstices. Embossed films can be used to increase the degree of impregnation, but may decrease dielectric strength.

Some mineral-oil-impregnated polycarbonated-film-insulated cables were manufactured on a trial basis. They showed a low tanδ as had been expected, but extremely low breakdown strengths. It is the present conclusion that this type of cable is less promising, so no further development has been made.

The next designs by which it was intended to improve performances of insulation were a combination of paper and polymers in the form of alternate lapping, laminate, or mixture. The objective was to reduce the dissipation factor by introducing polymer but at the same time to maintain the other advantages associated with cellulose paper. Some of the laminates show promise, but substantial improvement is necessary.

Alternate candidates for synthetic insulation may be found in the category of synthetic papers. Some of their properties have been shown already in Table 2.2.13. of Section 2.2.2. (Volume I). PP paper consists either of microfibers smaller than 10 μm in diameter or of a mixture of microfibers and oriented fibers. PAP paper is composed of PET fiber and a continuous micropore phase of polycarbonate. Tenax® is a synthetic paper made of poly(2,6-diphenyl-p-phenylene oxide) which is usually called P_3O. Intensive investigation and improvement of Tenax® encouraged expectations for its near-future use in UHV cables, but unfortunately it is no longer available. FHCP is claimed to be a new modified polyethylene paper which is highly oriented and crystallized. It has far greater compatibility with oil than ordinary polyethylene and also has high dielectric breakdown strength.

There is a special kind of paper named SSP or resin-impregnated paper. It is manufactured in such a manner than unwoven PET fabric is immersed in PC or PPO solution allowing PC or PPO to be fixed on the surface of PET fibers.

Table 1.2.10. is a summary of performances of synthetic fibrous papers developed so far.[76,78] Comparison with cellulose paper indicates that they all have lower permittivity, dielectric loss tangent, and water content than does cellulose paper. This originates from the small number of polar groups and the paper structure of the polymers. There are, however, some disadvantages; for example, care must be taken with respect to the tensile moduli of all the synthetic papers stated since these are three to five times smaller than that of cellulose paper.

The physical performance of a given tape must be such that creasing does not occur when a cable made of the tape is bent. A figure of merit for this purpose is usually represented by[51]

$$A = \frac{1}{\eta} (4E_1 E_2 G^2)^{1/4} \tag{1.2.12.}$$

where E_1 and E_2 are the elastic moduli of the tape in the longitudinal and transversal directions, G is the shear elastic modulus, and η is the static friction coefficient. The larger the value, the better the taping performance. Figure 1.2.13. shows a comparison among some typical materials of the critical interfacial pressure above which creasing would be anticipated.[51]

B. Compatibility with Impregnant Oil[78]

Compatibility includes swelling and crazing of a polymer in taped cable. Swelling of course results in dimensional change of the polymer and often in deterioration in mechanical properties. It has a strong influence on the dielectric loss, electric strength, and oil flow resistance, too. Much of the early interest in polymers focused on films, because of their availability. However, in recent years very thorough consideration has been given to synthetic papers and laminated structures.

Table 1.2.10.

TYPICAL CHARACTERISTICS OF SYNTHETIC PAPERS

Synthetic Paper

Performances	Polyolefine paper	PP nonwoven paper	PP paper	PAP	Resin imp. paper	Tenax® paper		FHCP (PE paper)	Low loss Kraft® paper
				Dry Paper					
Thickness (mm)	0.127	0.137 ± 0.013	0.117	0.125	0.090	0.095	0.080	0.128	0.125
Apparent density (g/cm)	0.75	0.65—0.75	0.75	0.75	0.64	0.75	0.65	0.69	0.63
Impermeability (Gurley sec/100 cc)	300—2000	200—800	1,300	3,000	1,600	3,000	1,000	3.7×10^5	1,560
Tensile strength (MD) (kg/cm)	210—350	140—210	286	690	520—590	280	200	37.8	509
Tensile strength (CD) (kg/cm)	140—210		224			230	200	40.9	290
Elongation to fracture (MD) (%)	4—6	5—6	10.4	25	16-18	4		10.8	3.0
Elongation to fracture (CD) (%)	3—5		11.3			2.8	2.3	9.8	5.3
Tensile modulus (MD) (kg/cm)	10,500—12,000	2,100	13,300	12,000	11,000	15,600	10,000	33,000	54,200
Tensile modulus (CD) (kg/cm)	9,840—12,000		11,300		11,000	14,500		34,000	34,400
Elastic limit (MD)(kg/cm)	98—154		100			100		250	250
Elastic limit (CD) (kg/cm)	70—98		95			95		200	200
Friction coefficient	0.30—0.33	0.28	0.30		0.2	0.55		0.32(0.20)	0.54
Water content (%)	0.1—0.2	0.1	0.3			0.055			7
Permittivity	1.7—1.9 (90°C)	1.94 (100°C)		2.5 (80°C)		1.7 (100°C)		2.0	1.99 (120°C)
Loss tangent (%)	0.02—0.04 (90°C)	0.031 (100°C)		0.04 (80°C)		0.010 (100°C)		0.01	0.286 (120°C)

Impregnated Paper

	Polybutene 2.2	Polybutene 2.16	DDB 2.25	DDB 2.65	DDB	Parafinic oil	DDB 2.4	DDB 2.2	DDB 3.26
Volume resistivity (Ω·cm)			3.61×10^{15} (100°C)				1.0×10^{16} (100°C)		4.55×10^{14} (120°C)
Impregnant oil Permittivity	2.2	2.16 (90°C)	2.25 (100°C)	2.65		2.5 (25°C 50 Hz)	2.4 (100°C)	2.2	3.26 (80°C)
Loss tangent (%)	0.03 ± 0.02 (100°C)	0.005 (90°C)	0.030 (100°C)	0.045 (80°C, 50 Hz)	0.041 (80°C)	0.025 (25°C, 50 Hz)	0.034 (100°C)	0.016(RT) 0.025 (80°C)	0.149 (80°C)
Volume resistivity (Ω·cm)			2.85×10^{15} (100°C)				1.3×10^{15} (100°C)	1×10^{15} (80°C)	
AC breakdown strength (RT) (kV/mm)	59 ± 4 (500 V/sec)	51 ± 12 (500 V/sec)	46 (80°C, 2 kV/mm/30 min)		65—75	78	72 (2 kV/mm/30 min)	104 kV/mm	54 (2 kV/mm/30 min)
Impulse strength (RT) (kV/mm)			100	100	120—130	147	145	229 kV/mm	127.5

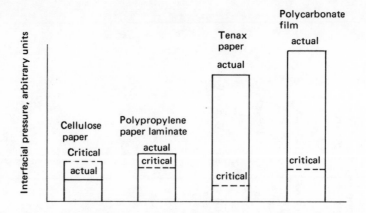

FIGURE 1.2.13.   Interfacial pressure in lapped cables (750 kV 2000 mm²
conductor). (From Miranda, F. J. and Gazzana-Priaroggia, P., *Proc. IEE*,
123(3), 231, 1976. With permission.)

a. Swelling

Swelling is now treated with the solubility parameters of the polymer and the oil. Oil
resistance is considered to be worse when the change in contact energy between the molecules
due to thermodynamic mixing ($\Delta\epsilon$) is at a minimum. The term $\Delta\epsilon$ is given by

$$\Delta\epsilon = V_0(SP_1 - SP_2) \qquad\qquad (1.2.13.)$$

where V_0 is the molecular volume of the solvent (oil), SP_1 is the solubility parameter of the
polymer, and SP_2 is the solubility parameter of the solvent. The solubility parameter is given
by the square root of the cohesive energy density (CED), i.e.,

$$CED = (SP)^2 = \Delta E/V \qquad\qquad (1.2.14.)$$

where ΔE is the latent heat of vaporization and V is the molecular volume.

The condition under which ΔE is a minimum occurs when $SP_1 = SP_2$. Hence, to avoid
swelling, the polymer and the oil should have widely different solubility parameters. Table
1.2.11.[78] gives the solubility parameters of candidate insulating oils and polymers. Experiments have verified the tendency for swelling to increase with decreasing solubility parameter
of the polymers. Swelling and solubility in glassy polymers generally increase with increasing
temperature. The above picture is not necessarily complete as far as swelling is concerned.
It is fortunate from the viewpoint of material improvement that other factors such as crys-
tallinity, orientation, and surface conditions tend to reduce swelling.

The paper called PAP is composed of a fiber material (PET) and a binding material (PC),
with high-solubility parameter, to achieve minimum swelling. Polyethylene exhibits appreciable swelling; nevertheless, it has been intensively investigated with an eye to its application
for UHV cable in the form of either a polypropylene paper or a laminate, where it is combined
as a film with cellulose paper. The swelling characteristics of polypropylene laminate (PPL)
are such that the paper restrains almost entirely any swelling in the plane of the laminate.
Swelling takes place in the perpendicular direction only, it is proportionately larger in this
direction than if the polymer were allowed to swell freely. Fortunately, PP is not susceptible
to crazing.

b. Crazing

Glassy polymers such as PPO, PC, and PS have an additional problem besides swelling

Table 1.2.11.
SOLUBILITY PARAMETERS OF INSULATION OILS AND POLYMERS

	Silicone oil	PB (LV-50)	Naphthenic oil	Sun XX	MO	DDB	SAS
SP	6.1	7.1	7.3	7.5	7.9	8.0	9.5

	PE	PP	PSt	PPO	PC	PS	P_3O	PET
SP	8.0	8.2	8.7	8.8	9.6	9.7	10.0	10.7

Table 1.2.12.
CRAZING OF PLASTICIZED POLYPHENYLENE OXIDE IN CABLE OILS

Oil	Viscosity cs (at temp.)	Temp. (°C)	Critical strain (%)
	Naphthenic-Base Mineral Oils		
55/100	7	100	<0.1 Cracked
XX	13	125	<0.1 Cracked
	White Oils (Aromatic-free Mineral Oils)		
70	6.7	125	<0.27
350	8.2	125	<0.27
70	7.4	100	<0.27
350	10	100	0.31
350, Antioxidant	10	100	<0.27
	Polybutenes		
15 H	18	125	0.29
15 SH	35	100	0.46—0.55
06 SH	14	100	0.30
0 SH	9	100	0.30
	Silicones		
200	1	100	0.87
	Dry		
		125	1.0
		100	1.5

in the presence of oils. Such glassy materials may develop small crazes or cracks when subjected to mechanical strain as may occur in a cable either in wrapping the film or in bending the finished cable. This occurs even in the absence of oil, but in the presence of oil, the phenomenon may be considerably enhanced. The critical strain (ϵ_{CR}) for crazing decreases with increasing temperature towards the glass transition temperature. With monoaxially oriented films, ϵ_{CR} for films with strain perpendicular to the direction of orientation was lower than with strain in the direction of orientation. It is preferable to use biaxially oriented film, if such film is available.

The influence of oils on ϵ_{CR} is shown in Table 1.2.12. Oils containing aromatics are clearly unsuited for application with PPO. PB fluids are the best of the hydrocarbons, especially those with high viscosities. Silicones have minimum effect upon crazing.

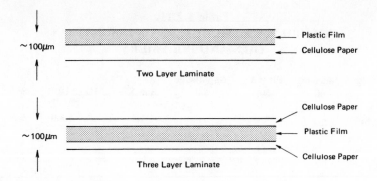

FIGURE 1.2.14. Plastic-paper laminate structure.

c. Mechanical Properties

With polymer films, the changes in mechanical properties on aging in insulating oils are to be expected frm the studies of swelling and crazing. If crazing occurs, mechanical failure is abrupt; if swelling alone occurs, deterioration in mechanical properties follows.

d. Dissipation Factors

Dielectric losses of candidate polymers are liable to be affected by extrinsic factors such as impurities, residual solvent, swelling, and electric stress.

Impurities can occur in polymers from processing residues, residual catalyst, or from impurities in the original chemicals. Impurities may also appear during the manufacture of synthetic paper; antistatic agent for P_3O paper making is one example. Residual solvent may be a factor if the polymer film is solvent-cast or, with synthetic papers, if a solvent has been used for producing the fiber and fibrillon material, assuming the latter is used in the paper-making process. Dissipation factors can increase substantially in impurity-rich polymers such as PPO, upon swelling, The effect is usually offset by the very low mobility of impurity in such glassy-structured materials before they swell. Electric stress is another factor which plays a significant role in the dielectric losses of polymer-fluid systems, or systems where we have alternate layers of polymers and fluid. Charges in the fluid become trapped at the polymer interface during a portion of a half cycle in a manner originally described in its simplest form by Garton; this causes the net losses to decrease. Then, at higher fields, enhanced dissociation of the ions in the liquid may take place, in a manner described originally by Wein; this causes the effective losses to increase again.

C. Most Promising Insulation Structures

An elegant and promising method for overcoming the difficulties associated with plastic tape insulation involves manufacturing a composite bonded laminar material[79] which comprises a layer of dielectric grade plastic material and one or two layers of high-purity Kraft® paper as sketched in Figure 1.2.14. The paper layers permit ready evacuation and impregnation and provide substantial accommodation and restraint for the swelling of the plastic. In addition, the laminate is mechanically robust. The use of cellulose paper in the composite does entail an increase in both permittivity and dissipation factor, so that part of the advantage of plastic material is sacrificed to obtain a practical construction. With three-layer laminates, the paper parts protect the plastic layer from the effects of ionization which might occur in butt gaps under surge voltage.[49]

Presently, polypropylene is one of the most attractive candidates for the polymer component, largely because of the excellent experience with this material in capacitors. Cellulose paper/polycarbonate film laminate (PPL or PPLP) is characterized by superior tensile strength

Table 1.2.13.
VARIOUS COMBINATIONS OF CELLULOSE AND POLYPROPYLENE FOR THREE-LAYER LAMINATE STRUCTURES

	1	2	3	Remarks
A	Cellulose paper	PP film	Cellulose paper	Easy to handle
B	Cellulose paper	PP film	PP paper	—
C	Cellulose paper	PP film	Biaxially oriented PP film (OP film)	—
D	PP paper	PP film	PP paper	Very low dielectric loss
E	PP paper	PP film	Biaxially oriented PP film	—
F	PP-cellulose mixed paper	PP film	PP-cellulose mixed paper	—

and high-impulse strength. The sandwich type seems to perform better than the two-layer type. To form the sandwich structure (called PPLP) polypropylene is usually extruded between two sheets of cellulose paper. There is good reason to believe that the adverse effects upon mechanical properties caused by swelling of the PP in PPLP will be cushioned considerably by the Kraft® paper. A variety of material combinations based on this structure are now under consideration as Table 1.2.13. shows. Performances of insulation systems A, C, and D are listed in Table 1.2.14.[81]

Such Kraft® paper/PP film laminates are, however, inferior in oil-flow readiness and swelling resistance, especially at higher temperatures. Silicone oil does not affect the laminates, but its cost is high. Development of novel structures (type F in Table 1.2.13.) is underway[85] to improve these unfavorable characteristics. This is in addition to the development of synthetic papers which tend to exhibit high oil-swelling and poor mechanical features. This is achieved by thermally bonding both sides of PP fiber paper with two sheets of mixed paper, using a hot calendar. A mixed paper to be bonded can be made from Kraft® pulp, PP fiber, and often PP pulp.

More recently, some new kinds of laminate paper have been under investigation in Japan; they are FEP laminate paper, SIOLAP, DCLP, PML paper and so on. EFP laminate paper is manufactured in such a way that cellulose paper is laminated on both sides of fluoroethylene-propylene film through the thermal calendering process. SIOLAP is a laminate of cellulose paper with silicone-grafted high-density polyethylene extruded from a T die. DCLP is composed of mixed paper (10% to 20% polypropylene fiber and 80 to 90% cellulose paper) and nonwoven polypropylene cloth. It is calendered thermally. PML paper is a laminate of cellulose paper with polymethylpentane film. Characteristics of these laminate papers are shown in Table 1.2.15. Any laminate using cellulose paper is generally for EHV cables. For UHV insulation, all-synthetic paper without cellulose is presumably required.

In spite of its superior dielectric properties, polyethylene has been virtually abandoned because of its inferior compatibility with insulating oils at higher temperatures. There has been some attempt to modify polyethylene paper for use in UHV cable insulation.[77] A new modified polyethylene paper called FHCP is already listed in Table 1.2.10. This design is predicated on the principle that the swelling and solubility of high-density polyethylene films

Table 1.2.14.
PROPERTIES OF CELLULOSE-POLYPROPYLENE LAMINATES[81]

Performances	Type						Cellulose paper[d]
	A-1	A-2	C-1	C-2	D-1	D-2	
Thickness (mm)	0.125	0.155	0.175	0.110	0.215	0.155	0.150
Density (g/cm³)	0.88	0.88	0.87	0.87	0.76	0.68	0.62
Impermeability (G sec)	—	—	—	—	—	—	1400
Tensile strength (kg/mm²)							
MD	6.0	5.9	4.6	5.1	2.8	1.3	5.1
CD	3.3	3.9	7.5	6.5	1.8	1.0	2.4
Elongation (%)							
MD	2.6	2.5	3.3	2.3	26	29	3.2
CD	7.5	8.1	11.4	14.4	18	28	8.4
Thickness of PP (%)[a]	40	40	60	60	—	—	—
Permittivity	2.8	2.8	2.5	2.5	2.2	2.3	3.1
Tangent delta (%) at 20 kV/mm							
80°C	0.08	0.08	0.05	0.05	0.01	0.01	0.17
100°C	0.09	0.09	0.05	0.05	0.03	0.03	0.20
Impulse strength (kV/mm)[b]	230	250	270/210[c]	280/240[c]	140	150	150
AC strength (kV/mm)[b]	120	135	150	140	90	80	80

a   (Mean thickness of PP layer/total thickness of PPLP) × 100.
b   Measured on impregnated single sheet.
c   Lower strength when PP faces positive electrode.
d   Extra-low loss paper.

Table 1.2.15.
CHARACTERISTICS OF DEVELOPED LAMINATE PAPERS

Insulation / Characteristics			Cellulose Paper for reference	Laminate Paper							
				PPLP			PPL	FEP	SIO-LAP	DCLP	PML
				CPC	CPF	SPS					
Dry Paper	Thickness (mm)		0.125	0.125	0.175	0.155	0.100	0.184	0.100	0.155	0.125
	Density (g/cm^3)		0.65	0.88	0.87	0.68	0.9	1.75	0.91	0.64	0.89
	Impermeability Air Resistance (sec/100cc)		1800	∞	∞	∞	∞	∞	∞	14,000	—
	Tensile Strength	MD	5.6	6.0	4.6	1.3	7.6	4.6	10.2[†]	4.0	11.3[†]
		CD	2.9	3.3	7.5	1.0	—	2.6	6.3[†]	—	7.3[†]
	Elongation to Fracture	MD	3.0	2.6	3.3	29	2.7	2.8	3.5	3.4	2.4
		CD	5.3	7.5	11.4	28	—	6.8	8.8	—	5.4
	Yound Modulus (10^4 kg/cm^2)	MD	5.42	5.0	3.0	—	—	3.0	3.9	2.6	2.95
		CD	3.44	2.7	2.6	—	—	1.8	1.8	—	2.39
	Static Friction Coefficient		0.54	0.58	0.31	—	—	0.53	—	0.43	0.65
Oil Impregnated	Permittivity		3.3	2.8	2.5	2.3	2.7	2.4	2.3	2.7	2.7
	tan δ (%)	80°C	0.14	0.08	0.05	0.01	0.08	0.06	0.073	0.073	0.07
		100°C	0.15	0.09	0.05	—	—	0.05	0.07	0.085	—
	Breakdown Strength (kV/mm)	AC	74	120	135	80	—	109	130	63	118
		Imp.	156	230	210	150	—	236	282	140	214

in hydrocarbon oils can be reduced dramatically by changing the morphology of the material by mono-axial stretching.[84] High-density polyethylene is made into a thin film, a few tens of micrometers thick by the inflation method, and then expanded in hot solvent to become high-crystalline polyethylene film comprising fibers with a "shish-kebab" or highly oriented molecular structure. Thus, fibrillated films are laminated without any binds, and rolled into paper form. It is claimed that this polyethylene paper possesses low solubility and low oil-flow resistance and will be applicable for future UHV cable following further intensive investigation.

Several prototype cables, insulated with synthetic paper or laminate paper, ranging from 132 kV to 750 kV, have been manufactured and have undergone extensive testing in various experimental projects. Some of the test results are listed in Table 1.2.16.

1.3. GAS-FILLED CABLES[87-95]

There are three kinds of gas-filled cables available besides compressed gas-insulated spacer cable. The first, which may be called a gas-compressed OF cable, is an oil-impregnated paper-insulated (OF) cable, pressurized externally by means of gas instead of oil. The second is a gas-filled cable with paper insulation impregnated with extra-high-viscosity oil. The third is literally a gas-filled cable with lapped plastic tape insulation. This has undergone intensive development, unfortunately with little success.

1.3.1. Gas Compression OF Cables
The compression-type cable is usually designed to be oval or semi-triangular in cross-

Table 1.2.16.
CHARACTERISTICS OF PROTOTYPE CABLES INSULATED BY COMPOSITE PAPER

Insulation	Voltage class (kV)	ε	tanδ (%)	εtanδ (%)	AC strength (kV)	AC strength (kV/mm)	Impulse strength (kV)	Impulse strength (kV/mm)	Remarks
Cellulose paper (DDB)	OF 275 (1200 mm²)	3.43ᵃ (20°C)	0.148 (80°C)	0.51ᵃ (80°C)	615	42.9	1420	99	
PPLP-CPC (DDB)	OF 275 (2000 mm²)	2.8	0.08	0.224ᵃ (80°C)	670	45	−1940	−130	Thermal resistivity 541°C · cm/W
PPLP-CPF (DDB)	OF 275 (2000 mm²)	2.5	0.045	0.113ᵃ (80°C)	—		—		ΔP = 0.8 kg/cm²
PPLP-CPC (polybutene) (1000 mm²)	POF 765 (17 kg/cm² G)	2.8 (90°C)	0.055 (90°C)	0.17ᵃ (90°C)	115nt	55.3	>2750	>132	ΔP = 4.4 kg/cm²
FEP laminate (DDB)	OF 275 (400 mm²)	2.47 (85°C)	0.058 (85°C)	0.143ᵃ (85°C)	560	61.6	−1610	−177	
SIOLAP (DDB)	OF 275 (2000 mm²)	2.75 (22°C)	0.085 (22°C)	0.234 (22°C)	660	49.3	−1710	−128	
DCLP (DDB)	OF 275 (800 mm²)	2.74	0.062 (60°C)	0.170ᵃ (60°C)	—		−1610	−128	
PML	Sheet	2.7ᵇ	0.05ᵇ (80°C)	0.135ᵃ	—		—		
FHCP (DDB)	OF model cable	2.2ᵇ	0.024 (80°C)	0.053ᵃ (80°C)	118	53	214	115	Thickness 0.8 mm
Tenax® (DDB)	OF 275 (1200 mm²)	2.4	0.02 (RT)	0.05ᵃ (RT)	149	69.5	1240	104	
PAP (mineral oil)	OF 66 (200 mm²)	2.65ᵇ	0.075 (80°C)	0.20ᵃ (80°C)	210	40.6	−450	−86.9	
PAP cellulose (DDB)	OF 154 (200 mm²)	2.56ᵃ	0.093 (80°C)	0.24ᵃ (80°C)	460	37.8	−1020 / +950	−83.6 / +77.9	
SSP (fluid paraffin)	OF 66 (600 mm²)	2.51	0.06 (80°C)	0.15ᵃ (80°C)	220	35.8	−480	−78.72	

ᵃ Calculated from capacitance in μF/km and tanδ in %.
ᵇ Values obtained from sheet experiments.

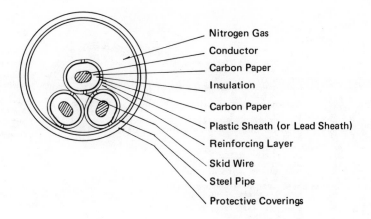

Nitrogen Gas
Conductor
Carbon Paper
Insulation
Carbon Paper
Plastic Sheath (or Lead Sheath)
Reinforcing Layer
Skid Wire
Steel Pipe
Protective Coverings

FIGURE 1.3.1.   Gas-compression cable.

sectional shape so as to be able to follow the thermal expansion and contraction caused by load change. Three insulated conductors are accommodated in a common steel pipe, each single cable core being covered with a lead or aluminum sheath. The pipe is filled with a gas at a pressure of 15 kg/cm^2. This acts on the lead sheath to pressurize the oil-impregnated paper insulation, thus preventing oil migration and void formation. An example of the pipe-type gas compression (PGC) cables is illustrated cross sectionally in Figure 1.3.1.

Because of the external pressure applied in this type of cable, higher voltage cables with heavier insulation may not respond immediately to conductor temperature change; buckling of the sheath may even occur. Therefore this type cable has an upper voltage limit of about 220 kV. It is usually operated in the range between 60 kV and 150 kV, 138 kV being a preferred level.[4]

1.3.2. Gas-Filled OF Cables

Lapped-paper insulation around the single conductor is impregnated with extra-high-viscosity oil, and pressurized to 15 kg/cm^2 by compressed nitrogen gas. Since there is no lead sheath provided around the insulation, the gas can diffuse into the insulation and fill any voids that may eventually form. Pressurization of the gas in the voids precludes ionization in service. A typical structure of this type of cable is sketched in Figure 1.3.2.[88] Cables of this type have been built for voltages up to 154 kV. If the nitrogen is replaced in part by SF$_6$ gas, a voltage of 230 kV is possible. Generally speaking, oil-filled cable is superior in dielectric properties to gas-filled cable. However, gas-filled cable is particularly advantageous for installation in places with large differences in altitude because the hydrostatic pressure problem is eliminated in this design.

Some interesting attempts are being made to increase the MVA rating of existing gas-filled cable systems by replacing the nitrogen with SF$_6$ gas. It is economically attractive because such cables are typically located in urban areas where construction costs are high. It was concluded from 1 trial, made on an installed 15-kV low-pressure gas-filled cable system, that the ionization starting potential of the system could be raised by factor of approximately 1.5 at 15 psig (10.4 × 10^4 N/m^2) after conversion to SF$_6$.[89]

1.3.3. Gas-Filled Plastic Tape Insulated Cables

The possible use of a very low-loss insulation has been intensively investigated.[90-95] The basic concepts of the development work on gas/plastic composite insulation are different from those of gas-filled paper-insulated cables described in Sections 1.3.1. and 1.3.2.; they follow more closely those of oil-impregnated synthetic tape-insulated cables described in

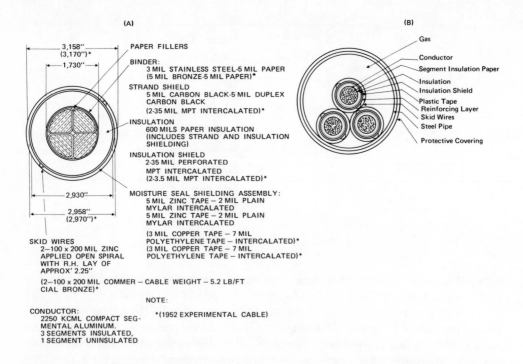

FIGURE 1.3.2.   138-kV Single-conductor 2250-kcmil aluminum compact segmental, high-pressure, gas-filled cable. (From Roughley, T. H., Corbett, J. T., Winkler, G. L., Eager, G. S., Jr., and Turner, S. E., *Design and Installation of a 138 kV High-Pressure, Gas-Filled Pipe Cable Utilizing Segmental Aluminum Conductors,* IEEE PES Summer Meeting T73 491-8, IEEE, Piscataway, New Jersey, 1973, 1 (Part A only). With permission.)

Section 1.2.5. The objective is to lap plastic-film tapes onto the conductor and to impregnate the core not with liquid but with an electro-negative gas under pressure.

This insulation system has an advantage in that the complicated problem of compatibility with oils, found in oil-impregnated plastic tape insulation, does not exist. Therefore, it is even possible to use polymers such as polyethylene which are susceptible to swelling and oil-assisted crazing. High-density polyethylene is now considered promising because of its low loss and low permittivity, as well as its low price. SF_6 gas is most suitable as the impregnant because of its high electrical strength and high thermal and electrical stability. Other gases such as nitrogen and SF_6/N_2 mixtures have also been investigated. Polycarbonate may be yet another candidate to improve the thermal resistance of this cable.

A. Prototype Design of a 275-kV Cable

A tentative specification for a 275-kV lapped-polyethylene cable with a conductor working stress of 15 kV/mm is shown in Table 1.3.1.[91] This cable was developed in 1965 and subjected to a series of tests. Unfortunately, nothing further appears to have evolved since that time. The 275-kV system voltage was selected, since this was considered to be the lowest value at which this type of cable was likely to show any appreciable economic advantage over oil-impregnated-paper cables.

The copolymer was selected because it has better stress cracking resistance than the high-density polyethylene, while it has a slightly lower tensile strength than the homopolymer. Tape thickness was determined through a compromise between electrical and mechanical requirements; maximum electric strength is obtained by use of the thinnest possible polyethylene tape, while the mechanical rigidity of tapes decreases sharply as the thickness is reduced. This suggested the possibility of using the thickness grading method. Conducting polyethylene was preferred to carbon paper as the conductor-screening material to prevent

Table 1.3.1.
275 kV GAS/PLASTIC CABLE

System voltage	275 kV
Design stress	150 kV/cm
Conductor size	1.0 in.^{2} (645 mm^2)
Conductor screen	Three layers of 6-mil conducting Regidex® 3 tape, 0.8-in. wide, applied to register with first 9 insulating tapes
Dielectric material	Rigidex® 3 copolymer
Dielectric construction	21 × 4-mil tapes, 0.8-in. wide; 30 × 5-mil tapes, 1.0-in. wide; remainder (≈50) 7-mil tapes, 1.0-in. wide
Reversal of lay	After first 9 and 21 4-mil tapes; every 10 layers with 5- and 7-mil tapes
Registration	65/35
Butt-gap width	0.050 in. (1.27 mm)
Lapping tension	150—450 lb*f*/in.^{2}
Dielectric screen	Two layers of 6-mil conducting Rigidex® 3 tape, 1.0-in. wide, applied to register with insulating tapes on final lapping head; third layer of conducting tape intercalated with 1 in.-wide 3-mil copper tape
Sheath	Plain aluminum sheath capable of withstanding internal gas pressure of 250 lb*f*/in.^{2}, applied with minimum practical clearance over insulated core

Note: 1 mil = 25.4 μm.

thermal wrinkling of the conductor-adjacent insulation tapes, thereby obtaining a favorable effect on impulse breakdown strength.

B. Problems to be Overcome

With SF_6 gas-impregnated, lapped, polyethylene-tape insulation, the loss tangent tends to increase rapidly once internal discharges take place. It is certainly possible to design a cable so as to preclude internal discharges for any probable voltage that may be anticipated on the cable system; but this requires an increase in gas pressure and in insulation thickness with attendant technical and economic disadvantages. The increase in the loss tangent is attributed to products created by the interaction of SF_6 discharges with the polyethylene. No tanδ increase is found in nitrogen-impregnated systems under the same condition. As a consequence, SF_6/N_2 mixtures have been investigated as impregnants. The formation of cavities by thermal expansion of the polyethylene will further add to the disadvantages of a decrease in the discharge inception stress.

Much more effort should be directed to increasing the impulse breakdown voltage and the discharge inception voltage. A satisfactory compromise has not yet been reached between the electrical and the mechanical requirements. In principle, the thinner the tape thickness, the better. Deliberate consideration and further investigation are needed on gas pressure, gas kind (N_2, SF_6, C_2F_6, C_4F_8, C_2F_5Cl, etc.), conductor shield, taping registration, and so on.

1.4. SPLICES AND TERMINATIONS

1.4.1. Kinds and Performances of Splices and Terminations

Limitations of manufacture, transportation, and installation require cables to be joined in

Table 1.4.1.
CLASSIFICATION OF SPLICES AND
TERMINATIONS

Joints	Remarks
Straight joints	
Normal joints	Taped cables
	Extruded cables
	Tape-wrap splices
	Field molded splices
	Prefabricated splices
Insulated joints	Shield-separated splices
Stop joints	Oil-filled cables — oil stop
Semistop joints	Pipe type oil-filled cables — oil stop
Gas-stop joints	Gas-filled cables
Terminal joints	
Termination in air	Taped cables
	Porcelain bushing
	Extruded cables
	Tape wrapping
	Elastomer mold bushing
	Porcelain bushing
	Epoxy-resin bushing (plug-in type)
Termination in oil	Porcelain bushing
Termination in SF_6	Epoxy-resin bushing
Direct connection	with circuit breakers, transformers, etc.
Hetero-joints	
SL-XLPE (PE)	
OF-XLPE (PE)	
OF-POF	
Branch joints	
Y or T branch	
X or + branch	

the field when the total length exceeds a certain value. Terminations are required where cables are connected to overhead lines or electrical apparatus. Obviously these splices and terminations should possess the same integrity as their associated cables. Table 1.4.1. indicates various types of joints and terminations which are currently used.

Straight joints splice two cables of the same kind in a straight line. Besides the normal joints which are commonly used, there are other kinds available such as insulation joints, stop joints, semi-stop joints, and gas-stop joints. Insulated joints are constructed such that the conductors are joined but the sheaths are insulated from each other. Joints of this type can be used for cross-bonding systems. Beyond a certain length, OF cable lines must be sectionalized for the purpose of oil-feeding and maintenance. In this regard, there is a certain optimum length. Stop joints are used in these cases; they connect cables electrically but block oil flow. The same principle is applied to gas-filled cables, except that the joints are referred to as gas stop joints. Semi-stop joints are utilized with POF cables. These enable oil to flow through a bypass in normal operation but the oil flow can be stopped when it becomes necessary.

The jointing of a single-conductor cable is the simplest and most easily understood. Figure 1.4.1. shows the basic concept, rather than the design of a normal joint. A cable joint is made up of three elemental components. First, there is the conductor compression sleeve (ferrule or connector) which joints together electrically and mechanically the conductors of the two cable ends. Another method of jointing conductors is called autogeneous welding. Secondly, there is the tapering down of the insulation on each cable to some predetermined

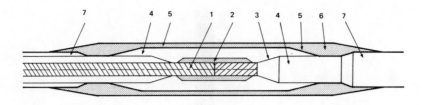

1　Conductor
2　Conductor Sleeve, (Ferrule or Connector)
3　Penciled or Stepped Part of Cable Insulation
4　Cable Core, Cable Insulation
5　Joint Insulation, Stress Control Cone (or Capacitive Graded)
6　Protective or Reinforcing Layer
7　Cable Sheath

FIGURE 1.4.1.　Basic concept of cable joint.

dimensions. This is called stepping of taped cables and penciling of extruded cables. Stepping consists of a set of steps having risers and treads from the level of the conductor surface to that of the cable insulation surface. Penciling, on the other hand, gives smooth surface. The insulation, as it is applied over the conductor sleeve, should be well blended into the cable insulation so as to make the overall insulation system as homogeneous as possible. Some degree of discontinuity inevitably occurs along the tapering at the ends of the conductor sleeve, resulting in the generation of longitudinal electric stress. The third, but not the least part, is the stress relief cone of the joint insulation. As is seen in Figure 1.4.1., the joint insulation is covered with a protective or reinforcing layer. It consists of a casing filled with oil or insulating compound in case of taped cables, and it is made of waterproof lapped tapes. Jointing is almost always carried out in the field. Considering this, together with the foregoing stress consideration with respect to the weakness of the tapes in the longitudinal direction and of the weakness of interfaces through penciling with molded insulation, it is usually necessary to build up the joint insulation to some diameter greater than that of the cable insulation. The penalty for this may be thermal discontinuities at joints along the cable system because thermal resistance tends to be higher at the joints than in any other part of the cable. At present it is inevitable and unfortunate that the quality of taping and molding are very dependent on the skill of jointers or splicers, and on jointing equipments. Furthermore, the design itself is most likely determined by rule of thumb, based on experience, rather than by any theory.

The three important elements of the joint which have been described, together with a kind of casing or housing, are essential, but of course they are not everything. Since the joint is part of the cable, at least it should have all the components of that cable, that is to say, conductor and insulation shields, sheath or protective covering, and so on. Besides the three main components, various other components are provided to complete and even improve the electrical, mechanical, thermal, and chemical integrity. With the electrical integrity, for example, insulation shields (wrapped lead tapes or helically wound, tinned copper wires) are extended from both cable ends toward the joint to cover the entire joint insulation. For cross-bonding, the shields are separated or isolated from each other. This type of joint is called an insulated joint. There are also various components peculiar to the type of cable, or simply taped cable or extruded cable. Taped cable is usually provided with a lead, copper, or steel tube casing, plumbers, grounding terminals, and so on. Stop joints, semi-stop joints, and oil-feeding joints are naturally more complicated than the normal joints, as will be described in due course. The joints or splices of extruded cable are much simpler than those of the taped cable as shown in Figure 1.4.1. This is one of the reasons why the extruded cable is preferred to taped cable whenever possible.

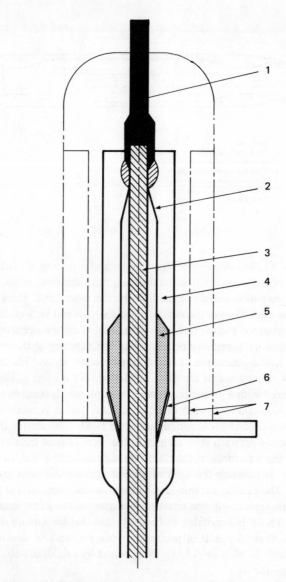

1   Conductor Lead-Out Rod
2   Stepping-down or Penciling
3   Cable conductor
4   Air, Gas, Oil, or Insulating Compound
5   Insulation Reinforcing Layer
6   Stress Relief Cone
7   External Insulation

FIGURE 1.4.2.   Basic concept of cable termination.

Designs for cable terminations are similar in concept to cable joints. Figure 1.4.2. shows the basic structure of a termination for a single-cable core. It consists of a conductor lead-out rod, an insulation reinforcing layer with a stress relief cone, and a casing (or a bushing in this case). The end of taped cable is usually encapsulated with a porcelain bushing which acts as external insulation. The space between cable core and the inner surface of the bushing may be filled with oil or other insulation compound. The external insulation of an extruded cable may be made of plastic tape, a porcelain bushing, an epoxy-resin bushing, or a rubber-mold bushing. Near sea shores or in industrial districts where salt or dust contamination is severe, a specially designed porcelain bushing is used. It has a longer surface current leakage path than any conventional bushing. This is also a bushing with lower pleats or beaks available for this purpose. Bushings may be coated with contamination-resistant silicone compound.

The end of the termination is often open to the air in order to link to an overhead line, but sometimes it is tied to some electrical apparatus in oil or SF_6 gas via a lead-wire, or directly, without any special oil or gas insulation. Terminations in oil are advantageous in that it can be installed in a limited space. Moreover, they are free from contamination and danger to men and beasts, since the high-voltage part is enclosed. They are commonly used for connections to transformers in elephant cases filled with oil. The top of a cable may be encapsulated and connected to a compact circuit breaking device in SF_6 gas. Should heavy corona occur, or an arc to be ignited, porcelain would be badly affected by the hydrogen fluoride solution formed by the chemical reaction of decomposed gases with any remaining water vapor. For this reason, epoxy-resin bushings are generally preferred.

It is occasionally necessary to join together two different kinds of cable. This can be achieved by heterogeneous joints, which include SL (or H) cable-extruded cable joints, OF cable-extruded cable joints, and OF cable-POF cable joints. Of course there are some other kinds available such as OF-SL (or H) cable joints, LPG cable-SL (or H) cable joints, and LPG cable-extruded cable joints, but they are rare. An SL-extruded cable joint should have some structure to prevent the flow of oil to the extruded cable. It is equipped with an epoxy-resin stop unit or a casing with an oil-stop structure for this purpose. Insulation reinforcing is accomplished by any method suitable to the respective cable. With OF-extruded cable joints, appropriate measures should be taken to suppress the outflow of pressurized oil. An epoxy-resin stop unit or a porcelain tube is a satisfactory solution.

Branch joints are used to connect one cable with two or more other cables. They may be classified according to their shape, such as T, Y, X, +, and π branch joints. Heterogeneous branch joints are also available. It has been a common practice to apply Y-shaped units made of epoxy resin for paper-insulated solid cables and extruded cables of the 22 ~ 33 kV class. More recently, extruded cables are spliced with compression or plug-in type premolded insulation. Oil-filled cables are usually jointed in a branch box filled with oil and provided with a stop unit of epoxy resin, thereby setting up a stop joint.

A. Performance Requirements for Joints

It goes without saying that joints should have the same performance as their companion cables. Joints and terminations in general should:

- Avoid ill effects on cable structural materials
- Withstand external forces and vibration generated during operation
- Withstand long-term operation below the maximum permitted temperature
- Carry the permissible short-time current of their companion cables
- Completely shut off the flow of insulant (oil or gas) in the case of hetero-joints
- Withstand mechanically any foreseeable wind loading, short-circuit forces, and seismic disturbances
- Coordinate with any electrical apparatus on which it may directly terminate

Conductor joints should:

- Produce no hotter spots than occur in adjacent cables
- Withstand the same permissible tensile forces as the cable conductor

Reinforcing insulation layers should exhibit a sufficiently high corona inception voltage, consistent with the detection sensitivity. The shielding layer or casing should:

- Support the same current as the cable sheath
- Withstand the pressure of insulant (oil or gas)
  Prevent ingress of water or moisture and egress of the insulating medium
- Hold off its specified contamination withstand voltage in the case of contamination-resistant terminations
- For terminations in air, satisfy its wet withstand voltage if required
- Withstand indefinitely sheath-induced voltages if of the insulated sheath type

Coverings for corrosion protection can also prevent permeation of water or moisture.

Various measures have been and will continue to be taken to pursue the technical integrity and economic viability of joints. These include simplification, reduction in size, improved reliability, and lower price. Prefabrication seems to be a promising approach to future jointing.

1.4.2. Insulation Design Principles

In the design of the reinforcing insulation layer (joint insulation), the radial electric stress may be chosen below that of the cable. One expects its dielectric strength to be smaller than that of the cable insulation since it is fabricated in the field and is liable to atmospheric exposure. The electric field should be so distributed among the insulation components as to partition the electric stresses according to their respective dielectric strengths. Furthermore, it should withstand repeatedly, and over a long period of time, the extraordinary transient voltages induced in cable systems.

A. Normal and Insulated Joints

The diameter of joint insulation is usually determined by the maximum electric stress on the surface of the conductor sleeve, which can be calculated from the following formula:

$$g_r = \frac{V}{r_s \ln \dfrac{R_j}{r_s}} \tag{1.4.1.}$$

where g_r = the maximum radial electric stress on the conductor sleeve, r_s = the outer diameter of the sleeve, R_j = the outer diameter of the joint insulation, and V = the voltage across the insulation. The value selected for the above maximum stress is less than half the maximum stress of the cable, alternatively it is determined by taking into consideration the breakdown strength of the specific material used for the joint insulation.

The edge of the reinforced joint insulation is likely to suffer some discontinuity of radial electric field distribution and some increases in longitudinal electric stress, which renders it liable to failure. The stress relief cone should be designed to improve the electric field distributions as far as possible. Approximate formulas given by Short[96] have been commonly used to determine the shape of the cone.

Half a joint, symmetrical about its center-line axis, is depicted in Figure 1.4.3. The electric field throughout the transitional curve A-B at the zero potential electrode (or ground electrode of the field), in passing from the cable insulation radius R_i, to the joint insulation radius R_j, is really three-dimensional. However, it is difficult to treat the problem with reasonable accuracy three dimensionally. A permissible approximation is to take the formula of the radial electric field distribution in the two-dimensional Cartesian coordinates as shown in Figure 1.4.3. The shape of the stress relief cone is then obtained as follows:[96]

$$x = \frac{V}{g_x} \ln \left\{ \frac{\ln(y/r)}{\ln(R_j/r)} \right\} \tag{1.4.2.}$$

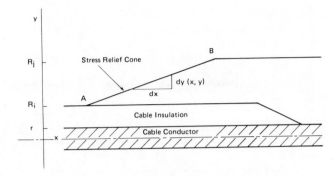

FIGURE 1.4.3. Cross-sectional picture of a stress relief cone.

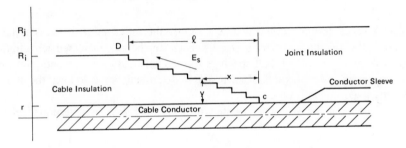

FIGURE 1.4.4. Cross-sectional view of stepping.

where g_x = the longitudinal electric stress, V = the applied voltage across the insulation, R_i = the radius of cable insulation, and r = the radius of cable conductor. It is obvious from the above formula that the combination of x and y, i.e., the shape of the cone, can be determined if V and g_x are given.

In the case of oil-impregnated paper insulation, the longitudinal electric stress is usually 15 to 20 times less than the radial stress. It may be either 1.0 to 2.0 kV/mm under conditions of anticipated cable failure, or 0.2 to 0.4 kV/mm in normal operation. Self-adhesive tape insulation is designed to hold 0.3 to 1 kV/mm for the prospective cable failure voltage. Another available design method keeps the longitudinal electric stress constant at y = R. The following formula then applies:

$$ x = \frac{V}{g_x} \left\{ 1 - \frac{\ell n(y/r)}{\ell n(R/r)} \right\} \qquad (1.4.3.) $$

The former formula is more popular, because the cone is longer with formula 1.4.2., resulting in a safer design.

Cable insulation stepping is generally designed so that the voltage gradient may be constant along the stepping as shown by the arrow in Figure 1.4.4. A constant difference of potential exists throughout the joint, between the conductor and the zero potential electrode (the ground potential of the field). With the surface of the joint insulation maintained at zero potential, in accordance with the more usual practice (commonly referred to as full shielding of the joint, or a fully shielded joint), the resultant formula is[96]

$$ x = \frac{V}{g_x} \cdot \frac{\ell n(y/r)}{\ell n(R_j/r)} \qquad (1.4.4.) $$

The stepping length can be obtained when y = R. The longitudinal electric stress g_x is chosen in the same way as for a stress relief cone; this seems to be a common design practice.

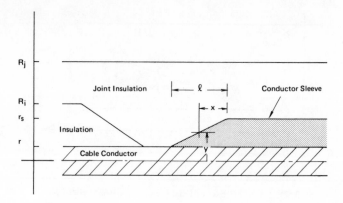

FIGURE 1.4.5.   Cross-sectional view of conductor sleeve.

There is another design method available. Only the radial electric stress is taken into consideration, the longitudinal electric stress in each step being considered zero. In this case, the radial voltage across the oil butt spacings formed between each step and joint insulation becomes important. The thickness of each step is determined so as to keep the above radial voltage within acceptable limits.

The shape of a conductor sleeve, as shown in Figure 1.4.5., is determined in a similar manner by:[96]

$$x = \frac{V}{g_x} \ell n \left\{ \frac{\ell n(R_i/y)}{\ell n(R_j/r_s)} \right\} \tag{1.4.5.}$$

The length of one tapered part is obtained by setting $y = r$. The value chosen for the longitudinal electric stress g_x, is the same as the g_x of the stress relief cone.

The length of interface between the cable insulation and the joint insulation may be another factor requiring consideration. This is called the internal insulation distance; it starts at the rising part of the stress relief cone and ends at the bare part of the conductor. To a first approximation it is represented by

$$L = \frac{V}{g_l} \tag{1.4.6.}$$

For oil-impregnated paper insulation, the longitudinal electric stress is chosen as 0.8 to 1.2 kV/mm when V is the prospective failure voltage, and as 0.2 to 0.3 kV/mm when V is normal service voltage. For self-adhesive tape insulation, g_l is in the range of 0.5 to 0.7 kV/mm for its prospective failure voltage.

B. Stop Joints

Stop joints of the bushing type are more complicated in their structure than the normal joints mentioned already. Much of the design is based on practical experience. In principle, the design method is similar to that for normal joints, certainly the stress relief cone can be designed in the same manner. A continuous oil path is provided along the inner surface of the bushing from the edge of a conductor sleeve to the edge of a stress relief cone. It is therefore logical to adopt the design longitudinal stress a little lower than that of the normal joints. The design in the radial direction is based principally on the stress on the shield ring in the center of the joint and follows the concept of normal joints design.

There is another type of stop joint with a stop unit made principally of epoxy resin. Its

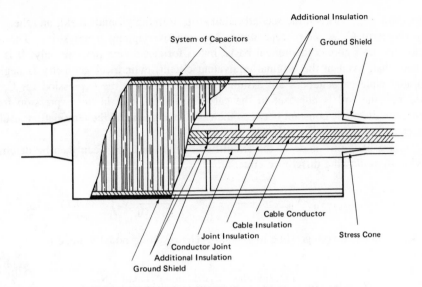

FIGURE 1.4.6. Basic construction of capacitively graded joint.

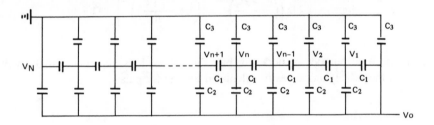

FIGURE 1.4.7. Simplified equivalent circuit for half of capacitively graded splice.

radial dimension is determined from formula 1.4.1. with the maximum radial stress 2 to 3 kV/mm for normal operation. The longitudinal design is made in a similar manner to the normal joints since no continuous oil paths exist. The shape of the unit is determined in the same manner with the stepping of the normal joints.

C. Capacitively Graded Joints

An alternative splice design concept employs capacitive grading over the length of the splice to control the electric stress distribution. The ground shield is removed from each end of the cable sections to be jointed, and a concentric system of cylindrical capacitors is connected between the conductor and ground shield of the cables as shown in Figure 1.4.6. The axial stress can then be controlled by selection of appropriate capacitor values.

Methods for calculating stress distribution in a capacitively graded splice may be either to solve rigorously for the currents flowing in circuit meshes formed by constituting capacitor elements, or to devise some suitable equivalent circuit for an approximate solution. Equations for mesh currents can be solved on a digital computer if necessary, although this solution in general needs a large-core storage in the computer and requires a long processing time.

A simplified equivalent circuit is proposed to calculate nodal voltages Vn, as shown in Figure 1.4.7. The capacitance C_2 is the average of two capacitances, the capacitance C_a between the first radial strip from the cable conductor and the cable conductor itself, and the capacitance C_b between the second radial strip from the cable conductor and the cable conductor. The capacitance C_3 is also the average of another pair of capacitances, the capacitance C_c between the first radial strip from the ground shield and the ground shield, and capacitance C_d between the second radial strip from the ground shield, and the ground shield. The term C_1 is series capacitance between overlapping metal strips. The above averaging has no rigorous theoretical basis, but is for calculation purpose only. It is to be noted here that, even in the original equivalent circuit, cylindrical geometry is neglected and parallel plate capacitances are assumed instead. Also, in the expression for C_a, it is assumed that the foil is adjacent to the cable insulation, and in the expression for C_b, differences in dielectric constant between the cable insulation and the termination insulation is neglected.

The voltage of the nth node of the simplified equivalent circuit can then be determined by solving the following difference equation:

$$V_{n+1} - 2V_n + V_{n-1} = \frac{C_2 + C_3}{C_1} \left(V_n - \frac{C_2}{C_2 + C_3} V_0 \right) \quad (1.4.7.)$$

where V_0 is the cable voltage (line to ground). The resulting nodal voltage is

$$V_n = \frac{V_0}{C_2 + C_3} \left\{ \frac{C_2 \sinh(\alpha N) - C_2 \sinh(\alpha n) - C_3 \sinh(n - N)\alpha}{\sinh(\alpha N)} \right\}$$

$$\text{with } \alpha = \pm 2 \ln \left\{ \left(\frac{C_2 + C_3}{4C_1} \right)^{1/2} + \left(1 + \frac{C_2 + C_3}{4C_1} \right)^{1/2} \right\} \quad (1.4.8.)$$

where N is the total number of nodes in half the splice. It is apparent that the voltage distribution in the splice can be easily calculated by this formula.

D. Terminations in Air

Like cable joints, cable terminations can use stress relief cones to mitigate stress concentration, especially in the neighborhood of the edge of the cable sheath. Practice confirms the effectiveness of this technique. For higher voltage, it is almost impossible to suppress the stress concentration sufficiently by this method. For this reason, a capacitively graded termination is usually preferred in these circumstances.

Asymmetrical capacitor bushings are in common use. A bushing of this type is made of cylindrical laminates of metal foils and insulating papers to form a set of cylindrical capacitances as depicted in Figure 1.4.8. They are designed so that voltage may be equally partitioned on individual capacitors. The two designs shown in Figure 1.4.8. are typical. In addition a capacitor-ring type of termination is available and common. It is made of multilayered doughnut-type capacitor elements in series connection.

We now discuss briefly one of the design methods designated (a) in Figure 1.4.8., which is often used in service. It is designed so that the thickness of insulation may be kept constant while the width of capacitances is changed. Figure 1.4.9. shows a line sketch of an asymmetrical capacitor cone and its equivalent circuit for design purposes.

If capacitances are chosen so as to make the voltages between adjacent metal foils all equal, we can derive the following equation from the consideration of currents flowing into a point s.

$$sC_{as} + C_{s-1} = C_s + (n - s)C_{bs} \quad (1.4.9.)$$

where $s = 1, 2, 3 \ldots (n - 1)$. We then obtain the electrode width 1_s of sth metal foil, i.e.,[6]

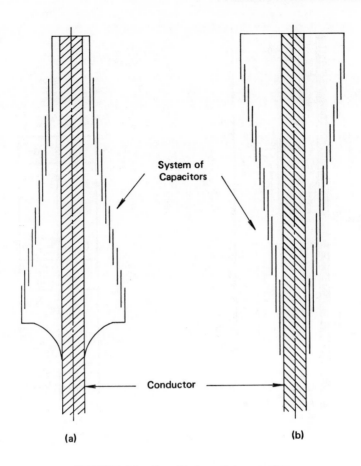

(a)   (b)

FIGURE 1.4.8.   Capacitively graded terminations.

$$L_s = (L_{s-1} - f) \frac{\dfrac{1}{\ln(d_s/d_{s-1})} - \dfrac{s}{\ln(d_s/d)}}{\dfrac{1}{\ln(d_{s+1}/d_s)} - \dfrac{s}{\ln(d_s/d)}}$$

$$+ f \frac{\dfrac{1}{\ln(d_{s+1}/d_s)} - \dfrac{k'}{k} \dfrac{n-s}{\ln(D/d_s)}}{\dfrac{1}{\ln(d_{s+1}/d_s)} - \dfrac{s}{\ln(d_s/d)}}$$

$$(1.4.10.)$$

where d = outer diameter of cable conductor, d_s = electrode diameter of sth metal foil, D = inner diameter of cylindrical ground metal surrounding the cable, f = staggered distance of external surfaces between the successive metal foil, g_s = staggered distance of internal surfaces between s^{th} and $(s-1)^{th}$ metal foils, L_s = width of sth metal foils, C_s = capacitance formed by s^{th} and $(s+1)^{th}$ metal foils, C_{as} = stray capacitance between the conductor and s^{th} metal foil, C_{bs} = stray capacitance between the surrounding ground (metal in a certain case) and s^{th} metal foil, and k, k' = constants containing permittivities. Once the length L_0 is predetermined, then $L_1, L_2...L_n$ can be successively calculated.

Since the stray capacitance to the surrounding ground can be virtually neglected for terminations in air, the width of sth metal foil is given by:[6]

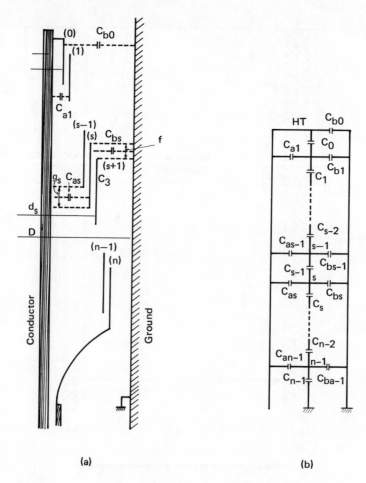

FIGURE 1.4.9.   Capacitive distribution of an asymmetrical capacitor cone and its equivalent circuit.

$$L_s = (L_{s-1} - f) \frac{\dfrac{1}{\ln(d/d_{s-1})} - \dfrac{s}{\ln(d_s/d)}}{\dfrac{1}{\ln(d_{s+1}/d_s)} - \dfrac{s}{\ln(d_s/d)}}$$

$$+ f \frac{\dfrac{1}{\ln(d_{s+1}/d_s)}}{\dfrac{1}{\ln(d_{s+1}/d_s)} - \dfrac{s}{\ln(d_s/d)}} \qquad (1.4.11.)$$

Internal insulation design is accomplished by coordinating bulk and interfacial breakdown voltages of capacitor elements with those of the cable of interest and the flashover voltage of its external insulation, especially for air terminations. An elementary capacitor is generally 1.0- to 1.5-mm thick. The length of the termination in air is determined by the flashover voltage of its bushing, and therefore gives a generous internal insulation design. The design stress used for capacitor elements is 12 to 15 kV/mm, each element being staggered with 20 to 25 mm space lengthwise from its neighbors.

Flashover voltage varies according to the relative position of internal reinforcing insulation

Table 1.4.2.
KIND OF EXTRUDED CABLE SPLICES

Types	Insulating materials
Tape wrap splices	Self-adhesive tape (EPR, PE, butyl rubber)
	Pressure-sensitive tape (XLPE)
	Heat-shrinkable tape (XLPE)
	EPR-Base tape
	PE-Base tape
	Irradiated PE-interposed EPR tape
	Often plus silicone oil
Field molded splices	Taping, heating, and injecting — cross-linking
	vulcanizable tape (EPR, PE)
	Injection mold (PE)
Prefabricated splices	EPR, silicone rubber
	Epoxy resin, insulating oil
Semi-prefab splices	EPR (mold or tape)
	Partial replacement of tape with mold

and the bushing. It arises from the difference in surface electric stress distribution caused by the relative position. Positive impulse and AC flashover voltages have a tendency to increase as the whole capacitor cone is set lower in position, while negative impulse flashover voltage has an inverse tendency.

E. Terminations in Oil

It is required to make the surface stress distribution on a bushing as uniform as possible. Hopefully, this will result in an increase in its flashover voltage. The oil-immersed portion bushing is much shorter than that of termination in air. Internal insulation design is basically the same as that for terminations in air, except that the stray capacitance to ground cannot be neglected in capacitor cone design because of the distance to the grounding shield box. Recently epoxy-resin stress relief cones have been introduced for voltage below 138 kV or 154 kV. Electric stress on the surface of the high-voltage shield ring is usually chosen as 2.5 to 3 kV/mm and 9 to 10 kV/mm for AC and impulse voltages, respectively, by taking into consideration long-gap breakdown strengths of insulating oil in a coaxial-cylindrical electrode configuration. The high-voltage shield may be covered with a solid insulation such as paper, especially for higher voltage cables.

F. Terminations in SF_6 Gas

The internal insulation design is almost the same as that for an oil termination. The design of an epoxy-resin bushing is determined by the gas pressure and the electric stress on the surface of internal high-voltage shield, because breakdown of an electro-negative gas is specified by the maximum electric stress as described in Volume I, Section 2.3.6. Impulse breakdown voltage is a determining factor in the design of gas insulation, since the impulse ratio is smaller for gas insulation than for OF cable insulation. Values of 9.5 to 11 kV/mm are usually chosen for the anticipated failure voltage when the pressure is 2.5 to 3.0 kg/cm². It is especially important for this type of termination that the shielding metal have as smooth a surface as possible and be as round as possible in its whole structure, so as to preclude extremely high local fields around the metal shield.

1.4.3. Designs for Extruded Cable Joints

Table 1.4.2. classifies typical types of extruded cable joints and comments on them.

A. Tape-Wrapped Splices

Taped splices have a long history of successful performance on 5-kV to 35-kV (or 3-kV

Table 1.4.3.
PERFORMANCES OF AN EPR-
BASED TAPE FOR SPLICE

Resistivity ρ (Ω-cm)	7.9×10^{15}
Loss tangent (%)	0.48
Permittivity	2.54
Breakdown strength (kV/mm)	56
Tensile strength (kg/mm^2)	0.45
Elongation (%)	1050
200% Md. (kg/mm^2)	0.14
Self-adhesiveness (kg/cm)	0.32
Thermal deformation (%)	
90°C	$3.9 \sim 4.8$
110°C	$3.8 \sim 5.0$

to 33-kV) cable systems.[97] Because of their adaptability, simple tape-wrap splices have been applied to newly developed higher voltage cables. Either a pressure-sensitive or self-amalgamating splicing tape has generally afforded a homogeneous build-up and provided good electrical strength and stability. At distribution voltages up to 27 to 35 kV, significant progress has been made in designing and making commercially available, snap-on and slip-on splice assemblies which can be installed in a relatively short time. As a consequence, these have been replacing tape-wrapped splices. At higher voltages, however, with attendant higher operating stress, it becomes very much more difficult and expensive to provide satisfactory accessory designs of this type. Tape-wrapped splices have recently given a good account of themselves in 46-kV and 69-kV (66-kV and 77-kV) systems and even in 138-kV (154-kV) systems.

Ethylene-propylene copolymer, polyethylene, and butyl rubber (plus silicone oil) are in use as tape materials. A variety of splicing tapes made from these materials is available. They include EPR self-amalgamating tape, polyethylene-based self-fusing tape, ethylene-propylene-based tape, pressure-sensitive cross-linked polyethylene tape, and heat-shrinkable cross-linked polyethylene tape. Yet others combine some type of splicing tape and a laminate structure comprising a layer of irradiated polyethylene tape interposed between layers of EPR insulating tapes. Among these, EPR selfamalgamating tape and EPR-based tape are the most popular.

A good tape insulation should have other properties besides having excellent insulation characteristics. Among these are

1. Excellent self-adhesiveness
2. An acceptable level for the modulus of tensile elasticity and stress relaxation resistance
3. Small thermal deformation

The self-amalgamating property is needed to bond one tape with another after taping, so as to reduce defects in the interfaces between successive tape laminations and prevent deformation of the splice. The splice is compressed by the inward radial force produced in tape-wrap structure by pulling and stretching insulation tape. This tensile force exerted on the tape is transformed into the compression force which would act to contact one tape with another closely and at the same time to reduce interfacial voids as far as possible. Thermal deformation is undesirable; it should be as small as possible. This is particularly true for the sloping parts of a splice. The performances of an EPR-based tape are tabulated in Table 1.4.3., which clearly demonstrate its excellent dimensional stability with temperature.

Tape wrapping can be executed either by hand or by machine. Generally speaking, better quality can be obtained by machine wrapping because it is easier to maintain tape tensile force and registration. In particular, it helps to reduce the size and number of voids produced in the rising part of the slope which is critical since it is subjected to comparatively high electric stress. Also, machine wrapping minimizes the surface roughness of external semiconducting layers.

In spite of their long history and economic advantage, tape-wrapped splices suffer from certain drawbacks. Some can be overcome, others may be inherent, in which case other alternatives should be considered. Problems are related to

- Interfacial voids
- Installation environment
- Workability

Where they exist, interfacial voids can cause early failures of finished splices. It is possible to reduce unfavorable voids to an allowable size and number by exerting a considerable degree of skill and care, but consistently fabricating perfect splices seems to be extremely difficult. In other words, it is very difficult indeed to avoid detrimental interfacial voids. Some experiments appear to demonstrate that no voids exist in wrapped-tape insulation near a ferrule, while voids of 20 to 40 µm in diameter are to be found near the outer layer. These are acceptable with present technology and regulation, and correspond with the permissible voids in cable. If wrapped-tape is to be applied for 138 kV and higher voltage cables, more and more skill and care will be needed.

The quality of the completed splice is a function of the ambient conditions in which the tape-wraps are formed. During installation, the splice may possibly be contaminated with dust and/or water (water, moisture, or even perspiration of splicers), resulting in degradation especially in electrical performance. One incident of splice failures is a matter of public record;[97] it occurred during DC proof tests, prior to placing a 69-kV cable in service. Tracking was found on the pencil surface of the splices which had failed. These failures were ascribed to water contamination. This experience was a warning to pay much more attention to the hygroscopic nature of XLPE surfaces when developing splicing procedures. This may be an extreme case, but more and more care of contamination should be taken because of the long splicing time.

Typically, the tape-wrapped splice is long (1.5 to 2.1 m) and up to twice the diameter of the cable. As a result, it is not unusual for a skilled splicing crew of 2 or 3 men to take up to 24 hr per phase in installing such a 138-kV cable splice in the field. Almost half of the time required for splicing may be consumed in hand-applied taping, thus, much improvement can be expected by the introduction of taping-by-machine. A suitable machine would also produce constant quality of splices.

A 69-kV taped XLPE cable splice is outlined in Figure 1.4.10., and the important design parameters are shown in Table 1.4.4.[98] For this joint, an electric planer, bladed file, and abrasive were used to shape the pencils. The abrasive used was 1-in. (25 mm) wide aluminum oxide tape. The shaped pencils were then wiped with cloths liberally soaked with trichlorethane. The next day an exothermic tape connector was installed, a final sanding with abrasive tape was carried out, and a second trichlorethane wash of pencils, connector, and conductor was undertaken. The stress levels listed in Table 1.4.4. are conservative except at the base of the belt. The recommended stress at this point is 6 V/mil (0.25 kV/mm) or less.

At the 138-kV level, conventional EPR self-amalgamating splicing tape affords an adequate joint insulation, although it provides only marginally acceptable impulse strength in the operating temperature range.[48] Excessive tape build-up is required to meet BIL impulse requirements at operating temperatures, since its dielectric strength is proportional to the square root of insulation thickness. It is possible that a pressure-sensitive cross-linked

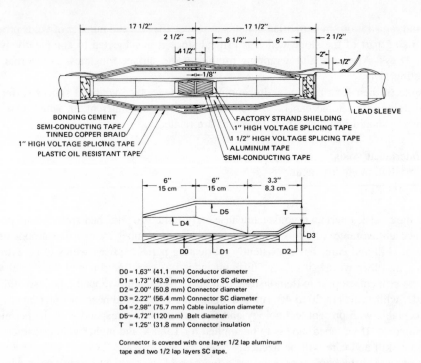

FIGURE 1.4.10. A 69-kV taped XLPE cable splice. (From Rittman, G. W. and Heyer, S. V., *IEEE Trans. Power Appar. Syst.*, 95(1), 303, 1976. With permission.)

Table 1.4.4.
STRESS PARAMETERS FOR 69-kV FIELD
JOINT

Creep stress	3.3 V/mil	(0.13 kV/mm)
Maximum radial stress at connector	47.6 V/mil	(1.87 kV/mm)
Maximum radial stress at base of pencil	45.9 V/mil	(1.81 kV/mm)
Maximum radial stress in cable	44.7 V/mil	(1.80 kV/mm)
Maximum stress at base of belt	6.7 V/mil	(0.26 kV/mm)
Maximum stress at base of pencil	4.7 V/mil	(0.19 kV/mm)

From Rittman, G. W. and Heyer, S. V., *IEEE Trans. Power Appar. Syst.*, 95(1), 306, 1976. With permission.

polyethylene tape could represent significant improvement in dielectric strength for tape-wrapped joints, but this is not presently available commercially. Another potentially attractive approach might well be heat-shrinkable cross-linked polyethylene tape, either alone or possibly in combination with some other type of splicing tape.

A 138-kV tape-wrapped joint construction is indicated in Figure 1.4.11.[99] The joint dimensions result in an average longitudinal stress of 3.5 V/mil, an average radial stress of 40 V/mil, and a maximum radial stress over the connector of 100 V/mil, all at rated voltage. The insulation of the joint was built up to approximately a 5-in diameter with consecutive applications of 2 layers of half-lapped EPR insulating tape and 1 layer of irradiated poly-

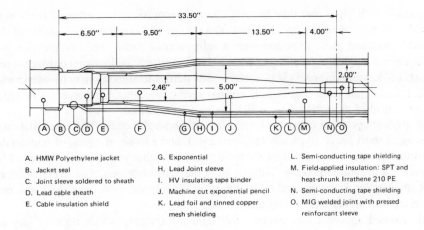

A. HMW Polyethylene jacket
B. Jacket seal
C. Joint sleeve soldered to sheath
D. Lead cable sheath
E. Cable insulation shield

G. Exponential
H. Lead Joint sleeve
I. HV insulating tape binder
J. Machine cut exponential pencil
K. Lead foil and tinned copper
mesh shielding

L. Semi-conducting tape shielding
M. Field-applied insulation: SPT and
heat-shrunk Irrathene 210 PE
N. Semi-conducting tape shielding
O. MIG welded joint with pressed
reinforcant sleeve

FIGURE 1.4.11.   A 138-kV tape-wrap splice design. (From Eager, G. S., Jr. and Silver, D. A., *IEEE Trans. Power Appar. Syst.*, 91(4), 1436, 1972. With permission.)

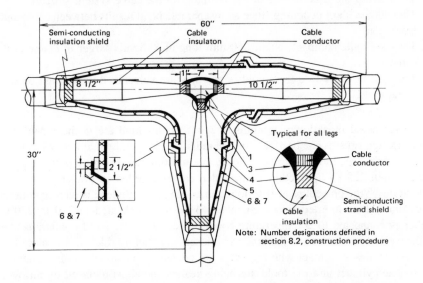

FIGURE 1.4.12.   A 69-kV tape-wrap tee joint. (After Balaska, T. A, Blais, L. D., and McBeath, R. H., *IEEE Trans. Power Appar. Syst.*, 93(3), 952, 1974. With permission.)

ethylene insulating tape, the latter in 8 mil thickness, applied with 1/4-in. open butt. The irradiated polyethylene tape was bedded into the underlying tapes by the application of heat. The shrinking of the irradiated polyethylene tapes by heat further served to place the hand-applied insulating tapes in compression around the cable and sharply reduce interlayer voids in the EPR tapes. It is interesting to note that the splice insulation can be stress-graded by using a specially formulated high dielectric constant (e.g., 4.3) ethylene-propylene base tape over the strand shield of the splice to a certain thickness (e.g., 0.125 in.).[100]

Hand-applied tapes can be applied to joints other than the straight joint as indicated for a field-installed 69-kV UN Tee joint in Figure 1.4.12.[101]

B. Field Mold Splices

One alternative to tape-wrapped splices is the field mold/cured insulation splice or hot splice,[97] which has gained increasing acceptance, particularly on 69-kV and the higher voltage cable systems. Such a field mold is usually made by applying a wrapping of cross-linking vulcanizable EPR tapes and then by heating and compressing the wrapped tapes together, the rubber being replenished by injection. The heating and compressing is accomplished by a portable molding press provided with a split metallic mold. The compression force is supplied by a wrapping of reinforcing tape (Mylar®, Teflon®, cellophane) or by gas pressure. It is light (20 kg) and compact (381 × 406 × 381 mm). Metal molds may be interchangeable to accommodate all cable sizes at all voltage levels. Vulcanizable polyethylene tapes can be used instead of the EPR tapes, which are half cross-linked.

The splicing operation consists of exposing only enough conductor for installation of the connector, a short pencil (typically less than 2 in.), and a loose wrapping of semiconducting and insulating tapes to the inside of the mold cavity, together with the injection of additional insulation during initial heating. Subsequently the splice is cured. The overall diameter of the finished splice is determined by maintaining the original cable wall thickness over the connector, with an additional tolerance for eccentricity.

Field-molded splices have several significant advantages, which become increasingly beneficial as the system voltage increases. Some of the advantages are as follows:

- They afford the least degree of discontinuity in the cable system structure
- They have a short penciling since an inseparable bond results between cable and splice insulation
- They are quite compact, about 350 mm long and slightly larger in diameter than the cable
- They are virtually free from voids since they are vulcanized under high pressure in a metal mold

The average radial breakdown strengths of joints of this kind are in the range of 9 to 15 kV/mm. Design stresses can be 10 kV/mm in radial and longitudinal directions, respectively.

Instead of the tape-mold described above, the injection mold of pure polyethylene is also applied to join 225-kV, polyethylene-insulated cables together in France.[102] According to the literature,[102] the procedure is as follows. Insulating polyethylene is packaged in the form of a cylinder. It is inserted into a chamber where it is preheated to 210 to 220°C. This chamber is connected to a temperature-controllable, two-part, mold containing the cable with the welded conductor and the inner rebuilt semiconducting shields. The mold is heated, but its extremities in contact with the cable are cooled by means of an air-radiator. As the polyethylene cylinder and the mold are being heated, the air is replaced by nitrogen so as to avoid oxidation of the polyethylene used. Polyethylene injection, which is in fact a transfer, is made possible by the pressure of nitrogen in the upper part of the chamber. The process takes approximately 2 hr.

Field molded splices also suffer from some significant shortcomings. While well engineered, the field molding unit is sophisticated, and the storage and handling of chemically reactive splicing compounds can be a problem. The skills of the installer seem to be comparable to that imposed by the tape-wrapped splice, bearing in mind that the splicer must also function as a molding technician. It is anticipated that the field-molding technique will be sensitive to ambient conditions. For the optimum achievable reliability, a controlled environment is considered essential. Turn-around time is 16 to 20 hr per joint for 138-kV XLPE cable, which is admittedly still too long.

C. Prefabricated Splices

A second alternative to tape-wrap splices is the factory molded, pretested slip-on splice assembly.[97] Premolded cable accessories gained wide acceptance at distribution voltages up to 25 kV and have been introduced for 35 kV applications. In spite of their innovative structure, because of their cost, they tend to be confined to special situations in which

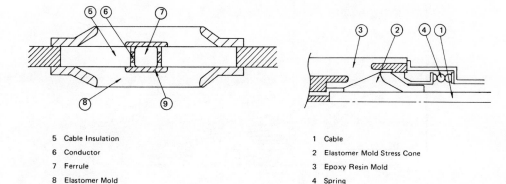

5   Cable Insulation
6   Conductor
7   Ferrule
8   Elastomer Mold
9   High Voltage Shield

1   Cable
2   Elastomer Mold Stress Cone
3   Epoxy Resin Mold
4   Spring

FIGURE 1.4.13.   Basic structures of prefab splice.

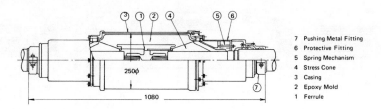

7   Pushing Metal Fitting
6   Protective Fitting
5   Spring Mechanism
4   Stress Cone
3   Casing
2   Epoxy Mold
1   Ferrule

FIGURE 1.4.14.   A 154-kV prefabricated splice.[104]

compactness, simplification in splicing, and shortening of installation time are extremely important. Because of their interchangeability and excellent insulation performance, it is to be expected that their use will expand to OF-XLPE cable hetero-joints, Y branch joints, joints with miniclad and in-air termination, and so on.

Basic designs are classified in two categories:

1.   Compression type
2.   Insert type (plug-in type)

The first type is provided with springs so as to transfer radial compressive force to the splice insulation, so making good contact. The second employs the elasticity of rubber (ethylene-propylene rubber or silicone rubber) to produce the compressive force. The distinctive features in their design are indicated in Figure 1.4.13.[103]

A prefabricated splice of the first type consists of main insulation and outer casing. The insulation is made from an assembly comprising an epoxy-resin-covered shielding metal tube and a rubber stress cone which reinforces the insulation at the edge of the cable insulation shield. The stress relief cone is provided with springs which act as a pressure compensating mechanism to retain good insulating performances in jointing interfaces and to make the splice follow the expansion and contraction of the cable insulation. The splice is housed in a metal case to prevent its being damaged and suffering water permeation after installation. Specially designed rubber packing is applied in contact with the cable sheath. Figure 1.4.14. outlines a prefab splice with basically the same structure as stated above.[104]

The main electrical problems in premolded parts are providing adequate wall thickness for radial electric strength, and seeing to it that the length of interference regions for electrical creep strength and electrical stress relief at diameter changes is sufficient to minimize stress concentration. Internal potential distribution can be calculated for a proposed splice design either by digital computer or by field mapping. Figure 1.4.15. shows the potential contours

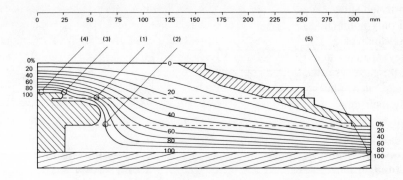

FIGURE 1.4.15. 138-kV Splice potential distribution.[97]

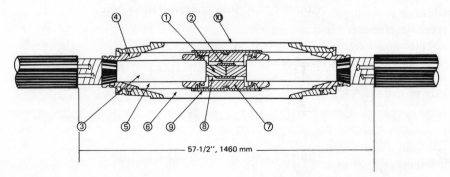

57-1/2'', 1460 mm

1. Cable Conductor. 1000 KCM compact aluminum conductor, diameter 29.26mm.

2. Conductor Connector. CADWELD connections have been used in the design qualification tests. A shielded cavity 140 mm long could accommodate other connectors.

3. Cable Insulation. Cross-linked polyethylene insulation 76.2mm O.D.

4. Cable Insulation Shield. Semi-conductive cross-linked poly-ethylene, diameter 94.49mm O.D.

5. Cable Adapters (2). The inside diameter of the adapters was 69.85mm. This size can accommodate cables with insulation diameter of 73.66 to 81.29mm.

6. Splice Sleeve Housing. The inside diameter wsa 111.76mm. The interference between the cable adapter and the splice sleeve housing was 127.mm.

7. Heat Transfer Sleeve. The inside diameter was 53.98mm. It maintains support for the adapter-housing interface in the event of cable insulation shrinkback up to 25.4mm each side.

8. Heat Transfer Medium. Semi-conductive tape built up over the conductor and the connector to an O.D. of 55.56mm.

9. Housing Conductive Insert.

10. Jacket. The electric field is terminated by a coating of conductive paint applied to the splice sleeve housing.

FIGURE 1.4.16. A 138-kV splice assembly.[97]

for a specified geometry of splice, generated by a computer program.[97] A numerical, iterative, finite difference solution of Laplace's equation was used to generate the potential at points on a grid. Some modification in design can be made, if necessary, such as the insertion of low resistivity or high dielectric constant materials in regions of high-stress concentration.

The conductor is an insulated splice tends to run hotter than the cable conductor because of the additional thickness of insulation; also some conductor connections have higher electrical resistance than an equivalent length of cable and therefore develop more heat. In this respect, a welded connection is preferred. A heat transfer sleeve can provide a path of low thermal resistance directly from the welded conductor connection to the inside of the splice sleeve housing.

A 138-kV splice assembly based on the above concept is presented in Figure 1.4.16.[99] The larger circumferential area of the housing yields a relatively low thermal resistance in spite of its considerable thickness. The overall effect is that heat is carried away not only from the connection inside the heat transfer sleeve but also from the adjacent cable where the cable insulation, cable adapter, and housing, form a very large thermal barrier. This may be effective in eliminating hot spots under the cable adapters.

Prefabricated splices seem to have the following features. They require short installation time and only nominal skills that are easily taught and learned. They are less sensitive to prevailing ambient conditions. They can provide for partially completed "jumpers" for emergency repairs, and it is claimed that it is possible to have the complete insulation system tested at the factory, prior to installation, thereby increasing the reliability of the system.

However, such splices possess certain disadvantages. Firstly their physical size, or more especially their circumferential area, is larger than that of the other types of splice. Their length might be reduced to that of tape-wrapped splices or less. The outer diameter of extruded cables varies within specified limits which may occasionally create trouble in interference fit designs. It is actually necessary to prepare several rubber stress cones with slightly different inner diameters for one splicing.

Semiconducting tape wrap or compound mold is used to electrically connect the cable insulation shield to the semiconducting layer of the rubber stress cone. The tape is liable to be affected in its insulation performance by the surface roughness of cable insulation to which it is wrapped. Greater care and further improvement are needed in the sanding procedure. Perfect sealing is required for cases that will be immersed in water during installation. Lastly, and admittedly this is the most significant disadvantage, there is the fact of cost which is much higher than for other splices. Hardware cost is probably 3 or 4 times higher, while installation cost can be reduced to 60 to 80%. As a consequence, overall cost may be twice as high as that of the other splices.

D. Semi-Prefabricated Splices

"Semi-prefab" splices are a possible modification of tape-wrapped splices dedicated to the simplification of installation, stabilization of electrically sound performance, and reduction in cost to the point of being comparable in cost with taped splices. Partial replacement of wrapped tapes by rubber stress cones must be the first challenge.

E. XLPE Cable Terminations

Terminations in air or directly connected to other electrical apparatus include rubber and epoxy-resin insulation elements, which can be called prefabricated accessories. Figure 1.4.17. outlines a 66- to 77-kV terminal.[105]

A drawing of the 138-kV extruded cable terminal, installed for cable testing at Waltz Mill is shown in Figure 1.4.18.[99] This is designated the Joslyn RD-148D terminal. The total creepage distance of this terminal is 130 in.; the dry arcing distance is 57 in; the weight is 500 lb, and the overall length from the bottom of the lower cover casting to the top of the hood casting is 87-7/8 in. The terminal, with inserts sized for the cable, was received fully assembled. The hood casing was removed prior to its assembly on the previously prepared end of the supported cable. The surface of the exposed insulation on the cable was coated with a light film of silicone compound and the terminal lowered onto the cable until the terminal seated on the base plate. The base plate was isolated from the mounting structure by small standoff insulators to permit power factor and corona discharge measurements to be made. The insulating and conducting elastomers were placed in compression around the cable by tightening down on the spring pressure plate at the top of the terminal. The terminal was then filled with silicone liquid, the hood casting replaced and tightened, and the weight of the cable transferred to the terminal by means of the connector stud nut.

F. Other Joint Designs

Rough sketches of several kinds of splices and terminations are shown in Figures 1.4.19.[103] and 1.4.20.[106] Figure 1.4.20. shows an injection-molded splice to connect 27-kV polyethylene to oil-paper cable.

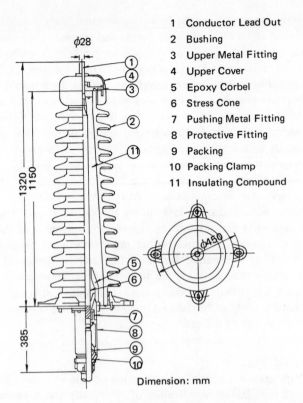

φ28

1 Conductor Lead Out
2 Bushing
3 Upper Metal Fitting
4 Upper Cover
5 Epoxy Corbel
6 Stress Cone
7 Pushing Metal Fitting
8 Protective Fitting
9 Packing
10 Packing Clamp
11 Insulating Compound

φ450

Dimension: mm

FIGURE 1.4.17.   A 66 ~ 77 kV XLPE cable termination.[105]

1.4.4. Designs of OF and POF Cable Joints

A. Straight Joints

Two designs of straight joints (normal and insulation) for OF cables are outlined in Figures 1.4.21.[6] and 1.4.22.[6] together with their dimensions according to voltage level. Asphalt-based compounds or some other insulating material can be applied to fill a space around the ferrule in oil-paper solid cables, thereby furnishing the main insulation of the joints. OF, PGF, and POF cable joints utilize the same insulation system as their companion cables, i.e., oil-impregnated paper and oil. Two kinds of papers are used: flat or hard paper or crepe paper. The flat paper is the same as paper applied to cable itself. The crepe paper is expected to reinforce the leakage path along the surface of the original cable insulation, which is normally the weakest point in splice designs.

Figure 1.4.23. shows normal and stop joints for 345-kV low-pressure oil-filled cable which were installed and tested together with their companion cable at Cornell University.[107] It was concluded after testing that carbon black crepe-paper might better be applied directly over the conductor for the stop joint.

Figure 1.4.24. shows construction details of a normal joint for a 500-kV HPOF cable tested at Waltz Mill.[67] A field-machine copper connector is secured in place with compression dies, then shaped by a special milling machine, and finally hand-finished with oiled aluminum oxide cloth in various grits. The creepage distance is 66 in. (168 cm), the pencil being 46.7 in. (119 cm) and the stress cone 19.3 in. (49 cm). The overall length of the joint from base-to-base of stress cones is 145 in. (370 cm) and the joint diameter is 6.20 in. (15.8 cm). The hand-applied insulation is straight paper matching the cable tapes.

Hand-taped joints can be thermal barriers under normal operating static conditions in high-power cable lines, and therefore require derating of the cable. To minimize such thermal hazards, a moderate insulation build up and/or a thermal smoothing by either oil oscillation,

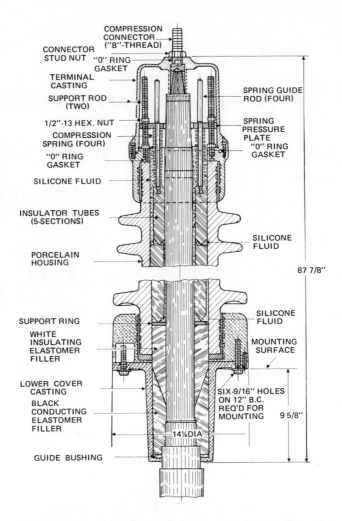

FIGURE 1.4.18.   A 138-kV extruded cable terminal. (From Eager, G. S., Jr. and Silver, D. A., *IEEE Trans. Power Appar. Syst.*, (9)4, 1436, 1972. With permission.)

or circulation in case of POF cable, should be employed. In this connection, hard paper insulation is preferred to crepe paper to achieve optimum heat transfer and improved dielectric strength. Furthermore, the reduced-wall joint also accommodates the concept of a dynamic cable system employing circulatory oil cooling for improved thermal control, with enhanced ampacity for the splice. If the return path is external to the pipe, the temperature at such locations is below the average oil temperature in the pipe and a net cooling occurs.

A semi-synthetic joint was designed and manufactured for Waltz Mill testing by using PP/paper laminates. It is characterized by low loss tangent ($8 - 10 \times 10^{-4}$ at 100°C), low dielectric constant (2.6 to 2.8), and high dielectric strength (approximately twice that of paper). This design is shown in Figure 1.4.25.[65] The impregnant was a medium-viscosity mineral oil.

Two other designs of joints for 500-kV HPOF cables appear in Figure 1.4.26.[108]

B. Capacitively-Graded Joints

As is often said, conventional hand-wrapped joints require highly paid skilled workmen for proper construction, and yet they are generally regarded as the weak links of a system.

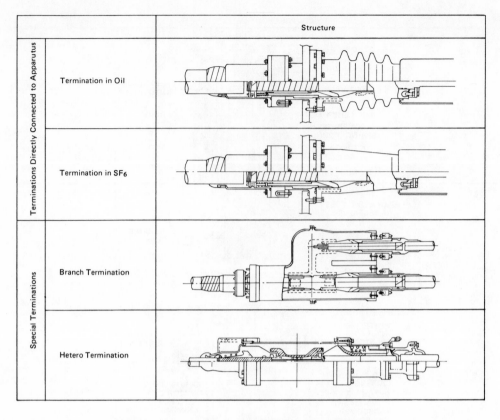

FIGURE 1.4.19.   Several designs of splice and terminations.[103]

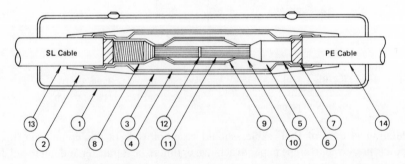

1. Lead Sleeve 3-1/2″ I.D.X20″ Length
2. Petrolatum
3. Copper Mesh Tape
4. Silicone Rubber Semicon Tape
5. Polyethylene Molding Compound-Voltage Stabilized
6. Petrolatum Seal
7. Cable Semicon Tape
8. Epoxy Impregnated Cotton Tape Oil Seal
9. Epoxy-Silicon Carbide Impregnated Cotton Tape Strand Shield
10. Elastomeric Conductive Tape
11. Soldered Copper Connector
12. Copper Diaphram Strand Oil Stop
13. Oil-Paper Insulated 27 kV 500 MCM Cable
14. Polyethylene Insulation Cable 27 kV 500 MCM

FIGURE 1.4.20.   Injection molded cable splice design. (hetero-joint). (From Brealey, R. H., Arone, N. F., Nakata, R., and Fischer, F. E., *Injection Molded Cable Splice for Connecting Oil-Paper Cable to Polyethylene Cable,* IEEE PES Winter Meeting Rec. C75 205-0, IEEE, Piscataway, New Jersey, 1975, 1. With permission.)

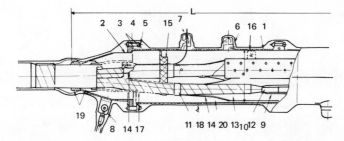

					Nominal Voltage	Nominal Cross Section [mm²]	Dimension [mm]	
							D	L
1.	Connecting Copper Pipe		11.	Insulating Tape	66kv	< 400	180	1450
2.	End Copper Pipe		12.	Cable Conductor	77kV	< 400	190	1500
3.	Flange		13.	Cable Core	110kV	< 400	250	2200
4.	Packing		14.	Tinned Annealed Copper	154kV	< 400	310	2600
5.	Clamping Bolt		15.	Core Holding Metal				
6.	Connector		16.	Core Holding Metal				
7.	Valve		17.	Trifurcate Lead Tube				
8.	Ground Terminal		18.	Anticorrosion Layer				
9.	Ferrule		19.	Plumber				
10.	Oil-Impregnated Paper		20.	Insulating Oil				

FIGURE 1.4.21. Normal joint for three-core OF cable. (From Iizuka, K., Ed., *Handbook of Power Cable Technology*, (in Japanese), Denki-Shoin, Tokyo, Japan, 1974, 304. With permission.)

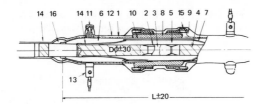

					Nominal Voltage	Nominal Cross Section [mm²]	Dimension [mm]	
							D	L
1.	Connecting Copper Pipe		9.	Clamping Bolt		< 1000	110	960
2.	Ferrule		10.	Packing	66kV	1200	120	1060
3.	Cable Conductor		11.	Connector	77kV	1500	130	1060
4.	Cable Core		12.	Anticorrosion Layer		1600~2000	140	1060
5.	Oil-Impregnated Paper		13.	Ground Terminal	110kV	< 800	140	1300
6.	Insulating Oil		14.	Tinned Copper		< 1000	160	1450
7.	Insulating Tape		15.	Flange	154kV	1200~2000	180	1550
8.	Insulating Sleeve		16.	Plumber	275kV	< 1500	240	1800

FIGURE 1.4.22. Insulation joint for OF cable. (From Iizuka, K., Ed., *Handbook of Power Cable Technology*, (in Japanese), Denki-Shoin, Tokyo, Japan, 1974, 305. With permission.)

The supply of such skilled technicians in this field has seriously declined in recent years. It has been apparent for some time that there is a need for a prefabricated joint for taped cables. Such a joint is now available. Figure 1.4.27. shows a printed circuit capacitive-graded joint for 138-kV cables.[109] Penciling is eliminated by taking advantage of capacitive grading. One of two basic design concepts control the axial voltage stress along the joint by printed circuit capacitors. The first concept, described in Section 1.4.2., employs uniform capacitive distribution. The second, more refined approach, is designed to achieve a uniform longitudinal stress by proper selection of the grading elements.

The main joint elements comprise (1) slip-on supporting sleeve, (2) graded capacitive sections, (3) electrostatic shield, (4) splice insulation, and (5) splice shield. According to the literature,[109] manufacturing details are as follows. The splice connector supports an electrostatic transfer link between the conductor and the equipotential ring located at the

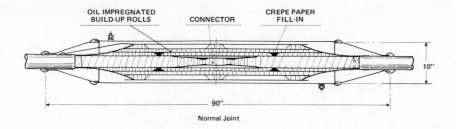

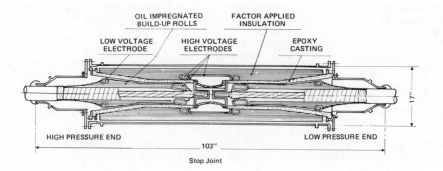

FIGURE 1.4.23.   Normal and stop joints for 345 kV low pressure oil-filled cable. (From Walker, J. J. and Juhlin, E. O., *IEEE Trans. Power Appar. Syst.*, 85(4), 344, 1966. With permission.)

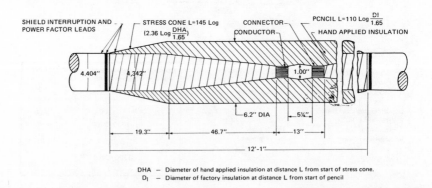

FIGURE 1.4.24.   A normal joint for 500-kV POF cable. (From Eager, G. S., Jr., Cortelyou, W. H., Bahder, G., and Turner, S. E., *IEEE Trans. Power Appar. Syst.*, 90(1), 248, 1971. With permission.)

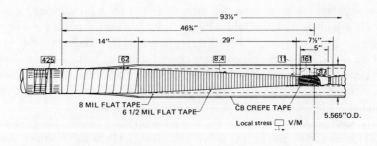

FIGURE 1.4.25.   A semi-synthetic prototype joint. (From Eich, E. D., *IEEE Trans. Power Appar. Syst.*, 90(1), 221, 1971. With permission.)

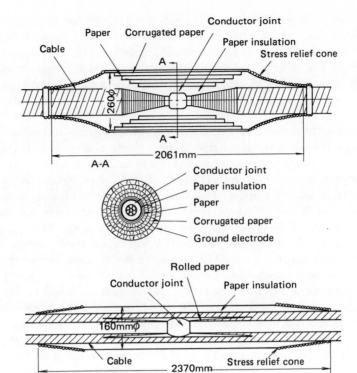

FIGURE 1.4.26.   Designs of 500-kV cable joints. (From Watanabe, T., Kikuchi, K., Matsuura, K., Yoshida, N., Haga, K., and Numajiri, F., *Higashi-Tokyo EHV Cable Test Project in Japan,* IEEE PES Winter Meeting Rec. C75 208-4, IEEE, Piscataway, New Jersey, 1975, 3. With permission.)

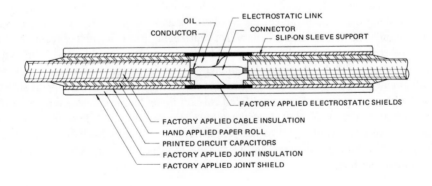

Typical Dimensions

I.D. Sleeve Support	3.75″	(9.5cm)
Thickness of Sleeve	0.125″	(0.3cm)
O.D. Capacitive Section	4.95″	(12.6cm)
Joint Insulation Thickness	0.7″	(1.8cm)
O.D. Joint Insulation	63.5″	(16.1cm)
O.D. Over Splice Shield	6.40	(16.3cm)
Overall Length of Joint	58.0″	(147.3cm)

FIGURE 1.4.27.   A capacitive graded joint for 138-kV HPOF cables. (From Allam, E. M. and McKean, A. L., *IEEE Trans. Power Appar. Syst.,* 96(1), 21, 1977. With permission.)

center of the slip-on sleeve. The controlled longitudinal distribution of stresses is achieved by the use of two identical capacitive graded rolls made of printed circuit sheets of suitable dielectric and provided with continuous, parallel, metallized strips. These two capacitive sections are located at either side of the equipotential ring, and are physically connected to it. To provide the necessary dielectric strength between adjacent layers of capacitors, each printed-circuit layer is alternated with a plain insulated sheet of the same dielectric. The capacitive cylinders are installed by wrapping them over the rigid sleeve, their end electrodes being connected to the cable shield and to the high-voltage electrostatic shield. The slip-on supporting sleeve is made of sturdy material. Its main functions are to provide a rigid element on which the splice components can be wound and to facilitate installation. Because of its excellent high-voltage insulating characteristics, phenolic resin-impregnated paper tubing was found suitable for this purpose. This sleeve is provided with slots at its midpoint, surrounded by a conductive cylindrical equipotential ring, to which the connector and the high-voltage terminals of the graded capacitors are connected. Crepe paper tape is used to fill the space between the printed-circuit elements, supporting sleeve, and the outer splice insulation. The space between the rigid slip-on sleeve and the cable insulation at each side of the connector is filled with insulating paper applied in the form of a wide roll. The entire length of the splice is insulated with high-purity, low-loss cellulose paper. In order to provide an equipotential conducting surface around the outside of the splice insulation for charging, eddy, circulating, and fault currents, a conductive shield is applied over the outer insulation of the splice. The splice assembly, comprising the components described above, is then factory-processed by placing it in a sealed vessel permitting removal of the moisture by heat and vacuum, after which the splice is oil-impregnated following established procedures normally employed for the impregnation of paper-insulated cables.

C. Flexible EHV Cable Connection

Transmission cable connections to an offshore floating plant present unique flexing problems due to motions of the plant during storm conditions even where the plant is moored in the basin of a protective breakwater.[110] Possible solutions are an overhead transmission wire, an underwater cable catenary, or an overhead cable. The overhead transmission wire has a very significant disadvantage, the insulators may be exposed to salt water or salt ice contamination. The underwater catenary concept seems to be feasible except in shallow basins. The structure-supported overhead cable design appears feasible, too. The matter of support presents quite a developmental challenge.[110] The supporting structure so far envisaged for each circuit is basically a three-hinged bridge consisting of two vertical support structures, one on the floating plant and one on a fixed caisson in the basin, connected by a horizontal bridge structure. The three-hinged design allows the structure to follow plant motion. In this design, the cable is clamped to the support system structure but is required to flex at the hinges. Distribution of strains due to flexing is achieved by means of stainless steel arches around the hinges.

D. Terminations

Figure 1.4.28. shows some designs of cable heads for 500-kV class cables.[108] With outdoor sealing end boxes, 8-m-long insulators are employed to withstand very heavy salt contamination. The surface of the stress relief cone is oil-cooled. In some cases, the concentric cylindrical capacitor is divided into several parts to facilitate handling. Some designs use FRP insulation tubes to supply sufficient resistance against internal pressure. Metal-enclosed potheads are oil-cooled on the surface of their stress-relief cones, and insulated with SF_6 gas outside the insulator. Oil-immersed sealing end boxes are also oil-cooled on the surface of their stress cones; insulation outside the insulator is provided by heavy alkylbenzene.

Figure 1.4.29. shows a cross section of a pothead for a 500-kV cable, installed and tested

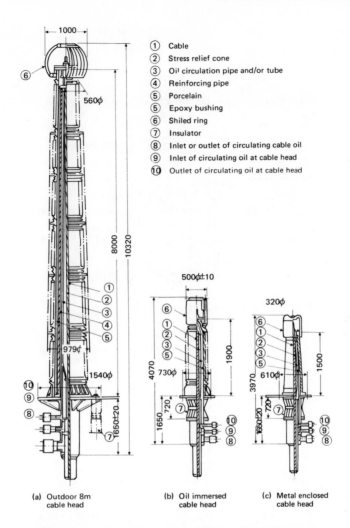

① Cable
② Stress relief cone
③ Oil circulation pipe and/or tube
④ Reinforcing pipe
⑤ Porcelain
⑤ Epoxy bushing
⑥ Shiled ring
⑦ Insulator
⑧ Inlet or outlet of circulating cable oil
⑨ Inlet of circulating oil at cable head
⑩ Outlet of circulating oil at cable head

(a) Outdoor 8m cable head

(b) Oil immersed cable head

(c) Metal enclosed cable head

FIGURE 1.4.28.   Construction of 500-kV class sealing end boxes. (From Watanabe, T., Kikuchi, K., Matsuura, K., Yoshida, N., Haga, K., and Numajiri, F., *Higashi-Tokyo EHV Cable Test Project in Japan,* IEEE PES Winter Meeting Rec. C75 208-4, IEEE, Piscataway, New Jersey, 1975, 3. With permission.)

at Waltz Mill,[111] which is somewhat different in design from that just described. Internal and external field control of the longitudinal electric stresses are accomplished by a graded stack of 50 doughnut-shaped capacitor elements. Physical support for the stack of capacitor elements is provided by a centrifugally cast epoxy tube. Use of the cast epoxy support tube, in place of a phenolic support tube, permits operation of the conductor within the pothead at appreciably higher temperatures because of the greater thermal stability of the capacitor support. Transition of the ground potential electrode from the cable-shielding diameter to the bore diameter of the epoxy capacitor support tube is accomplished by a preperforated paper roll stress relief cone.

It is claimed that both stress cone-type and capacitor-type sealing ends will be feasible for 750-kV OF cable.[112] For still higher voltages, i.e., 1000 to 1200 kV, the former seems no longer possible because of the intrinsic limitations of its electrical performance and the higher oil pressure required. The latter will probably be put into service by taking advantage of its better withstand voltage performance and its smaller radial dimensions. The withstand voltage increases linearly with length.

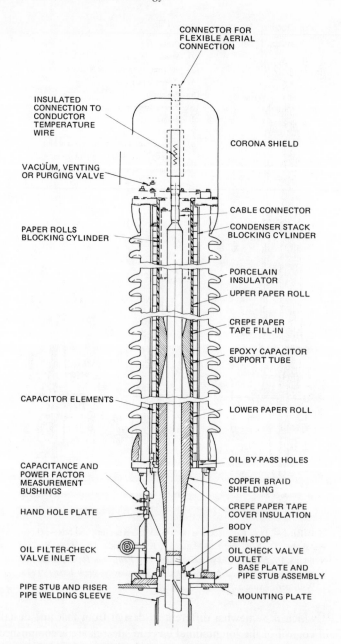

FIGURE 1.4.29.   A 500 kV pothead design. (From Gear, R. B., Heppner, D. R., Lusk, G. E., and Nicholas, J. H., *IEEE Trans. Power Appar. Syst.*, 90(1), 201, 1971. With permission.)

1.4.5. Improvement of Splices and Terminations

A. Thermal Problems

One limitation to current up-rating is excessive temperature in accessories, which are often referred to as thermal bottlenecks. The most critical condition is found in straight-through joints. Joints are more heavily insulated than their companion cables, resulting in an increase in thermal resistivity at these spots in a cable system. Improvements can be obtained by several methods, but as a prerequisite it is important to develop suitable methods for calculating the temperature distribution in and around a joint.

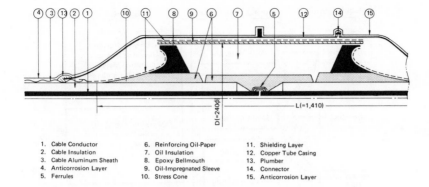

1. Cable Conductor	6. Reinforcing Oil-Paper	11. Shielding Layer
2. Cable Insulation	7. Oil Insulation	12. Copper Tube Casing
3. Cable Aluminum Sheath	8. Epoxy Bellmouth	13. Plumber
4. Anticorrosion Layer	9. Oil-Impregnated Sleeve	14. Connector
5. Ferrules	10. Stress Cone	15. Anticorrosion Layer

FIGURE 1.4.30.   A normal joint with low thermal resistivity. (From Maeno, T., Gomi, Y., Ando, N., and Ide, S., *Hitachi Hyoron*, 57(8), 78, 1975. With permission.)

Several methods of reducing the rise in joint temperature are available.

- Utilization of backfills with low thermal resistivity.
- External cooling: in water-cooled cable systems, the problem can be tackled by cooling the joint casing by water circulation in a jacket around the joint. In spite of this, the joint still tends to run at a higher temperature than the cable conductor.
- Internal cooling: thermal smoothing by circulation in one direction or oscillation of the cable oil in alternate directions has been successfully used, the latter method being essential where efficient cooling of the terminations of integral sheath-cooled installations is required.
- Utilization of insulating materials with low thermal resistivity. Use of epoxy-resin and insulating oil with low thermal resistivity as reinforcing insulation can reduce the temperature rise.
- Reduction in thermal resistivity as a whole. This can sometimes be achieved by reducing insulation thickness.
- Reduction in ambient temperature. This is possible by optimizing joint bay layout or by reducing temperature in manholes or tunnels.

Figure 1.4.30. illustrates one design for a normal joint in a 275-kV OF cable. The objective is to lower the thermal resistivity of the electrical insulation.[113] Instead of oil-impregnated paper as joint reinforcing insulation, insulating oil is used. It will be noted that it is surrounded by a pair of epoxy bell-mouths and an oil-impregnated insulation tube. Convection of the oil increases heat transfer and removal to the outside of the joint. The low thermal resistivity of epoxy resin is an additional factor. It is interesting to note that this design eliminates an elaborate tape-wrapping operation, thereby meeting the requirements for simplification and reliability, too. The use of low-permittivity oil can relax the electric stress on the oil-impregnated paper-covered conductor and on the ferrule. Molded epoxy-resin is homogeneous with respect to breakdown path; it has no preferentially weak directions. In this regard, it is unlike wrapped paper which will withstand higher longitudinal stress than oil-impregnated paper.

In the development of 138-kV XLPE cable splices described previously in Figure 1.4.16., the devised heat transfer sleeve provides a path of low thermal resistance directly from the welded conductor connection to the inside of the splice sleeve housing.

B. Optimization

In order to achieve optimum designs for cable joints, one must aim to improve reliability,

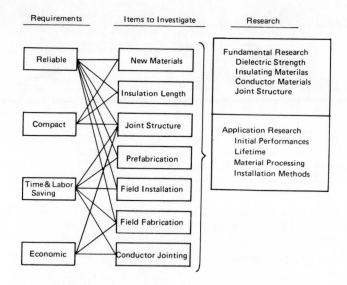

FIGURE 1.4.31. Optimization strategy for cable joints.

compaction, simplification of field installation procedures, and overall cost, both from the standpoint of the joints themselves and the whole cable system, including its companion cables and other apparatus to which it may be connected. One research and development strategy is illustrated in Figure 1.4.31., which is self-explanatory.

1.5. FORCED-COOLED CABLE SYSTEMS

As discussed in Section 1.2.1., forced cooling is the last but most efficient method of increasing the ampacity of conventional cables with oil-impregnated paper, extruded plastic, or rubber insulation. Broadly speaking, there are two methods for artificially cooling the cables, direct cooling and indirect cooling, as shown in Table 1.5.1. Integral or lateral external cooling is simple technically, but the ampacity is limited by the thermal resistivity of the electrical insulation. For this reason, external cooling is not very effective for UHV cables with heavy insulation. However, this cooling system provides a good way of extending the limits of conventional cables, especially self-contained, oil-filled cables. Moreover, it can be achieved without major technical and economic risks. The simplest method, the use of cooling by water flow in parallel tubes, can enhance the ampacity by about 50% to 60%.[4] It is obvious that pipe-type, oil-filled cables are advantageous because of their inherent adaptability to forced cooling. Oil that bathes the conductors can be refrigerated and circulated through a number of cooling loops comprising the entire cable length. Such direct cooling may increase steady-state rating of a circuit by over 100%.[61]

1.5.1. Analysis of Cooling

A. General Treatment

Thermal behavior of a family of cables installed with forced cooling systems are governed by the following basic equations:[6]

$$T_i - T_o = [R_{ij}][W_i']$$ (1.5.1.)

$$W_i'' = W_i' + W_i$$ (1.5.2.)

$$W_i = k_i C_i Q_i \frac{dT_i}{dz}$$ (1.5.3.)

Table 1.5.1.
REPRESENTATIVE COOLING METHODS

Cooling Method	Medium	Cable	Arrangement	Remarks
Direct Cooling Internal Cooling	Oil	OF Cable	C Cooling Medium	Effective to increase power for relatively short length of cable.
	Water	Extruded Cable		
External Cooling (Integral Cooling)	Air in Tunnel	Any Cable	Air or Water C C C	Easy to apply for multiple cable installation.
	Water			Effective for long distance power transmission.
	Oil	POF Cable	C C C	
	Water	Any Cable	c c c	
Indirect Cooling (Lateral Cooling) (External Cooling)	Water	Any Cable	c c c c c	Easy to apply for long distance cable.

Note: Symbol C represents a cable.

where T_i = temperature of a cable pipe or ambient medium (°C), R_{ij} = self (i = j) and mutual (i ≠ j) thermal resistance of cable pipe or cooling pipe (°C − cm/W), W_i = heat dissipated from cable pipe or cooling pipe toward ground (W/cm) (negative for the cooling pipe), W''_i = heat generated inside cable pipe or cooling pipe (W/cm) (zero for the cooling pipe in case of the indirect cooling), W_i = heat absorbed by coolant, C_i = the specific heat of coolant (W·sec/cm³·°C), Q = flow rate of coolant, (cm³/sec), k_i = constant determined by the direction of coolant flow, k_i = +1 (direct cooling), −1 (indirect cooling) in case of positive flow direction with respect to z, and k_i = −1 (direct cooling), +1 (indirect cooling) in case of negative flow direction with respect to z, T_0 = base temperature, T_i = oil temperature of ith pipe, $[T_i − T_0]$ = the matrix of $(T_i − T_0)$, $[R_{ij}]$ = the matrix of thermal resistances R_{ij}, $[W'_i]$ = the matrix of heats W'_i, and z = distance along cable lines installed.

From the above three equations, we obtain

$$\left[W''_i − k_i C_i Q_i \frac{dT_i}{dz} \right] = [R_{ij}]^{-1} [T_i − T_0] \qquad (1.5.4.)$$

where $[R_{ij}]^{-1}$ is the inverse matrix of $[R_{ij}]$. This can be changed in the form of:

$$C_j Q_j \left[\frac{dT_i}{dz} \right] = [a_{ij}] [T_i] − [b_j] \qquad (1.5.5.)$$

and also

$$C_j Q_j \left[\frac{dT_i}{dz} \right] = [a_{ij}] [t_i]$$

$$\text{with } [a_{ij}][K_i] = [b_i] \text{ and } T_i = t_i + k_i \qquad (1.5.6.)$$

General solution of Equation 1.5.6. is

$$t_i = \sum_{k=1}^{n} P_k C_{ki} \exp \left(\frac{\lambda_k}{C_j Q_j} z \right) \qquad (1.5.7.)$$

where λ_k = the eigenvalues of $[a_{ij}]$, C_{ki} = eigenvectors of $[a_{ij}]$ and P_k = undetermined multipliers. Finally we obtain

$$T_i = K_i + \sum_{k=1}^{n} P_k C_{ki} \exp \left(\frac{\lambda_k}{C_j Q_j} z \right) \qquad (1.5.8.)$$

The undetermined multipliers P_k can be calculated by solving the first order multidimensional allied equations under given boundary conditions at $z = 0$ or 1. We thereby obtain the oil temperature T_i at any distance z. For systems including different cooling directions, Equation 1.5.6. must be solved at each section with the same cooling direction. Some modification of the equation is needed for POF cable, because heat is generated in the pipe, also.[114] A flow chart to solve the equation by computer is shown in Figure 1.5.1.

B. Special Cases

For simple cases, it would be convenient to express solutions explicitly. Solutions are given below for some specified cable installation conditions.

a. Cooling by Wind in Tunnel Installation

The atmospheric temperature at a distance z from the air inlet is given by[115]

$$T_z = T_f + (WR_e + T_0 - T_f) \left(1 - \exp \left(- \frac{z}{CQR_e} \right) \right) \qquad (1.5.9.)$$

where W = heat generated in the tunnel (W/cm), R_e = external thermal resistance of the tunnel [°C − cm/W], Q = air flow rate, C = the specific heat of air at constant pressure [W·sec/cm^3 °C][1.1×10^{-3}], T_0 = base temperature, and T_f = air temperature at the air inlet (°C).

b. Indirect Cooling in Conduit Installation

Consider a cable system comprising a cooling pipe with unidirectional coolant flow and several cables installed in pipes as shown in Figure 1.5.2(a). For simplicity, it is assumed that all the cables here have the same characteristics and suffer the same current flow. Then the following formulas hold.[116-118]

$$\left. \begin{array}{l} W'' = W + W' \\[2mm] (T_s - T_0) = R_s W' \\[2mm] (T_s - T_p) = R_6' W \\[2mm] dT_p/dz = W/CQ \end{array} \right\} \qquad (1.5.10.)$$

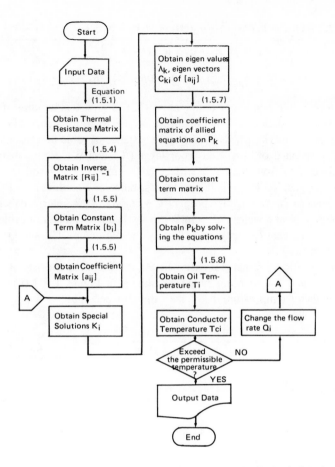

FIGURE 1.5.1.   Flow chart to calculate temperature distribution in forced cooled conditions.

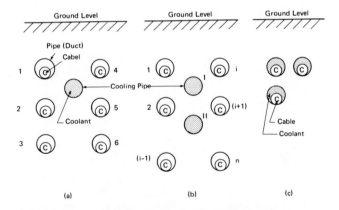

FIGURE 1.5.2.   Several types of cable cooling arrangement.

where T_p = temperature of cooling water (°C), T_s = temperature of pipes (°C), W″ = heat generated by cable in a pipe, W = heat absorbed by cooling water out of W″ (W/cm), W′ = heat dissipated from pipes toward ground out of W″ (W/cm), R_5 = equivalent soil thermal resistance (°C − cm/W), and R'_6 = thermal resistance of soil between the cooling pipe and pipe ducts. From these equations and a similar equation to 2.1.9. (Volume I), we can deduce the permissible current I (ampere) as follows:

$$I = \frac{1}{r_c} \cdot \frac{(T_c - T_0) - (T_{po} - T_0) Ae^{-B} - W_d((1/2)R_1 + R_2 + R_3 + R_5)(1 - Ae^{-B})}{R_1 + (1 + P)(R_2 + R_3 + R_5)(1 - Ae^{-B})}$$

with $A = R_5/(R_5 + R_6 + R_p + R_w)$ and $B = L/CQ(R_5 + R_6 + R_p + R_w)$ (1.5.11.)

where T_c = permissible conductor temperature (°C), T_{po} = temperature of cooling water at its inlet (°C), W_d = dielectric loss (W/cm), r_c = electrical resistance of conductor at the permissible temperature (Ω/cm), R_1 = thermal resistance of cable insulation (°C − cm/W), R_2 = thermal resistance of corrosion-proof layer (°C − cm/W), R_3 = thermal resistance against heat dissipation from the cable surface (°C − cm/W), R_5 = soil thermal resistance of cable pipes (°C − cm/W) R_6 = soil thermal resistance of a cooling pipe (°C − cm/W), R_p = thermal resistance of a cooling pipe proper (°C − cm/W), R_w = interface thermal resistance between a cooling pipe and water (°C − cm/W), p = sheath loss factor of cable, L = cooling length (cm), C = the heat capacity of cooling water (W·sec/cm³°C), and Q = flow rate of cooling water (cm³/sec).

Next consider a little more general case where a cable system consists of multiple cables of the same kind and two cooling pipes, as shown in Figure 1.5.2.(b). As a result of calculation, conductor temperature T_{ci} of the i th cable becomes

$$T_{ci} = T_0 + R_{int}W + \frac{R_A W + \lambda_1 CQ(\delta_1 A_1 R_{iI} + \delta_{II} A_2 R_{iII}) e^{\lambda_1 z}}{\lambda_2 CQ(\delta_1 B_1 R_{iI} + \delta_{II} B_2 R_{iII}) e^{\lambda_2 z}} \quad \text{with } R_A = \sum_{k=1}^{n} R_{ik}$$ (1.5.12.)

R_A = the sum of soil thermal resistance of a pipe,

$$A_1 = \frac{a_1}{b_1} \frac{\lambda_1 \lambda_2}{\lambda_1 - \lambda_2} \left\{ \frac{b_1 + \lambda_2}{a_1 \lambda_1} (T_{poI} - T_0) + (T_{poII} - T_0) - \frac{c_1}{b_1} \frac{b_1 + \lambda_2}{a_1 \lambda_2} - \frac{c_2}{b_2} \right\}$$

$$B_1 = \frac{a_1}{b_1} \frac{\lambda_1 \lambda_2}{\lambda_1 - \lambda_2} \left\{ \frac{b_1 + \lambda_2}{a_1 \lambda_1} (T_{poI} - T_0) + (T_{poII} - T_0) - \frac{c_I}{b_1} \frac{b_1 + \lambda_2}{a_1 \lambda_2} - \frac{c_2}{b_2} \right\}$$

$$A = -\frac{b_1 + \lambda_1}{a_1 \lambda_1} A_1, \quad B_2 = -\frac{b_1 + \lambda_2}{a_1 \lambda_2} B_1$$

$$a_1 = \frac{\delta_{II} R_{I\,II}}{\delta_I R_{pI}}, \quad a_2 = \frac{\delta_I R_{I\,II}}{\delta_{II} R_{pII}}$$

$$b_1 = -\frac{1}{\delta_I COR_{pI}}, \quad b_2 = -\frac{1}{\delta_{II} COR_{pII}}$$

$$c_1 = -\frac{R_I W}{\delta_I COR_{pI}}, \quad c_2 = -\frac{R_{II} W}{\delta_{II} COR_{pII}}$$

$$\lambda_1 = \frac{-(b_1 + b_2) + \sqrt{(b_1 + b_2)^2 - 4b_1 b_2(1 - a_1 a_2)}}{2(1 - a_1 a_2)}$$

$$\lambda_2 = \frac{-(b_1 + b_2) - \sqrt{(b_1 + b_2)^2 - 4b_1 b_2(1 - a_1 a_2)}}{2(1 - a_1 a_2)}$$

where T_0 = base temperature (°C), R_{int} = equivalent thermal resistance of a cable (°C − cm/W), W = loss generated in a pipe (W/cm), R_{ij} = soil thermal resistance of a pipe (°C − cm/

W), $R_{I\,II}$ = mutual soil thermal resistance between the two cooling pipes I and II (°C − cm/ W), $R_{pI} = R_{II} + R_p + R_w$, $R_{pII} = R_{II\,II} + R_p + R_w$, R_{II} = selfsoil thermal resistance of the pipe I, $R_{II\,II}$ = selfsoil thermal resistance of the pipe II, R_p thermal resistance of a cooling pipe proper, R_w = interfacial thermal resistance between a cooling pipe and water.

$$R_I = \sum_{k=1}^{n} R_{ik}, \quad R_{II} = \sum_{k=1}^{n} R_{IIk}, \quad R_{Ik}, \quad R_{IIk}$$

equals mutual soil thermal resistance between a cooling pipe and a cable pipe duct k, δ_I, δ_{II} = +1 or −1, T_{poI}, and T_{poII} = temperatures of cooling water at its inlets.

When a system with only one cooling pipe is to be investigated, it is sufficient to neglect all the terms with respect to either of the cooling pipes. For return circulation, $\delta_I = -1$ and $\delta_{II} = 1$ are the necessary conditions. Note that it is impossible to treat a case of two-end cooling directly with Equation 1.5.12. Unfortunately, no established formula is available for interfacial thermal resistance between a cooling pipe and coolant water, but the following formula is often utilized in the condition of Reynold's number exceeding 10^4:

$$R_w = 1.218 \times 10^8 \, \frac{1}{\lambda} \left(\frac{dv\rho}{\mu}\right)^{-0.8} \left(\frac{C_p\mu}{\lambda}\right)^{-0.4} \tag{1.5.13.}$$

where d = the inner diameter of a given pipe (m), λ = the average thermal conductivity of a coolant medium (kcal/m·hr·°C) (0.514 (20°C), 0.540 (40°C), 0.560 (60°C) for water), v = the velocity of the coolant (m/hr), ρ = the density of the coolant (kg/m³) (1000 for water), μ = the dynamic viscosity of the coolant (kg/m·hr) 3.6 (20°C), 2.3 (40°C), 2.0 (60°C) for water), C_p = the specific heat of the coolant at constant pressure (kcal/kg·°C) (1.0 for water), R_w = interfacial thermal resistance (°C·cm/W). For indirect pipe cooling, the interfacial resistance R_w can be neglected, because it is smaller than the thermal resistance between a cooling pipe and a cable.

Pressure loss can be estimated from the cooling formula in the case of indirect cooling

$$\Delta P = \rho f \left(\frac{L}{d}\right) \left(\frac{v^2}{2g}\right) \tag{1.5.14.}$$

where ΔP = pressure drop in distance L (g/cm²), ρ = the specific gravity of a coolant, L = cooling distance (cm), d = inner diameter of a cooling pipe (cm), v = flow velocity of a coolant, g = acceleration of gravity (980 g·cm/sec²), and f = friction factor dependent on Reynold's number R.

$$f = \frac{64}{R_e} \qquad \text{for } R_e < 2000 \text{ (laminar flow)}$$

$$f = 0.186 \, R_e^{-0.2} \qquad \text{for } R_e > 4000 \text{ (turbulant flow)}$$

$$\text{with } R_e = \frac{d\rho v}{\mu}, \quad \mu = \text{viscosity (poise)} \tag{1.5.15.}$$

c. Direct Cooling in a Pipe

It is almost impossible to express a general solution by a simple formula.[113] Now consider the simplest case in which the flow of coolant is unidirectional as shown in Figure 1.5.2.(c), and assume that all the cables and pipes are identical, then we obtain the flow rate of coolant required:

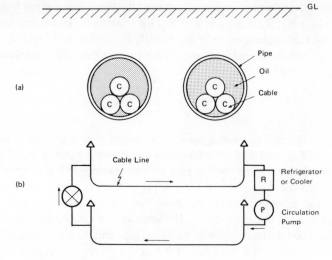

FIGURE 1.5.3.   Two-circuit system of POF cable for cooling.

$$Q = \frac{\dfrac{L}{CR_s}}{\ln\left(\dfrac{T_0 + WR_s - T_{po}}{T_0 + WR_s - T_{pL}}\right)} \qquad (1.5.16.)$$

where R_s = soil thermal resistance of a pipe (°C·cm/W), W = heat generated in a pipe (W/cm), L = cooling length (cm), C = the heat capacity of coolant (W·sec/cm³ °C), T_0 = base temperature (°C), T_{p0} = temperature of coolant at its inlet (°C), T_{pL} = the maximum allowable temperature of coolant at the end of the cooling system, and Q = coolant flow rate (cm³/sec).

Pressure drop on this case is calculated from Formula 1.5.14. but the inner diameter of a cooling pipe must be replaced by the effective diameter expressed by d_e = 4SF, where S = the actual cross section through which coolant can flow (cm²) and F = circumferential length adjacent to coolant (cm).

d. Oil Circulation Cooling of POF Cables[114,120]

Consider a system of two circuits of POF cable installed as shown in Figure 1.5.3.(a), which are cooled in the way indicated in Figure 1.5.3.(b), and let $T_{01}(z)$ and $T_{02}(z)$ be oil temperature in the "go" pipe and "return" pipe at a distance z from the oil inlet and at the cooling position, respectively, then we obtain

$$T_{01}(z) = T_0 + \frac{ab}{\sqrt{1 - b^2}}\left\{A\exp\left(\frac{az}{\sqrt{1 - b^2}}\right) + B\exp\left(\frac{-az}{\sqrt{1 - b^2}}\right)\right\} + \frac{c'}{a} \qquad (1.5.17.)$$

$$T_{02}(z) = T_0 + \left(\frac{a}{\sqrt{1 - b^2}} + a\right) A\exp\left(\frac{az}{\sqrt{1 - b^2}}\right)$$

$$+ \left(\frac{a}{\sqrt{1 - b^2}} - a\right) B\exp\left(\frac{-az}{\sqrt{1 - b^2}}\right) + \frac{c'}{a}$$

with $a = \dfrac{1}{(R_3 + R_4 + R_5)\,CQ}$, $b = \dfrac{R_m}{R_3 + R_4 + R_5}$

$$c' = \frac{(R_3 + R_4 + R_5 + R_m)\,W_0 + (R_4 + R_5 + R_m)\,W_p}{(R_3 + R_4 + R_5)\,CQ} \qquad (1.5.18.)$$

where A,B = constants of integration, C = the heat capacity of oil(W·sec/cm³ °C) (usually 1.9), Q = oil flow rate(cm³/sec), W_0 = generated loss inside a pipe (W/cm), W_p = generated loss in a pipe itself (W/cm), R_3 = interfacial thermal resistance between a pipe and oil (°C·cm/W), R_4 = thermal resistance of pipe-corrosion-protective layer (°C·cm/W), R_5 = selfsoil thermal resistance (°C·cm/W), R_m = mutual soil thermal resistance (°C·cm/W), and T_0 = base temperature (°C). In the case of one-end cooling, $T_{01}(0)$ is given at z = 0, so A and B can be determined from the condition that $T_{01}(L) = T_{02}(L)$ at z = L. With double-end cooling, they can be obtained also, since $T_{02}(L)$ is given at z = L. It is, therefore, obvious that oil temperature at any distance z can be computed.

In order to determine conductor temperature from the oil temperature estimated according to the above procedure, it is necessary to define the interfacial thermal resistance R_2(°C − cm/ W) between the surface of the cable and the oil. Again, it remains still unsolved. The next formula can be used in conservative design:

$$R_2 = \frac{56.4}{2.16 d_2}$$ (1.5.19.)

where d_2(cm) is the cable diameter excluding skid wire. No established formula exists for R_3, either. For convenience, the following expression is often used:

$$R_3 = \frac{56.4}{d_p}$$ (1.5.20.)

where d_p is the inner diameter of a pipe.

Pressure drop, also, has no established formula available, but may be expressed approximately by

$$\Delta P = \lambda \rho \left(\frac{1}{d_e}\right)\left(\frac{v^2}{2g}\right)$$ (1.5.21.)

where ΔP = loss of pressure in a distance l, ρ = density of oil (g/cm³) (0.86 to 0.9), l = cooling distance, g = acceleration due to gravity (g·cm/sec²), v = oil flow rate (cm/ sec), d_e = diameter of equivalent circle (when the cross section is not a circle, then d_e = $4S_2/F$ holds, where S_2 = the cross section through which fluid can flow, F = the circumferential length adjacent to fluid), and λ = the friction coefficient of pipe. The last item is determined as follows:

$$\lambda = \frac{64}{R_e} \qquad \text{for } R_e < 2000 \text{ (laminar flow)}$$

$$\lambda = 0.3164 \, R_e^{-0.25} \qquad \text{for } 300 < R_e < 10^5 \text{ (turbulant flow)}$$

$$R_e = \frac{d_e l v}{\mu} \qquad \mu = \text{viscosity of oil (poise)}$$

e. Internal Cooling

This is represented by a single core OF cable with an oil duct large enough to circulate oil through and by an extruded cable provided with a closed duct through which cooling water can flow and carry the generated heat away.

Consider a single-core cable so designed as to permit coolant (oil or water) flow through the duct provided in the center of its conductor; we obtain the formula for temperature as follows:[121,122]

$$T_c = T_0 + R_{i_1} W_c \left\{ 1 - \frac{R_{i_2}}{R_{i_2} + R_w} \exp\left(- \frac{z}{CQ(R_{i_2} + R_w)} \right) \right\}$$

$$+ \frac{R_{i_2}}{R_{i_2} + R_w} \left(T_a - T_0 \right) \exp\left(- \frac{z}{CQ(R_{i_2} + R_w)} \right) \qquad (1.5.22.)$$

where T_c = conductor temperature at a distance z form the coolant inlet, T_0 = base temperature, T_a = temperature of coolant at its inlet

$$R_{i_1} = \frac{R_i + (1 + p)R_a}{1 - \alpha W_c \{ R_i + (1 + p)R_a \}} \quad , \qquad R_{i_2} = \frac{R_i + R_a}{1 - \alpha W_c \{ R_i + (1 + p)R_a \}}$$

R_i = thermal resistance between conductor and sheath, R_a = thermal resistance outside cable insulation, R_w = interfacial thermal resistance between conductor and coolant

$$W_c = I^2 r_{ac}$$

I = conductor current, r_{ac} = AC conductor resistance, p = sheath loss factor

$$\alpha = \frac{\alpha_{20}}{1 + \alpha_{20}(T_c - 20)}$$

α = temperature coefficient of conductor resistance, α_{20} = α at 20°C, C = the heat capacity of coolant oil, and Q = oil flow rate.

The contribution of dielectric loss should be taken into consideration for EHV and UHV cables. The friction factor f may be estimated from the following:

$$f = \frac{64}{R_e} \qquad \text{for } R_e < 2000 \text{ (laminar flow)}$$

For turbulent flow ($R_e > 2000$), in the case of a smooth surface

$$f = 0.3164 R_e^{-0.25} \qquad \text{for } R_e < 8 \times 10^4$$

$$f = 0.0032 + 0.221 R_e^{-0.257} \qquad \text{for } R_e > 8 \times 10^4$$

in cases where the inner surface is rough enough to cause turbulent flow

$$\frac{1}{\sqrt{f}} = \frac{1}{\sqrt{f_t}} = 2 \log_{10}\left(1 + \frac{d}{\epsilon} \right) + 1.14$$

$$\text{for } \left(1 + \frac{d}{\epsilon} \right) \frac{1}{R_e \sqrt{f}} < 0.005$$

$$\frac{1}{\sqrt{f}} = \frac{1}{\sqrt{f_t}} = 2 \log_{10}\left\{ 1 + 9.31 \left(1 + \frac{d}{\epsilon} \right) \frac{1}{R_e \sqrt{f}} \right\}$$

$$\text{for } \left(1 + \frac{d}{\epsilon} \right) \frac{1}{R_e \sqrt{f}} > 0.005$$

where d = the average diameter cross sectional area of the coolant flowing and ϵ = roughness of inner surface.

All the equations and formulas obtained above are only for the cables proper. Therefore, in a cable system thermally designed on the basis of the above criteria, their joints may occasionally attain a higher temperature than their permissible temperature. In such cases, it is required either to cool the joints with some other facility or to run the cables at temperatures below the permissible conductor temperature.

1.5.2. Forced Cooled OF Cables

When OF cables are submitted to continuous maximum current loading, 24 hr a day, 7 days a week, the ground behaves differently from when the cable is loaded cyclically. Such continuous operation has become more and more common in practice. Continuous operation of a cable, with its surface temperature at 60 or 70°C, will vaporize the moisture in the direct vicinity of the cable causing the vapor to migrate outwards through the ground, leaving behind dried-out ground with high thermal resistivity of the order of 300 thermal ohm-cm (°C − cm/W).[53] The reduction in current carrying capacity caused by this increase in ground thermal resistance can be overcome by increasing the conductor size, except when the largest practicable conductor size is already in use. In this event, some heat must be removed by artificial means if the cable rating is to be maintained. This can be achieved by several methods such as[49]

1. Cooling water pipes laid alongside the cables
2. Placing the cables in water troughs
3. Enclosing each cable in a cooling water pipe
4. Circulating cable oil as a coolant through the conductor

Such artificial cooling has produced favorable results to the extent that it can overcome the thermal limitations of the OF cable system at higher transmission voltages. This solution parallels or even takes preference over foreseeable contribution from the search for a synthetic insulation with low dielectric loss.

A. Indirect Cooling by Adjacent Water Pipes

Four basic components are necessary for an indirectly cooled cable system. They are (1) a pipe to carry the coolant close to the cable, (2) a pump to circulate the coolant, (3) a heat sink which is usually the atmosphere but may be river water or underground water, and (4) a heat exchanger, unless the coolant is directly discharged and not recirculated.

This cooling system is favored, particularly in the U.K., because no variation from standard cable design is needed. It therefore offers the simplest and most economic solution, although it is inferior to the integral sheath cooling from the thermal efficiency point of view. One example installed in England[123] comprises a double circuit system (760 MVA per circuit) of 275-kV cables. Polyethylene-over-sheathed, corrugated, seamless aluminum (CSA) pipes are buried alongside the single-core 2.5- and 3-in.[2] (1600- and 1940-mm^2) oil-filled cables for these 1600-A circuits. The cooling pipes are arranged for closed water circulation via a "go" pipe of the largest practicable bore (2.6 in. or 66 mmφ) and a pair of slightly smaller (2 in. or 50.8 mmφ) "return" pipes, the diameters being such that go and return pumping pressure gradients are substantially identical. Water flows down one pipe and returns along the other two. The cables and pipes are in flat formation, with cable center spacing averaging about 10 in. (254 mm). To minimize the effects of asymmetry, the array is transposed at each joint bay, i.e., every 300 yd (275 m). CSA pipes obviously have an advantage over the high-density polyethylene pipes as far as their permissible operating pressures are concerned. They are particularly suitable for installation on hilly routes involving high-static

Table 1.5.2.
HYDRAULIC PARAMETERS FOR 275 kV OF CABLES INSTALLED

		Beddington to Addington	Tottanham to St. John's Wood
Circuits		2	2
Conductors (in.^{2})		2.5 and 3.0 (mixed)	3.0
Maximum heat	in cables	128	115
generation per	in pipes	27	27
circuit @ 760 MVA			
(kW/1,000 yd)	total	155	142
Flow rate (gal/min/1,000 yd/circuit)		11.0	10.5
Water inlet temperature @ 760 MVA (°C)		15	30
Circuit lengths			
End-to-end (yd)		11,700 and 10,600	18,150 and 71,940
Water-cooled (yd)		11,700 and 10,600	16,400 (each)[a]
Cooling sections (each served by one cooling station)			
Number		4 per circuit	5 per circuit
Longest (yd)		3,600	4,350
Shortest (yd)		1,860	2,000
Mean (yd)		2,790	3,280
Cooling subsection			
Number		7 per circuit	10 per circuit
Longest (yd)		1,860	2,280
Shortest (yd)		1,300	1,000
Mean (yd)		1,590	1,640

[a]   Remaining lengths, naturally cooled, in trough alongside the Regents Canal.

heads. The normal design maximum pressure for CSA pipes is 125 lb/in.^{2} (8.8 kg/cm^2); whereas it is 50 lb/in.^{2} (3.5 kg/cm^2) at 50°C for polyethylene. The main disadvantages of CSA pipes are that their corrugations entail hydraulic "roughness" and that eddy currents electromagnetically induced in them cause heat generation. Table 1.5.2. indicates values associated with hydraulic parameters of the systems designed.[123] These systems require quite complicated heat exchanger equipment as well as a water pumping plant, which must be housed in buildings provided every 2 mi (3 to 4 km) or so along the cable route.

The lateral pipe cooling method appears to be applicable to 750-kV cable systems.[124] Figure 1.5.4. gives an idea of the power rating that can be transmitted by either 3 single-core cables with a conductor cross section of 1100 mm^2 (2200 kcmil), or by 3 cables with a conductor cross section of 2000 mm^2 (4000 kcmil). Four high-density polyethylene water pipes are arranged around the cables which are laid in flat formation.

B. Cables in Troughs

Cables in water-filled concrete troughs can be laid in flat or trefoil formation. The cooling water runs from one end, following the slope of the trough, and weirs are placed at intervals in order to maintain the water level above the top of the cables. Small cable movements at weirs are allowed by leaving a sufficient clearance and having a flexible seal to the cables passing through.[74]

Some modifications of this type of cooling are to be found. Figure 1.5.5. indicates a double circuit system comprising 6 3-in.^{2} (1940 mm^2) 275-kV OF cables, laid in a trough, with a central water channel to ensure a waterlogged condition under all circumstances. The trough can be filled therefore with water-logged sand with a thermal resistivity as low as

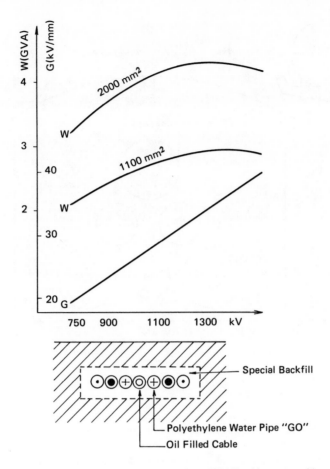

FIGURE 1.5.4. Power transmissible by a 750-kV cable system with lateral pipe forced cooling.

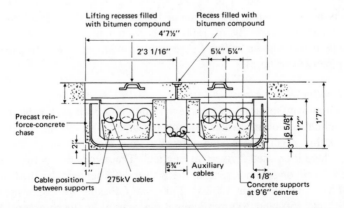

FIGURE 1.5.5. Double circuit of 275-kV OF cables in trough with waterlogged sand. (From Watson, E.P.C., Brooks, E. J., and Gosling, C. H., *Proc. Inst. Electr. Eng.*, 114(4), 516, 1967. With permission.)

18° C-cm/W. These cables were installed in canal towpaths, and could therefore operate even without water cooling. For this purpose, the trough was filled with a sand chosen for its inherently low thermal resistivity, an attribute determined by the range of grain sizes. In

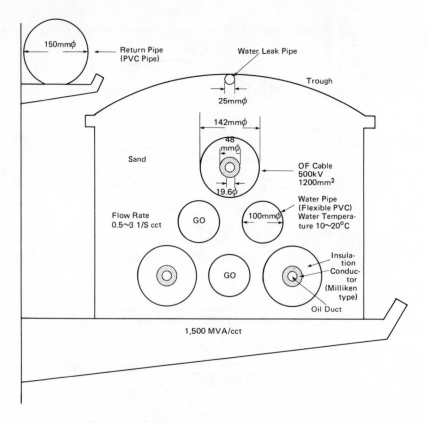

FIGURE 1.5.6.   500-kV 1500-MVA OF cables in a trough in a tunnel.

this condition, the installation could achieve a 760 MVA rating in winter and a 620 MVA rating in summer. The installation was approximately 1500-yd (1370-m) long, and contained a total of 42 joints.

Another modification is shown in Figure 1.5.6. which basically illustrates lateral pipe cooling. This is an experimental installation of 1200 mm², 500-kV OF cables, which should have a current carrying capacity of 1800 A, corresponding to a power rating of 1500 MVA.[126,127] The trough shown in this figure is filled with sand, waterlogged by means of a water leak pipe (27 cm³/min).

C. Cables in Cooling Water Pipes

One way to improve the rating of a cable system is to install each cable of the circuit in a pipe and to circulate cooling water through the pipe to absorb the cable heat. A cable design based on this principle has been utilized to take underground a heavy-duty, 400-kV overhead line where it crossed 2 rivers in England. The details are as follows.[74] The cable is installed in a 4-km long tunnel beneath the 2 rivers. To achieve the final required winter rating of 2600 MVA (3780 A), at 400 kV, 2600 mm² OF cable with a corrugated aluminum sheath is pulled into 250- and 300-mm internal diameter glass fiber-reinforced plastic pipes through which cooling water is to be circulated.

The next example is a German design for a 380-kV system, with a power rating of 1120 MVA, which is to be integrally cooled.[129] After a thorough testing of all the well-known pipe materials, asbestos-cement and polyvinyl chloride (PVC) were considered most suitable for the water-carrying pipes enclosing the cable, particularly with respect to their pulling-in characteristic. Asbestos-cement pipes have a very good manufacturing uniformity and proven connecting techniques (Reka-couplings). However they are not flexible and are likely

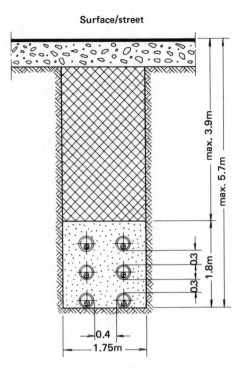

Surface/street

FIGURE 1.5.7.   Conduit design for surface-water-cooled 380-kV OF cables. (From Müller, U., Peschke, E. F., and Hahn, W., *The First 380 kV Cable Bulk Power Transmission in Germany,* CIGRE Paper 21-08, CIGRE, Paris, 1976, 12. With permission.

to release calcium ions, thereby disturbing the lime/carbonic equilibrium and causing the precipitation of calcium carbonate ($CaCO_3$). On the other hand, PVC tubes are lighter in weight, and more elastic except at low temperatures. Unfortunately, no service experience is yet available with respect to the strength at elevated temperatures over a period of 30 years. In any event asbestos-cement pipes were selected if only for economical reasons. A diameter ratio between cable and pipe of around 1:2 is considered optimum. The normal pipe route consists of the 2 adjacent systems each with 3 pipes in vertical arrangement, as shown in Figure 1.5.7.;[128] the width is 1.75 m and height 1.8 m. To assure a minimum covering layer of 1.5 m required for public road areas, a minimum depth of 3.3 m for the trench bottom is necessary. The lowest pipes lie on a stone-free, compacted bottom of sand which is obtained partly by hardening the bed with cement mortar and by the application of a back-fill material of specified quality.

An experimental installation of a 500-kV 1200 mm² OF cable system is shown in its cross-sectional view in Figure 1.5.8.; it is designed for a power rating of 1500 MVA.[126] Some new problems were experienced, specifically associated with forced cooling. As the frictional force opposing cable movement decreases due to buoyancy from cooling water, the cables are likely to move 2 to 2.5 times as much as they would without water under the same load condition. This may necessitate either the construction of larger manholes to offset the thrust produced, or control of the water temperature. Water hammer phenomenon should also be taken care of.

In Italy, an attempt is being made to use this cooling system for 750-kV class OF cables.[124] The 750-kV, 1100 mm² cable can transmit about 2200 MVA, with a distance between heat

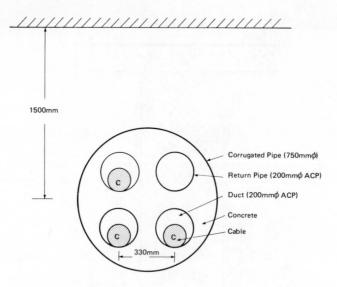

FIGURE 1.5.8.   500 kV 1 × 1200 mm² OF Cables with surface cooling.

exchangers of 2 to 3 km, by circulating 60 m³/hr (16.7 ℓ/sec, 264 gpm) of water in a 250-mm (10 in.) bore pipe. It is interesting to note that the same cable with internal oil circulation can carry the same power if the heat exchangers are no more than 250 to 300 m apart. The main difficulty with this method arises from the uncontrollable cable movements which occur in the pipe during thermal cycles. These may possibly result in sheath damage or fatigue. Corrugated aluminum sheaths are better candidates than lead sheaths. Another possible problem is the development of "hot spots" in joints and terminations. This problem may be overcome by using an "oil oscillation" technique.

D. Internal Water or Oil Circulation

Cables are cooled by circulation of coolant through a central duct in their conductor. Due to its high thermal capacity and low viscosity, water would give the lowest coolant flow rates and pumping pressures. However, cooling water passed through the conductor would have to be segregated from the insulation and would also pose a problem at the points of entry to and exit from the high-voltage conductor of the cable.[129] Oil circulation is a better alternative in this respect, as the problem of coolant entry and exit is much less severe. A central duct of comparatively large diameter is required. It is claimed that for 400-kV 3200 A duty the most economic system is obtained with a duct diameter of about 50 mm, reducing the pressure to about 1/1000 of that for a normal 12-mm duct serving the same duty. Designs are shown in Figure 1.5.9.[129] This method is very effective in principle but it has two major disadvantages.[124] First, the limited dimension of the oil duct naturally leads to the limited flow of oil, and the oil temperature increases dramatically along the length of the cable as a consequence. Because of this, the distance between adjacent heat exchangers cannot exceed a few hundred meters. Second, the thermal capacity of the cable is relatively small so that a failure of the cooling system produces a rapid rise in conductor temperature. This requires corrective action to be taken very quickly.

Figure 1.5.10. shows a German design of a 100-kV, single-core, oil-filled cable with internal water cooling.[130] A watertight pipe surrounds the water channel through which water at a temperature of 30°C is pumped using a pressure of 31 bars (31.6 kg/cm²). By the time the water reaches the end of the cable its temperature may be 80°C and the pressure may have fallen to 1 bar. Calculation shows that with a water channel diameter of 70 mm and

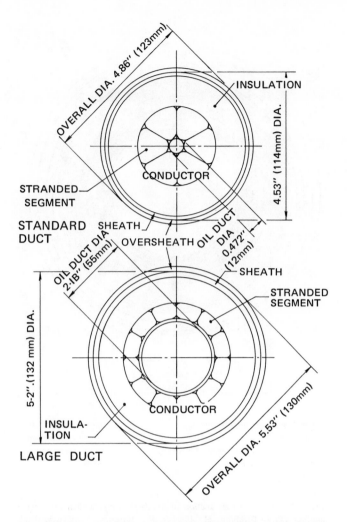

FIGURE 1.5.9. Comparison of standard and large duct — 3 in.^{2} (1935 mm^2) — copper conductor oil-filled cable. (From Beale, H. K., Hughes, K. E. L., Endacott, J. D., Miranda, F. J., Flack, H. W., and Nicholson, J. A., *The Application of Intensive Oil-Filled Cables and Accessories for Heavy Duty Transmission Circuits,* CIGRE Paper 21-09, CIGRE, Paris, 1968, 5. With permission.)

a conductor diameter of 100 mm, the distance between cooling stations can be 10 km if 600 MVA are to be transmitted, and that even with a transmission power of 900 MVA this distance may still be 6 km for this cable.

1.5.3. Forced Cooled Extruded Cables

The logical solution to increase the power rating of extruded cables appears to be in water cooling. This can be achieved either by internal cooling or by surface cooling. However, internal water cooling may be limited to comparatively low voltage because of difficulties in structuring the water inlet and outlet.

A. Internally Water-Cooled XLPE Cables

A 22-kV, 15,000 A, XLPE cable system has been developed, which is internally cooled by pure water; this is shown in Figure 1.5.11.[131] The conductor contains a smooth, watertight copper pipe of 2.6 mm inner diameter and a wall thickness of 2 mm. This water channel

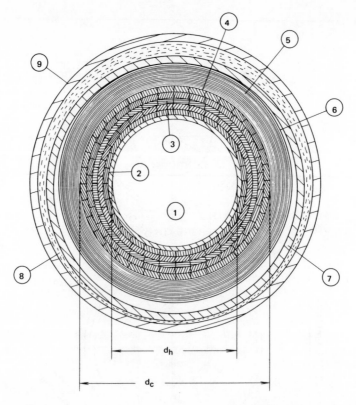

1 Water channel, channel diameter d_h
2 Water-tight pipe
3 Stranded conductor, aluminum, diameter d_c
4 Semi-conducting papers
5 Insulation
6 Shielding tapes
7 Corrugated aluminum sheath
8 Anti-corrosion protection
9 PVC jacket

FIGURE 1.5.10. Cross section of a 110-kV cable with internal water cooling. (From Birnbreir, H., Fischer, W., Rasquin, W., Grosse-Plankermann, G., and Schuppe, W. D., *High Power Cable with Internal and External Water Cooling,* CIGRE Paper 21-09, CIGRE, Paris, 1974, 4. With permission.)

avoids insulation deterioration due to water and assures less pressure drop. Six segmented copper conductor elements, bound with stainless steel tapes, are applied over the copper pipe. A conductor with such segments has the same flexibility as conventional power cables; it also has a low radial thermal resistance. The conductor is 2000 mm² in cross section, which is sufficient for its 15-kA rating. A cable terminal box design, developed for this type of cable, is shown in Figure 1.5.12.[131] The box comprises the cable terminal, the water terminal, and a hollow porcelain insulator. Cooling water in the cable conductor at working voltage flows through the hollow porcelain insulator into a cooling plant pipe at ground potential. Water purity is maintained by means of an ion exchanger, which assures that its electrical conductivity does not exceed 1 μmho/cm(μsiemens).

B. Surface-Cooled Extruded Cables

A German design for a 100-kV, 300 MVA, XLPE cable, with integral cooling, is shown in Figure 1.5.13.[132] The conductors are 630 mm² in cross section and their electrical insulation, i.e., the XLPE, is 18 mm thick.

A French design for a 225-kV, 600 MVA, pure PE cable, with integral cooling, is

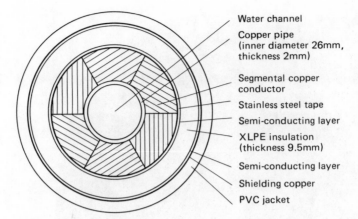

FIGURE 1.5.11. Cross section of a 22-kV internally water cooled XLPE cable. (From Hayashi, H., Kamiyama, T., Torigoe, Y., Ichino, T., and Tanaka, N., *22 kV Internally Water-Cooled 15 kV Capacity XLPE Cable,* IEEE PES Winter Meeting Paper A76 167-7, IEEE, Piscataway, New Jersey, 1976, 2. With permission.)

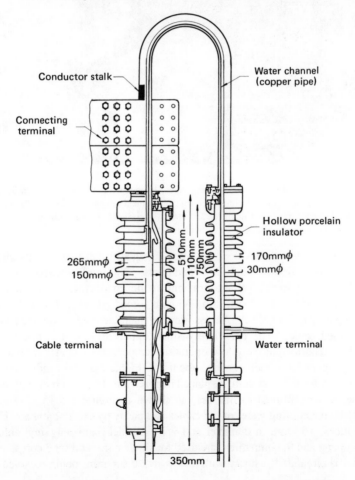

FIGURE 1.5.12. Cable terminal box for 22-kV internally water cooled cable. (From Hayashi, H., Kamiyama, T., Torigoe, Y., Ichino, T., and Tanaka, N., *22 kV Internally Water-Cooled 15 kV Capacity XLPE Cable,* IEEE PES Winter Meeting Paper A76 167-7, IEEE, Piscataway, New Jersey, 1976, 4. With permission.)

FIGURE 1.5.13.   XLPE cable in cooling pipe: (1) conductor, (2) conductor shield, (3) XLPE insulation, (4) semiconducting layer, (5) copper shield, (6) plastic covering, (7) cooling pipe made of hard pvc (outer diameter 250 mm). (From Falke, H. and Schlang, P., *Electrotech. Z. Ausg. B*, 25(20), 557, 1973. With permission.)

available.[133] A 200-m length has been installed for quality verification. Its characteristics are as follows: cross section of aluminum conductors: 1200 mm², thickness of polyethylene insulation: 22 mm, maximum voltage gradient at the conductor: 8.3 kV/mm, and outer diameter over the lead sheath: 103.8 mm. Without forced cooling, this cable system has a transmission capacity of 300 MVA. Accessories are manufactured by molding as in the case at naturally cooled cables. The particularly small dimensions of the splices as shown in Figure 1.5.14.[133] lead to relatively low operating temperatures. Under normal conditions, the maximum temperature at their center is about 5 °C higher than the conductor temperature. The terminations of the cable system are placed in porcelains filled with insulating compound as in the case of conventional techniques, as shown in Figure 1.5.15.[133] The cooling is effected partly by the cooling water of the cables and partly by the ambient air. High-density polyethylene tubes, 160-mm in diameter and 9.1-mm thick, and polyvinyl chloride tubes, 315-mm in diameter and 9.1-mm thick, are used for single cores and for 3 cores, respectively. Ordinary water is circulated, usually with a flow rate 0.6 ℓ/min, being recycled in a closed system.

1.5.4. Forced Cooled POF Cables

Recognizing that forced cooling permits a significant increase in the ampacity ratings of

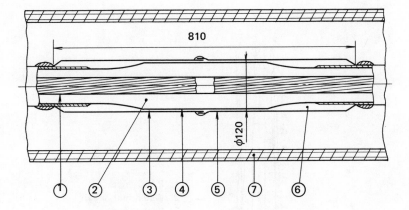

1. Conducting sheath
2. Insulation remade by moulding
3. Conducting layer
4. Metal screen
5. Metal shell
6. Filling compound
7. Pipe

FIGURE 1.5.14.   Splice with external cooling for a service voltage 225 kV. (From Lacoste, A., Royere, A., Lepers, J., and Benard, P., *Experimental Construction Prospects for the Use of 225 kV—600 MVA Links Using Polyethylene Insulated Cables with Forced External Water Cooling*, CIGRE Paper 21-12, CIGRE, Paris, 1974, 4. With permission.)

pipe-type cables, development and installations have been made by raising thermally limited power-transfer capability. One logical approach is to utilize the mechanism of "surface cooling" by the circulation and cooling of the dielectric oil surrounding the one or more cables contained in the pipe. Forced cooling of pipe-type cables has a number of advantages:[134]

1. The current-carrying capacity, or ampacity, of existing lines can generally be significantly increased with the possibility that new systems, specially designed for forced cooling, can have current ratings exceeding twice those of self-cooled lines.
2. The "critical length", that is, the length at which the line-charging current is equal to the rated cable current, is substantially increased.
3. The power rating of a line can be increased in steps, thereby saving on investment-carrying charges.
4. The installation of new self-cooled transmission lines may be deferred with attendant economic savings.
5. More power can be transmitted over a given right-of-way.
6. Forced cooling may permit the efficient use of extra-high-voltage cable systems, where dielectric loss becomes a very serious consideration.
7. With forced cooling, the cable system becomes more thermally independent of its environment, allowing installation without special attention to backfill materials.
8. The cost per MVA-mile of transmission is usually reduced by the application of forced cooling.

A. Service Installations

Figure 1.5.16. indicates the cross section of a trench containing a double circuit, 345-kV, forced cooled, POF cable with a total transfer capability of 960 MVA,[135] installed in New York. The conductors are 2000 and 2500 kcmil (4000 and 5000 mm²) copper. The 1.025-in. (26 mm) paper insulation is enveloped with a 5-mil (127 μm) metallic ground

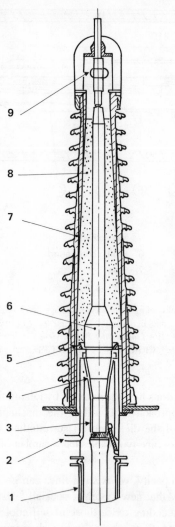

1. P.V.C. pipe
2. Water inlet or exit
3. Sealing cone
4. Deflector
5. Metal ring
6. Insulation reinforcement
7. Insulation
8. Filling compound
9. Flexible connector

FIGURE 1.5.15.   Termination of a cable with external cooling for
a service voltage of 225 kV. (From Lacoste, A., Royere, A., Lepers,
J., and Benard, P., *Experimental Construction Prospects for the Use
of 225 kV—600 MVA Links Using Polyethylene Insulated Cables with
Forced External Water Cooling*, CIGRE Paper 21-12, CIGRE, Paris,
1974, 4. With permission.)

sheath and reinforced by a helically wound D-shaped stainless steel skid wires. The forced
cooling loop consists of the series hydraulic connection of a portion of the feeder pipe and
the coolant supply pipe, in conjunction with a high-flow circulating pump and heat exchanger,
as schematically shown in Figure 1.5.17.[135] The direction of fluid flow in all loops is through
a circulating pump, sending line, cable pipe inlet diffusion chamber, cable pipe, cable pipe
exit diffusion chamber, heat exchanger, and back to the circulating pump. In this manner,
the heat exchangers and the potheads are subjected to minimum operating pressures. Feeder
cooling loop lengths range from 9000 ft (2.7 km) to 2000 ft (6.1 km), being determined by
the limiting pressures of 400 psi (28 kg/cm²) at potheads, 600 psi (42 kg/cm²) on the 10-
in. (254 mm) cable pipes, and 800 psi (56 kg/cm²) on the 5-in. (127-mm) dielectric pipes
of the sending line.

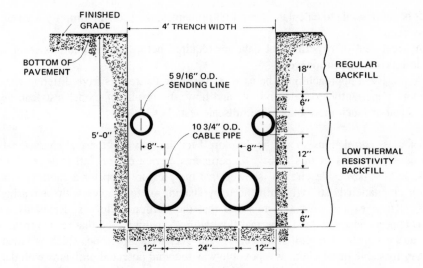

FIGURE 1.5.16.   Cross section of trench for a double forced cooled circuit. (From Buckweitz, M. D. and Pennell, D. B., *Transmission Distribution,* April 1976, p. 51. With permission.)

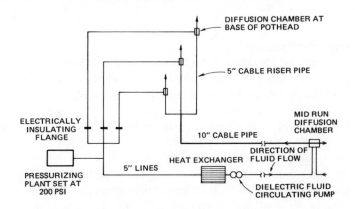

FIGURE 1.5.17.   Hydraulic connections on forced cooled feeders at station ends. (After Buckweitz, M. D. and Pennell, D. B., *Transmission Distribution,* April 1976, p. 54. With permission.)

B. Experimental Installations

Slow oil circulation, to distribute the heat generated by hot spots over a wider area, is common practice today.[136] Oil flow rates are typically 5 to 10 gpm (19 to 38 ℓ/min). For laminar flow produced by either moderate cooling or slow circulation, pressure drop calculations made by using the AIEE guideline[137] or the formula described in Section 1.5.1. are sufficiently accurate. Recent theoretical and laboratory evaluation studies, necessitated by the prospect for intensive forced cooling, have revealed that it is not possible to translate the analytical methods of classical fluid mechanics to the complex geometry of a pipe-type cable system, where both the cables and the pipes are heat sources, and the cables change configuration with change in load.[138]

There seems to be a dilemma for the cable engineer designing a forced cooling system in which the cable configuration (semiopen birdcaged, full birdcaged closed triangular, or cradled) changes with both location and time (at different loads). This probably arises as a consequence of the facts that[138]

1. It is not possible to calculate, from first principles, the pressure drop for a given cable system.
2. A great deal of experimental data are required before the hydraulic system can be designed with confidence.
3. A consistent approach must be used, to correlate the results from different tests, and more importantly, to determine if, and how, the results of model tests can be used instead of much more expensive fullscale tests.

Dimensionless correlations should be developed for the flow and thermal characteristics of forced cooled systems of this type.[139] Experiments carried out on full-scale, 600-ft long, 230-kV and 345-kV cable circuits, showed that the pressure drop for a cable system with rapid oil circulation is strongly influenced by snaking of the cables with increasing load currents. These experiments also revealed that the friction factor vs. Reynolds number analysis appears adequate for isothermal flow with no snaking, and that results from model cable can be translated to full-scale cables. An oversized pipe, one which is larger than necessary for cable installation, is a possible way to avoid potential problems with delicate, very-low-viscosity pressurizing oils.

Experiments were also carried out in Japan to obtain a current capacity of 1800 A with a 500-kV, 3×2000 mm^2 higher viscosity oil-filled pipe-type cable system and a 500-kV, 3×1400 mm^2 lower viscosity oil-filled pipe-type cable system.[126] Their performances are listed in Table 1.5.3.

1.5.5. High-Ampacity Potheads

The ampacity ratings of conventional high-voltage potheads have in general been more than adequate to meet the demands of self-cooled, high-pressure oil, pipe-type cable systems. If the constraint of the pipe-earth interface temperature is reduced, or eliminated, by forced cooling the pipe cable, the conventional pothead may allow an increase in ampacity.

Three different designs of 230-kV high ampacity potheads are indicated in Figures 1.5.18., 1.5.19., and 1.5.20.[140] According to the literature,[140] the description of each pothead is as follows. The first is intended for existing installations where it is desirable to avoid pothead disassembly and where only a relatively small increase in pothead ampacity is required to match that of the forced cooled cable. It incorporates a closed loop hydraulic system which permits circulation of the insulating oil upward and within the pothead porcelain insulator. The return path for the oil is through an adjacent insulator column, which incorporates internal thermal insulation to decrease the heat gain from the ambient air, and then to a heat exchanger and circulating pump. The increase in pothead ampacity is an inverse function of the circulating oil temperature and therefore is limited by the allowable negative temperature gradient across the wall thickness of the pothead porcelain insulator, since this temperature gradient adds mechanical stress to the stress developed by the pipe cable hydraulic oil pressure (normally 200 psi or 14 kg/cm^2).

The second design of Figure 1.5.19. is intended for new or existing cable systems where a modified pothead construction can be used. In the case of existing installations, modified potheads can be installed without replacing the paper roll insulation or stress relief cone termination. It is based on the same principles as the first design but utilizes a larger diameter porcelain insulator to permit the placement of a thermal insulation liner adjacent to the insulator bore. An internal oil return path is also incorporated into the design to eliminate the necessity for an external return insulator column. Plastic tubing is used for the return path as shown in Figure 1.2.19. The increase in pothead ampacity possible with this design can be significantly greater than with Figure 1.5.18., because the temperature gradient between the ambient air and the cooled oil is shared by the porcelain and the thermal insulation liner, thereby permitting the fluid to circulate at a lower temperature.

Table 1.5.3.
PERFORMANCES OF FORCED COOLED POF CABLES

Item		500kV POF Cable #1 Lower Viscosity Oil-Filled	500kV POF Cable #2 Higher Viscosity Oil-Filled
Route Length		3φ 460 m	3φ 230 m
Installation System		Direct Buried & Tunnel	Tunnel
Conductor Size	Normal	1,400 mm²	2,000 mm²
	Riser	1,400 mm²	2,000 mm²
Sealing Ends	Outdoor Type	3 sets	3 sets
	Oil-Immersed	3 sets	3 sets
	Metal-Enclosed Gas	3 sets	3 sets
Joints	Normal	3 sets	3 sets
	Oil Stop	3 sets	3 sets
Oil Feeding System		Pumping plant	Pumping plant
Cooling Medium		Lower Viscosity Oil	Higher Viscosity Oil
Refrigerator		2°C/5°C: 120 Refrigeration Ton / 5°C/10°C: 160 "	
Cooling System		Oil Circulation Cooling	

Item		Cable #1	Cable #2
Conductor	Size (mm²)	1,400	2,000
	Shape	6 Segmental Stranding	4 and 7 Segmental
	Oil Duct (mm)	—	—
	O. D. (mm)	46	56
Insulation	Thickness (mm)	30	30
	tan (%)	0.15 or less (at 80°C, 290kV)	
	Permittivity	3.2 or less	
	Paper	Low Loss Kraft Paper	
Insulation Oil	Type	Pipe Oil / Alkylbenzene	Pipe Oil / Alkylbenzene
	Viscosity C.S. 30°C	5.8 / 1,500	80 / 1,500
	Viscosity C.S. 80°C	1.7 / 62	10.5 / 62
Metal Sheath etc.		Stainless Steel Pipe & Skid Wire	
Protective Covering			
Cable Diameter (mm)		115	125
Cable Weight (kg/m)		25	33
I.D. of Steel Pipe (mm)		300	350

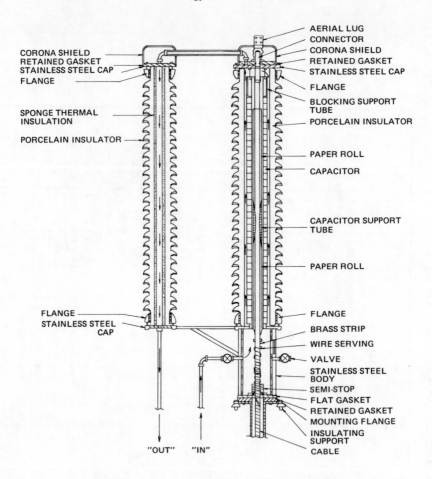

FIGURE 1.5.18.   A high ampacity pothead design #1.[140]

The third design of Figure 1.5.20. is intended only for new systems, or where major modifications to an existing cable system are acceptable. It requires the use of hollow-core conductor cables connected to the main pipe system by a trifurcating joint and extending through the riser pipes to the pothead terminations. It is the most efficient of the three designs as it transfers the primary heat energy developed by the cable conductor ohmic loss through the hollow-core passage to an air-cooled heat exchanger mounted at the top of the pothead and does not require any refrigeration. An oil-injector tube extends from the top of the pothead downward through the hollow-core conductor to a hydraulic restriction located at a level near the base of the pothead. The oil returns upward outside the injector tube but still within the hollow-core conductor to the circulating pump and heat exchanger mounted at the top of the pothead. Electrical energy for the motor-driven circulating pump is obtained from a window-type current transformer surrounding the pothead connector and the aerial lug.

The most common dielectric oils presently employed in the self-cooled pipe cable systems are the mineral and polybutene oils having viscosities in the range of 550 to 750 SUS at 100 °F. Use of these oils in forced cooled pipe cable systems at high flow rates and low temperatures would increase the hydraulic problems. Oils having viscosities less than 100 Saybolt Universal Seconds (SUS) at 100 °F are preferred, or necessary, for those systems requiring low circulating oil temperatures. It is interesting to note that axial heat flow complicates the analysis of thermal problems associated with pothead design.

A number of modifications are necessary for pipe-type pothead constructions to achieve

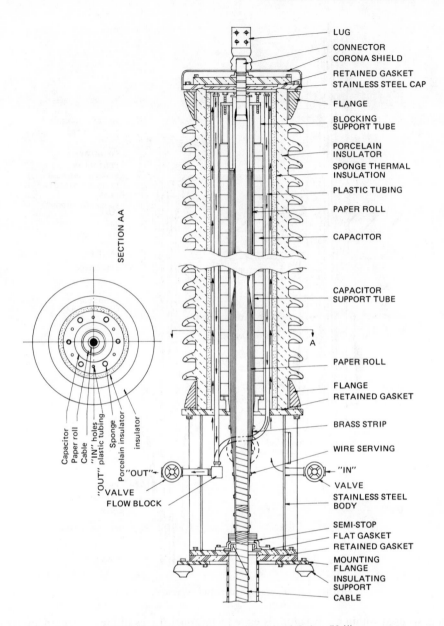

FIGURE 1.5.19. A high ampacity pothead design #2.[141]

an appreciable heat sink zone at one or both ends of the pothead and thereby materially increasing the ampacity ratings required for forced-cooled pipe cable systems. They are concerned with massive heat sink pothead connector design, pothead connector incorporating a "heat pipe" heat transfer and radiator element, forced cooling of pothead body zone, and hot pothead zone suppression by axial internal fluid circulation.

1.6. COMPRESSED GAS INSULATED CABLES[141-166]

1.6.1. Characteristics of CGI Cables

A. General Features

Compressed gas insulated cables (CGI cables) evolved from the technologies of metal-enclosed bus duct and gas-insulated substations. In accord with previous practice, a short

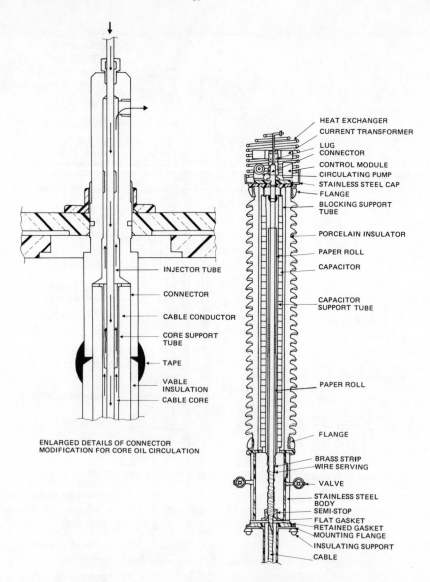

Labels on left diagram:
INJECTOR TUBE
CONNECTOR
CABLE CONDUCTOR
CORE SUPPORT TUBE
TAPE
VABLE INSULATION
CABLE CORE

ENLARGED DETAILS OF CONNECTOR
MODIFICATION FOR CORE OIL CIRCULATION

Labels on right diagram:
HEAT EXCHANGER
CURRENT TRANSFORMER
LUG
CONNECTOR
CONTROL MODULE
CIRCULATING PUMP
STAINLESS STEEL CAP
FLANGE
BLOCKING SUPPORT TUBE
PORCELAIN INSULATOR
PAPER ROLL
CAPACITOR
CAPACITOR SUPPORT TUBE
PAPER ROLL
FLANGE
BRASS STRIP
WIRE SERVING
VALVE
STAINLESS STEEL BODY
SEMI-STOP
FLAT GASKET
RETAINED GASKET
MOUNTING FLANGE
INSULATING SUPPORT
CABLE

FIGURE 1.5.20.   A high ampacity pothead design #3.[140]

section, as used within a substation, is called a bus duct. Lines of greater length are termed compressed gas insulated transmission lines or cables.[148] One of the most significant features of such a system is its ability to coordinate with overhead lines with respect to transmission voltage and ampacity. It is especially useful where conditions mandate undergrounding of an overhead line system for bulk power transmission. Sulfur hexafluoride (SF_6) is normally used to insulate this type of cable. CGI cables have been used in buried installations, getaways in substations, for risers inside transmission towers, through tunnels, and as critical links in urban areas.

Among the useful characteristics of CGI cables are[141]

1.   An inherently high insulating strength as described in Volume I, Section 2.3.6. — Under practical conditions, the voltage and field strength for certain densities and types of gas (SF_6 at 27 psi; 1.9 kg/cm²) can exceed the volume strength of hydrocarbon oils and ceramics, and can withstand the highest voltages encountered by, or envisioned for, electric power systems.

2. A dielectric constant of virtual unity, even at the highest practical pressure — This property of insulating gases is important in underground AC power transmission because of the diminished charging current and longer length of line which is permissible without compensation.

3. Excellent heat-transfer characteristics — The convective process of heat transfer reduces the temperature drop developed in transferring ohmic losses from the central conductor to the surrounding pressure sheath to relatively small value. The heat transfer rate is about twice that of conventional paper-oil cable insulation.

4. Low dielectric loss — Since it is desirable and feasible to operate compressed gas insulation below the corona onset level, the dielectric loss in the gas volume is negligible.

5. Compatibility with high temperature operation — In principle, neither the gas molecules themselves nor selected solid dielectrics used for support and termination functions are significantly affected by temperatures up to 150°C. Thermal run-away, a limitation of oil-paper cables, may be therefore avoided.

6. Power ratings that can equal and exceed the AC power ratings of overhead lines of the same voltage.

7. An enclosed system which prevents the high voltage conductors and the external environment from influencing each other.

8. Nonflammable insulating medium.

9. No increase in static pressure due to major difference in level.

In spite of their superiority as stated above, CGI cables have some significant disadvantages. Since they are usually rigid, their unit length is short (up to 60 ft or 18 m), resulting in a number of joints in field installation. Special precautions are required for making such joints at the time of installation especially as regards the environment, because it is essential to prevent gas insulation from being contaminated. Furthermore, special joints are necessary to accommodate the thermal expansion and contraction of their conductors. Therefore, installation cost seems to be rather independent of their size. They may be economically suitable for ampacities greater than 3000 A.

CGI cables are either rigid or flexible. Flexible cables are now under development. Rigid CGI cables can be classified into two designs: one comprises three identical isolated cables, the other has three conductors enclosed together in one pipe. Commercial CGI cables have been of the isolated phase design. Three conductor designs have so far been used only for short bus runs in European and Japanese substations. This design concept is being evaluated in the U.S. for transmission applications.

The isolated phase design simply consists of two coaxial aluminum tubes (or stainless steel-aluminum or stainless steel-copper assemblies), the inner conductor at line potential being supported by solid insulating spacers. The outer tube serves not only as enclosure, containing SF_6 gas at a moderate pressure above atmospheric (50 psig for example), but also as electrical sheath. Modular components, which should be readily assembled into permanent systems, allow flexibility in application. These are (1) straight shipping sections; (2) terminations to air, oil, or SF_6 insulated equipment; (3) miscellaneous components such as elbows, T-joints, switches, lightning arresters, and the like; and (4) auxiliary and protective equipment.

B. Cable Section

A typical straight shipping section is outlined in Figure 1.6.1. Expansion joints in the conductor at tulip-type joints can accommodate differential thermal expansion of the conductor relative to the sheath. Insulating spacers must be designed for a minimum distortion of the electrical field distribution, while they perform the mechanical functions of keeping the conductor centered within the enclosure. They must occasionally act as gas barriers,

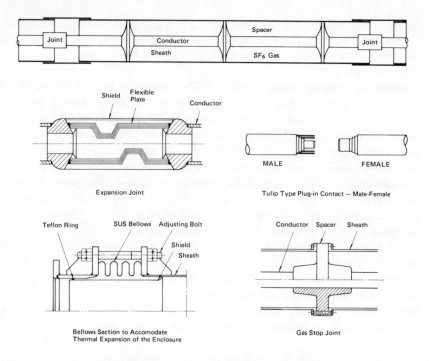

FIGURE 1.6.1.   General features of CGI cables.

too. The material is usually an epoxy resin, preferably a filled cycloaliphatic epoxy, vacuum cast and selected for arc tracking resistance, so that flashover which may occur during conditioning and testing will not harm the cable. There are three kinds of spacers in use; disc, conical, and post-type spacers. Disc-type spacers have operated successfully in both transmission and substation applications. The advantages of cone-shaped spacers is that they have the lowest stress in the solid dielectric and also an increased surface creepage path.

Compressed gas insulation is easily influenced by contamination, especially due to metallic particles. To achieve dielectric reliability, some contamination control measures are therefore often taken at various stages in design, manufacture, and installation. Chemical machining and single-weld joints are examples which minimize the ingress of contamination in the form of dirt, moisture, and metallic particles. Conducting particles that might remain after assembly can be removed from the dielectrically active area by contamination control devices by a process of electrical conditioning which is described in the next section.

C. Installation

Because of the small temperature difference between the conductor and the sheath due to the excellent heat transfer characteristic of the gas, the power transmission capacity may be limited by the outer thermal resistance. The temperature limit of 40°C at which the soil begins to dry is reached at rather low transmission power. Therefore, thermally stabilized backfill should be used when the cable is directly buried. In buried installations, the equipment is well-protected and environmental impact is minimal. Conversely, corrosion protection is required and the cost of installation has been relatively high.

The covered trench also achieves a small environmental impact, but it is a rather expensive mode of installation. The cables are maintained in an air environment where they are able to achieve a high ampacity with the help of ventilators. Tunnel installation may be considered where a utility tunnel already exists within a city, or to carry power from a hydro-generator through a mountain. Again, the cable is able to achieve high ampacity when aided by air blowers.

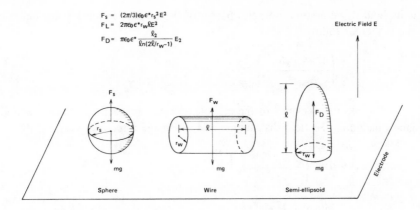

FIGURE 1.6.2. Modeled conducting particles and forces under electric field.

Above-ground installations have to consider the effect of solar heat flux on the ampacity rating. An improvement in the radiative heat transfer is obtained with special paints of controlled emissivity. In a protected area, the CGI cable can be placed on simple concrete foundations; otherwise, it may be placed on elevated structures that can provide all-around protection against access. The available place for installation might be on the lower part of existing transmission towers, median strips of highways, along railroad lines, under bridges for river crossing, etc.

1.6.2. Various Factors in Insulation Design

Breakdown processes in SF_6 gas, as described in Volume I, Section 2.3.6. are well documented, thus the suitability of SF_6 gas for CGI cables is widely recognized. This is based on the experimentally confirmed, superior electrical insulation characteristics, as described in the previous section. Extrinsic factors influence breakdown of SF_6 gas. These are contamination, especially free-conducting particles, metal surface represented by the phenomenon called conditioning, and the shape of spacers to modify the electric field distribution.

A. Free Conducting Particles and the Firefly Phenomenon

When a conducting particle is resting on an electrode, it acquires a charge Q from the existing local electric field E, which is dependent on the shape, orientation, and size of the particle, as shown in Figure 1.6.2. This produces an electrostatic force on the particle of magnitude qE. When this force exceeds the gravitational force, the particle is levitated. Once this occurs, the attractive force on the particle due to the image charge decreases, the resultant force increases, and the particle moves more quickly. It then traverses the gas gap, reverses the polarity of its charge on contact with the opposite conductor, and is repelled back to the original conductor. This fundamental behavior is the same for both AC and DC voltages.

It is worthwhile to discuss a few characteristic particle shapes, and typical formulas for the particle charge q and field E_L when levitation occurs.[150,153,154,158] The motion of a particle which has acquired a charge q at t = 0 can be described by the equation

$$m \frac{d^2 x}{dt^2} = qE(x,t) - mg \qquad (1.6.1.)$$

Here viscosity effects are neglected for the particle sizes of interest in compressed gas insulation. It is obvious that the critical condition for levitation is qE = mg.

For a spherical particle of radius r_s, the formulas $q = 4\pi\epsilon_d/\epsilon^* r_s^2 E$ and $m = (4/3)\pi r_s^3 (\rho_p - \rho_g)$

in SI units hold if we assume no perturbation of the electric field. The levitation electric field, E_L, is then as follows

$$E_L = \left\{ \frac{(\rho_p - \rho_g)g}{3\epsilon_0 \epsilon^*} r_s \right\}^{1/2} \tag{1.6.2.}$$

where ρ_p and ρ_g are the densities of the particle and gas, respectively. Exact solutions for the equation including the field distribution and the image force are given by[150,158]

$$E_L = 0.49 \left\{ \frac{(\rho_p - \rho_g)g}{\epsilon_0 \epsilon^*} r_s \right\}^{1/2}$$

$$\text{with } q = 2\pi \epsilon_0 \epsilon^* r_s^2 E/3 \tag{1.6.3.}$$

With a wire particle of radius r_w and length L lying on an electrode, which approximates a practical contamination, the levitating field becomes

$$E_L = 0.836 \left\{ \frac{(\rho_p - \rho_g)g}{\epsilon_0 \epsilon^*} r_w \right\}^{1/2}$$

$$\text{with } q = 2\pi \epsilon_0 \epsilon^* r_w LE \tag{1.6.4.}$$

Once a wire particle is levitated in the field, it aligns with the field because of the dipole effect and stands vertically on the electrode. In this position it is reasonable to approximate the wire by a semiellipsoid of base radius r_w and vertical length L. For this case, the levitating field is

$$E_L = [\ln(2L/r_w) - 1] \left\{ \frac{(\rho_p - \rho_g)g}{\epsilon_0 \epsilon^*} \cdot \frac{r_w^2}{L(\ln(L/r_w) - 0.5)} \right\}^{1/2}$$

$$\text{where } q = \pi\epsilon_0 \epsilon^* L^2 E/(\ln(2L/r_w) - 1) \tag{1.6.5.}$$

Generally speaking, the induced charge is greater when the wire is vertical than when it is lying horizontally. In other words, once the wire is levitated from the horizontal to the vertical position it acquires more charge and so becomes more active in the field. The levitation field is smaller in the former case than in the latter case. Once a particle has moved into the vertical position on an electrode, the field required for the particle to fall over and stop moving may be lower by a factor of two or three, depending on the particle length and radius. This means that after the particle leaves the conductor, it can return and possibly cause microdischarges.

Modes of particle movement can be derived from Equation 1.6.1. With alternating voltages in parallel-plane gaps, the velocity u(t) of a spherical particle of radius r and density ρ, in the nth excursion, at time t measured from the last contact with the lower electrode, is then given by the equation[153]

$$u(t) = A_n[\cos\phi_n - \cos(\omega t + \phi_n)] - gt - RU_n \tag{1.6.6.}$$

and the distance h(t) traveled from the lower electrode is

$$h(t) = A_n[(\sin\phi_n - \sin(\omega t + \phi_n))/\omega + t \cos\phi_n] - gt^2/2 - RU_n t$$

$$\tag{1.6.7.}$$

where $A_n = \pi^2 \epsilon_0 \epsilon^* E_n E / 2r\rho\omega$, E_n is the field at the lower electrode at the beginning of the nth excursion, E is the peak value of the applied field, $\phi_n = \sin^{-1}(E_n/E)$, U_n is the particle impact approach velocity at the electrode just before the nth excursion, and R is the coefficient of restitution for the impact. Computer calculation and experimental results reveal that following elevation, particles do not immediately cross the gap, but the particle activity increases with increasing voltage; only at substantially higher voltages will they get across. In AC fields, they can remain in the midgap region for a long period of time and can take several cycles of voltage to cross the gap.[150]

In DC fields, they are first lifted and then cross directly to the opposite electrode. As the field increases, they will move back and forth between electrodes with increasing velocity. However, after the particles have crossed the gap many times, they will sometimes remain hovering at the negative electrode, even at high field stresses, with intense corona and emitting visible light. This is known as the firefly phenomenon in that it appears as if a "firefly" is flitting along and around the inner conductor.

Microdischarges are considered to be the principle contribution to the reduction of breakdown strength of SF_6 gas in coaxial electrode system, which is typically in the range of 2 to 10 times. In other words, this reduction in both corona and spark over levels results from particles which

1. Rapidly bounce to and fro between the electrodes
2. Become attached to an electrode and act as corona sources
3. Remain in dynamic equilibrium on or near the surface of the most highly stressed electrode, emitting corona current ("fireflies")

Lightweight, twisted metallic fragments of aluminum (10^{-5} to 10^{-2} cm in size) have the most deleterious effect on insulation stability of such systems.

B. Scavenging Metallic Particles

Certainly, the best way to eliminate "fireflies" is to fabricate the CGI cable system completely free from any conducting particles. Every effort should be exerted to make contamination-free systems as far as they are economically feasible. Unfortunately, a few metallic contaminants cannot be avoided during fabrication and installation.

There are basically two measures that can be taken to mitigate their effects. One is to make a dielectric modification around the most highly stressed electrode surface where free-conducting particles are most likely to gather and stay. The other is to let the free conducting particles migrate to a special place where, once they are trapped, they can never be lifted to electrically active regions. Both solutions, but especially the second, are standard practice in the U.S.

The first concept is realized by way of dielectric electrode coatings. The influence of free conducting particles lessens when the conductor surface is coated with a dielectric paint. Insulation strength increases by 20% or more and can reach a level close to the ionization limit (the ultimate level for SF_6 gas). The type of dielectric used and the coating thickness makes no noticeable difference; durability and long-term performance should be the important considerations.

The second defense against free-conducting particles has resulted in the development of "particle trapping devices". Inactivation of particles by trapping can be accomplished in two existing ways; both make use of induced particle motion under the action of an applied electric field. In the first method, regions of low field are introduced by such means as raised perforated screens which form the outer electrode and which allow particles to enter but keep them from leaving. Figure 1.6.3. shows one such design.[156] The Tri-Trap® particle entrapment system consists of a three-post epoxy spacer and a perforated plate. This system

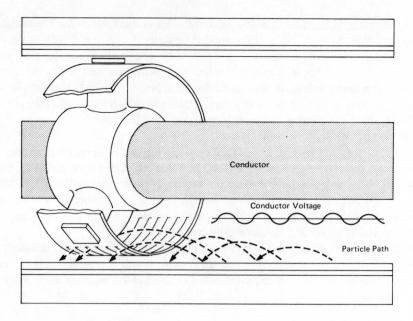

FIGURE 1.6.3. A particle trap design and particle movement.[156]

works in two ways to prevent flashover from any conductive particle that might remain in the system after field installation. Such a particle moves because of the charge induced on it by the high AC voltage on the conductor. This AC voltage creates high AC voltage gradients which alternately lift and drop the particle, causing random longitudinal motion along the sheath. For this reason, particles will move along the bottom of the sheath. The particle passes between the two lower posts of the spacer and drops between the trap and the sheath. Since both the trap and the sheath are at the same potential, the particle stays in the trap, permanently neutralized.

Some other designs are indicated in Figure 1.6.4.[157] They are equally effective in scavenging particles. For DC use, different types of trapping devices are being developed as shown in Figure 1.6.5.[157] It is to be noted that all three types are deep relative to the minimum (shielding) width of the opening. When any of them is placed at the lowest point of an incline, it traps particles efficiently, quickly, and with essentially no hesitation.

In the second method of trapping particles, a sticky surface is applied to the outer electrode. This can also hold particles and thus keep them away from the high field center conductor. Greatly reduced prebreakdown currents and about 20% higher breakdown voltages result from the application of either particle trapping method.

After the CGI cable has been filled with SF_6 to the specified value of density and with a sufficiently low moisture content, a process of electrical conditioning can be implemented. This conditioning removes conducting and semiconducting particles by transferring them into the particle trap or the contamination control devices. A slowly rising conditioning voltage, applied from an AC source with relatively high internal impedance, is able to effect the transfer of the mobile particles to the trap. The net effect of voltage application to a CGI cable with a trap is a gradual increase in the dielectric withstand. Low-energy breakdown through the high-impedance power source will hopefully not damage the cable insulation and conductor. The spacer-insulators should be designed to be track-resistant. Care should be taken with aluminum conductors since they are susceptible to surface damage due to sparking.

As a complementary scavenging technique, mechanical vibration may be effective for an enclosure so designed as to incline toward the trap. It is recognized, however, that particles

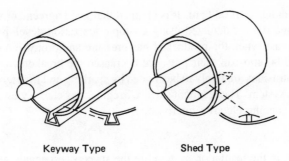

Keyway Type   Shed Type

FIGURE 1.6.4. Designs of particle trapping devices. (From Nakata, R., *Controlled Particle Scavenging Technique For Use in HVDC SE₆ Gas Bus,* IEEE PES Summer Meeting Paper A76 410-1, IEEE, Piscataway, New Jersey, 1976, 4. With permission.)

Single Pipe   Multiple
SUMP TYPE   SUMP TYPE

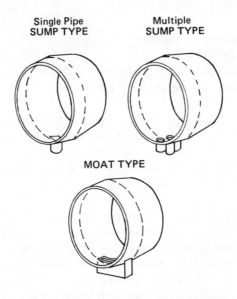

MOAT TYPE

FIGURE 1.6.5. Particle traps for HVDC use. (After Nakata, R., *Controlled Particle Scavenging Technique For Use in HVDC SF₆ Gas Bus,* IEEE PES Summer Meeting Paper A76 410-1, IEEE, Piscataway, New Jersey, 1976, 5. With permission.)

often gather at unexpected nodal points within the enclosure, such as points of support, or locations stiffened by intersecting enclosure branches.

C. Electrode Surface

Aluminum and stainless steel electrodes with similarly finished surfaces exhibit almost the same performance as far as their electrical behavior is concerned. However, aluminum is more susceptible to surface damage; its spark marks being about 10 times larger than those on stainless steel.

Large surface scratches on the electrodes can be breakdown sites. This has been confirmed by experiments on 0.25-mm wire surface irregularity wrapped on a 76-mm center conductor, and on protrusions of steel or aluminum spheres with 0.079-cm radius or a rod 10 times higher than its tip radius of 0.039 cm. It is claimed that good agreement was found between measured values and those calculated using a simple ionization development for discharge initiation. In each case when the product of gas pressure and protrusion height above a flat electrode exceeded 80 atm-μm their presence decreased the breakdown strength.

Microscopic protrusions are likely to be created by sparking and to deteriorate the electrical performance of the system. The resultant surface irregularities can be viewed and investigated with the aid of a scanning electron microscope.

D. Support Spacers

The weak points in the insulation system are the spacers, especially at their triple points of insulator, conductor, and gas. Spacers are selected for minimum distortion of the original electric field and for mechanical integrity as supports. It is, of course, imperative to fabricate spacers with intimate void-free electrode contact and strong mechanical connections at both ends. Metal inserts are utilized for these necessary functions and also to modify the normal electric field to a more favorable distribution. The electric field in the presence of a spacer is controlled by[145]

1. The electrode geometry surrounding the spacer
2. The geometry of the inserts
3. The geometry of the dielectric material
4. The dielectric constant of the dielectric material

Figure 1.6.6. gives a comparison between "optimum" and worse post-type spacers with respect to shape.[145] The spacer is fabricated from vacuum cast bisphenol-A epoxy, heavily loaded with Al_2O_3, and possesses a dielectric constant of 6 only for experimental purposes. A computer-generated equipotential distribution of the optimum spacer indicates the field to be relatively uniform along the dielectric-gas interface with a maximum value of 0.76 times the conductor field, decreasing somewhat near the metal ends. Corrugated post-type spacers are one modification designed to increase the creepage path along the surface. This is a favorable shape in most cases where no particle traps are used. Care should be taken lest the corrugation reduces the insulating ability of the CGI cable. This can come about if the electric field increases near the protruded regions of the corrugation.

As regards the selection of the type of spacers, it has been fairly conclusively established that, by careful and knowledgeable design, approximately equal performance level can be achieved from the three basic shapes: disc, conical, and post.

As for three-phase CGI cables, both electrical and structural designs are among the major problems. Electrically, the chief difficulty resides in selection of a spacer configuration which minimizes the field enhancement at the spacer boundary with the gas. Structural problems are concerned with the determination of the electrodynamic forces acting on the conductors and the selection of spacers that can withstand the resulting mechanical stresses. A systematic approach, demonstrated in the literature,[160,161] defines an optimum spacer as one having the minimum volume of material which enables it to fulfill its structural function without adversely affecting the electrical integrity of the cable system.

1.6.3. Designs and Performances

A. Isolated Phase CGI Cables

A typical design of a single-core CGI cable has already been shown in Figure 1.6.1. Table 1.6.1.[152] and Table 1.6.2.[148] present some design and performance details of CGI cables of 138 kV to 765 kV in the U.S. The conductor is a rigid tube of high conductivity aluminum alloy 6101, with a wall thickness of 0.5 in. (12.7 mm) at 60 Hz, and is relatively independent of operating temperature. This wall thickness is sufficient to give nearly max-

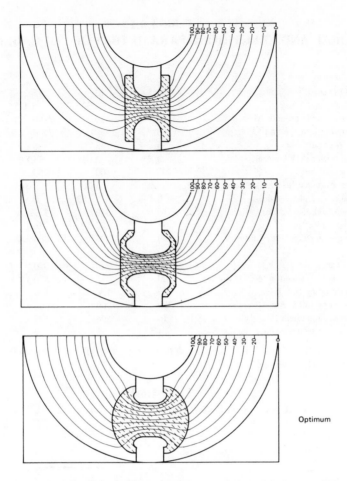

FIGURE 1.6.6. Equipotential plots for candidate spacer and their comparison. (From Cooke, C. M. and Trump, J. G., *IEEE Trans. Power Appar. Syst.*, 92(5), 1442, 1973. With permission.)

imum conductivity. The conductor expansion joint is designed to allow an axial motion of ± 0.50 in. (± 12.7 mm) which accommodates a temperature differential of 90°C. The sheath is a rigid tube of aluminum alloy 6061 (T6), which combines high strength and good electrical conductivity. The wall thickness ranges from 0.25 in. (6.35 mm) to 0.5 in. (12.7 mm) depending on the voltage and power rating of the line. When buried in long lengths, the backfill prevents movement of the sheath. Under these conditions, a temperature rise of 60°C will cause a compressive stress of about 14,000 psi (984 kg/cm², 96 MPa), which is well within the yield stress of 6061 (T6) aluminum. An operating SF_6 gas pressure of 50 psig (3.5 kg/cm², 344 kPa) is chosen at 20°C, resulting in a density of 1.75 lb/ft³ (0.028 g/cm³). Consistent with industry practice, the moisture content of the gas is held below 30 ppm; this prevents moisture condensation down to −40°C. Stop joints can be used to divide long runs of CGI cable into sections of about 1/2 mi (800 m), depending on system needs.

Disc-type spacers are used in these designs. They are placed in the sheath during factory assembly and the outer stress relief ring is securely locked to the sheath by means of pins inserted and welded from the outside. A flexible metal strap connects the inner ring electrically to the conductor. The conductor is free to slide on Teflon® bearing surfaces on the inner ring, except, of course, in the case of a fixed spacer or stop joint. A design of pothead termination to air is illustrated in Figure 1.6.7.,[152] which shows it gas-insulated with electrical grading provided by internal and external electrodes.

Table 1.6.1
PHYSICAL AND ELECTRICAL PARAMETERS OF CGI CABLE DESIGNS

System voltage (kV)	138	230	345	500
Conductor O.D. (in.)	3	4	5	6.75
Sheath O.D.(in.)	8.5	12	15	20
Sheath wall (in.)	0.25	0.30	0.375	0.375
Power rating underground (MVA)	300	620	1200	2200
Current rating underground (A)	1230	1550	2000	2530
Losses (W/circuit foot)	36	38	46	54
Charging power (MVA/circuit-mile)	0.76	2.05	4.8	10.9
Critical length (mi)	390	300	250	200
Effective resistance (μ-Ω/phase foot)	7.93	5.26	3.84	2.80
Capacitance (μF/phase mile)	0.106	0.103	0.107	0.115
Inductance (μH/phase mile)	316	340	337	337
Surge impedance (ohms)	54.5	57.2	56.1	54.1
Withstand voltage (60 Hz, 1 min-kV (RMS)	240	395	460	690
BIL (kV-crest)	550	900	1050	1550
Switching surge insulation level (kV-crest)	460	745	870	1290
Momentary rating (kA-1 sec)	80	100	100	100
Open air rating (30°C rise-MVA)	750	1800	3400	6500
Peak operating gradient (V/mil)	76	90	108	115
Phase separation (centers-in.)	14.5	18	22	28

From Pederson, B. O., Doepkin, H. C., Jr., and Bolin, P. C., *IEEE Trans. Power Appar. Syst.*, 91(6), 2632, 1971. With permission.

Table 1.6.2.
TYPICAL SYSTEM PARAMETERS OF CGI CABLES

Max. rated system voltage (kV)	145	242	362		550		800
Nominal system voltage (kV)	138	230	345		500		765
BIL (kV)	650	900	900	1050	1300	1500	1800
Example standard design	9A4	12A4	14A6	14A10	18A10	20B12	25B18
Enclosure O.D. (in.)	9	12	14	14	18	20	25
Enclosure thickness (in.)	0.25	0.25	0.25	0.25	0.25	0.375	0.375
Conductor cross section area (in.²)	4	4	6	10	10	12	18
Capacitance (pF/phase ft)	22.4	17.0	19.1	19.1	15.8	17.7	18.0
Inductance (μH/phase ft)	0.050	0.064	0.058	0.058	0.069	0.065	0.061
Surge impedance (Ω)	47.0	61.5	55.0	55.0	66.0	62.4	58.0
Charging current (A/mi)	3.5	4.5	7.6	7.6	9.1	9.6	15.8
Underground installation							
Minimum trench width (in.)	62	72	78	78	90	96	112
Nominal rated current (kA)	1.6	1.7	2.1	2.3	2.4	3.0	3.5
Nominal rated power (GVA)	0.38	0.68	1.25	1.37	2.08	2.60	4.64
Phase resistance ($\mu\Omega$/phase ft)	8.11	7.23	5.26	4.30	4.06	3.03	2.20
Losses at rated current (W/circuit ft)	62	63	70	68	70	82	81
Critical length (mi)	460	380	280	300	260	310	220
Tunnel installation							
Nominal rated current (kA)	2.4	2.7	3.7	4.2	4.6	5.5	7.3
Nominal rated power (GVA)	0.57	1.08	2.21	2.51	3.98	4.76	9.67
Phase resistance ($\mu\Omega$/phase ft)	7.59	6.90	5.01	4.04	4.10	3.01	2.22
Losses at rated current (W/circuit ft)	130	151	205	213	260	273	355
Critical length (mi)	690	600	490	550	500	570	460

From Cronin, J. C. and Dethlefsen, R., *Guide to the Use of Gas Cable Systems*, NTIS PB-247 898, Nat. Tech. Inf. Serv., Springfield, Va., 1975, 17. With permission.

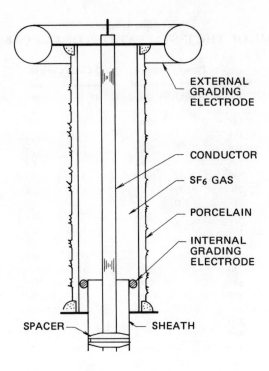

FIGURE 1.6.7. Typical CGI bushing. (From Pederson, B. O., Doepkin, H. C., Jr., and Bolin, P. C., *IEEE Trans. Power Appar. Syst.*, 90(6), 2633, 1971. With permission.)

The sheaths of the three phases are usually solidly bonded together and grounded at the terminations, and in the case of long runs, at each manhole also (about every 1/2 mi [800 m]). Cross bonding may be used to reduce sheath currents.

Some representative installations of CGI cables in North America are listed in Table 1.6.3. Table 1.6.4. indicates some of the Japanese designs of bulk power CGI cables. As far as gas is concerned, SF_6 is standard practice, but some other gases, such as freons and mixture gases, are now under investigation.

B. Three-Conductor CGI Cables

Only one three-conductor CGI cable design appears in Table 1.6.3. It is the prototype cable shown in Figure 1.6.8.[167] Three-conductor CGI cables have the following advantages when compared with isolated-phase CGI cables:[167]

1. Much less sheath materials.
2. Lower sheath losses.
3. Even though only 1 30-in. (76.2-cm) sheath is used in place of 3 15-in. (38.1-cm) sheaths at 345 kV, the 3-conductor CGI cable ampacity can be higher than the isolated phase CGI cable ampacity for the same conditions.
4. Fewer field welds resulting in faster and less costly installation. With a totally welded isolated phase CGI cable there are six circumferential field welds per three-phase joint. For a totally welded three-conductor system, only four welds are needed. On the other hand, if a reliable plug-in conductor joint is available, there are three welds in the isolated phase CGI cable and only one in the three-conductor CGI cable. By employing plug-in conductor joints and a three-conductor system, the number of field welds can be reduced by almost 85%.

Table 1.6.3.
SOME OF THE INSTALLATIONS OF CGI CABLES

Operating company	Voltage (kV)	Power rating (MVA)	Rated current (A)	3-Circuit length ft	3-Circuit length m	Installation[a]	Year energized
Cleveland Electr. Illum.	345		2000	450	137	A	1971
Cleveland Electr. Illum	345		2000	450	137	A	1972
Con. Ed.	345		3350	520	158	U	1972
PEPCO	138		1200	210	64	A	1972
PSE & G of N.J.	230	600	1600	600	183	U	1973
South Cal. Ed.	230	600		450	137	A/U	1973
New Orleans P.S.	138	300		150	46	A	1973
PEPCO	138		1200	520	158	A	1973
AEP	138	300		500	152	U	1974
AEP	138	300		1,100	335	U	1974
Duke Power	230	1,500		600	183	O	1974
Arizona P.S.	230		2000	370	113	A	1974
Commonwealth Ed.	345		3000	1,900	579	A	1974
Boston Ed.	345		1100	630	192	U	1974
BPA	500	3,500		700	213	U	1975
Arizona O.S.	230	600		250	76	O	1975
Hydro Quebec	345					A	1975
	800		3000	100	30	A	1975
Central Illum. Ltd.	345		2000	1,100	335	A	1975
Hawaiian Electr.	138		1600	150	46	A	1975
Ontario Hydro	230		3000	2,300	701	A	1976
AEP	138		1600	2,000	610	U	1976
P.S. of N.H.	345		3000	3,300	1,006	A	1976
C. Hydro	500		3000	1,760	536	T	1976
AEP (2 circuits)	138	480		350	107	U	1977
Ontario Hydro (10 circuits)	230	1,200		1,600	488	A	1977
AEP (3 circuits)	138	480		1,800	549	U	1977
Ontario Hydro	230	1,200		3,900	1,188	A	1978
Duke Power	230	1,500		1,000	305	A	1978
EPRI	345	1,200	30-ft 3 conductors			U	—
ERDA	1200	10,000		40	12	U	—

[a] A: aboveground, U: underground, O: open trench, and T: tunnel.

Table 1.6.4.
TYPICAL DIMENSIONS OF BULK POWER CGI CABLES

Rated voltage (kV)	Ampacity (A)	Transmission power (MVA)	Single conductors Conductor O.D. (mm)	Single conductors Sheath O.D. (mm)	3ϕ Conductors Conductor O.D. (mm)	3ϕ Conductors Sheath O.D. (mm)
154	4,000	1,100	100	380	150	1,200
275	4,000	2,000	150	480	180	1,400
	8,000	4,000	230	560	230[a]	1,500
500	4,000	3,500	230	815	230	2,000
	8,000	7,000	230	815	230[a]	2,000
	12,000	10,000	350	970		

Note: SF$_6$ Gas Pressure z.7 Kg/cm^{2}G (0.36 MPa, 53 psi).

[a] Forced cooling is needed.

From Furukawa Electr. Ind., *Outline of CGITLs*, Ichihara-shi, Japan, 1975. With permission.

FIGURE 1.6.8. Prototype three-phase CGI cable. (345 kV, 1200 MVA).[167]

GAS-INSULATED CABLE TRENCH REQUIREMENTS
(1200 MVA at 345 kV)

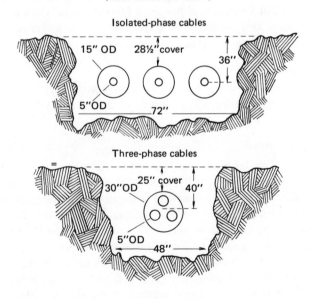

FIGURE 1.6.9. Excavation reduced by using a three-phase CGI cable.[167]

5. Narrower trenches required. Typically a conventional 345-kV isolated CGI cable would require a more than 72-in (180-cm) wide trench while a 345-kV, three-conductor CGI cable would require a 48-in. (122-cm) trench, as demonstrated in Figure 1.6.9.
6. Lower installed cost on a $/MVA-mile basis.

The design for the 1050-kV BIL, 345-kV, 1210-MVA, buried CGI cable with 40-kA short-circuit capability consists of 3 5-in. outer diameter × 4-in. inner diameter aluminum con-

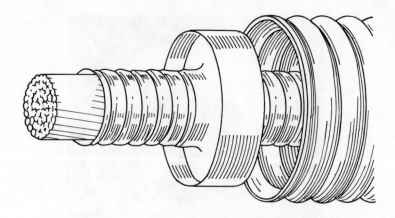

FIGURE 1.6.10.   Flexible CGI cable design.[4]

ductors located equilaterally within a 30-in. O.D. × 0.312-in. wall aluminum sheath. The insulating gas is SF$_6$ at 50 psig 3.5 kg/cm^{2}G and the spacers are cast with a filled epoxy (an alumina-filled bisphenol-A epoxy with a M-phenylenediamine hardener). The spacers are solidly mounted on a ring shield with one ring shield every 10 to 12 ft. Differential thermal expansion requires that the ring shields move relative to the sheath and at least two of the conductors move relative to the ring shields.

Short circuit forces seem to be more important in designing spacers for three-conductor cables than for isolated-phase cables, although no mechanical behavior of a three-conductor configuration has been well documented. Cantilever (electromagnetic) force on spacers should be investigated in order to optimize their design.

C. Flexible CGI Cables

For long-distance transmission, flexible CGI cables have obvious advantages over the rigid CGI cables so far described. One example of a flexible CGI cable with corrugated tubes is illustrated in Figure 1.6.10.

1.6.4. Forced Cooling[165,166]

The ampacity of CGI cables is presently limited by the maximum sheath temperature. In buried CGI cable systems, this is usually set at 60°C or less to prevent moisture migration in the soil or backfill, which might otherwise result in thermal runaway and failure of the cables. Under these conditions, the conductor temperature is usually around 80°C. Higher operating temperatures are possible by careful selection of the spacer material.

The internal thermal resistance of CGI cables is much (two to six times) smaller than that of solid cables. This high-heat transfer characteristic opens up possibilities for external forced cooling with concomitant higher efficiency. This will probably be realized by circulating a cooling fluid, preferably water, in pipes attached to the sheath. Other methods propose cooling the gas directly and the conductor by fluid circulation, and the use of heat pipes or an evaporative method has also been suggested.

Conceptual designs are shown in Figure 1.6.11.(A) and (B) for isolated-phase and three-conductor systems, respectively.[167] The basic notion is to water-cool the sheath, and remove the heat from the recirculating water by an air-cooled heat exchanger. The cooling pipes can be integrally extruded with the sheath. These have less thermal impedance between the sheath and coolant than ''nonintegral'' pipes which are welded or attached to the sheath. For single-core cables, water leaving the cooling station enters pipes attached to the central sheath, and then returns with half of the flow cooling each outside sheath. An alternative scheme would be to cool each phase in parallel and provide a combined return pipe that is not actively used in cooling.

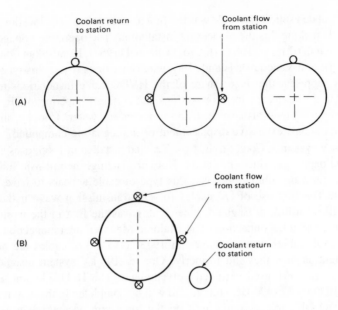

FIGURE 1.6.11.   Cooling pipe arrangement for isolated-phase and three conductor CGI cable systems.

Selection of a forced-cooling configuration for the three-conductor system is not as clear cut as the isolated phase. Cooling and return pipes could both be thermally integral with the sheath. In one option, shown in Figure 1.6.11.(B), the water leaves the cooling station and enters pipes on the sheath. Reverse mode is possible; the cooled water enters the separate pipe, flows to the end of the line, and then cools the sheath on its return to the station.

1.7. DC CABLES

1.7.1. Technical Evolution and Installation

Compared to AC, DC constitutes a much better method of transmission of electric power.[169] DC power transmission cables, unlike AC cables, bear no charging or capacitive current and, in principle, no dielectric loss. This permits DC power transmission to be considered over long distances without compensation and with reduced losses. For a given insulation, the transmissible DC power corresponds to the peak voltage which this insulation can withstand. The transmissible power is thus, in principle, $\sqrt{2}$ times as high as the effective AC power. Further, the design electric stress can be selected higher for DC than for AC insulation. Thus the increase in transmissible power by a factor of more than $\sqrt{2}$ is to be expected.

With submarine cables, for which DC cables have actually been developed so far, weight creates mechanical problems during installation. In general, one can say the lighter the better. For a given power transmitted weight is much higher for AC than for DC cables. It is a difficult task to lay submarine cables even without reactors. With this penalty, submarine AC links are limited in length to less than around 100 km.[169]

All types of cables developed for AC power transmission can be candidates for DC transmission. Three types have been used for high-voltage DC transmission during the more than 20 years of their history:

1.   Impregnated-paper insulated "solid type" cables
2.   Gas-filled pre-impregnated paper-insulated cables
3.   Oil-filled impregnated-paper insulated cables

The paper-insulated solid-type cable was the first to be installed. The location was Gotland, Sweden in 1954, making this the pioneering installation. The operating voltage was 150 kV and the circuit length 61 mi (~100 km). Cross Channel (1961), Konti-Skan (1965), Sardinia-Mainland (1965), and Vancouver Island (1969) are other links where this type of cable was installed. They represent the major share of the HVDC cable installations in present-day operation.[170] All are submarine interconnections and certainly superior to OF cables, which require oil feeding for long-distance installation under sea water. Cooling minimizes temperature rise and avoids excessive displacement of impregnating compound.

One ± 250-kV system at Cook Strait, New Zealand installed in 1965, uses high-pressure, oil-impregnated paper, gas-filled insulation. Notwithstanding, the intrinsic suitability of the gas-filled cable for long submarine links, this type of cable appears to have been ignored in recent years. The first use of OF cables for DC transmission was for the Kingsnorth-Beddington-Willesden link in England; indeed, this was the first in the world to use only land cables for a DC interconnection. The Majorca-Menorca interconnection by means of ± 200-kV DC OF cables was completed in 1973. HVDC OF cables are now receiving much more attention than they did formerly. One ± 100-kV system installed across the English Channel in 1961 used extruded polyethylene. The Hokkaido link in Japan was established in 1979 by 125-kV DC voltage and will be completed in the near future by using a ± 250-kV DC OF cable system based on the long-term acceptance tests which were carried out on ± 500 kV DC OF cables.[171]

Interest in the U.S. centers naturally around the use of HPOF systems in the voltage range ± 400 ~ 750-kV DC. A completed 1/4-mi experimental underground cable transmission system was installed in New York City in 1975, which included converter and inverter terminals and cable splices and terminals to operate at a voltage stress equivalent to ± 600 kV DC, on a 2400-MW system. High-voltage (400- to 750-kV) DC power cables, insulated with extruded dielectric, are not developed as yet, mainly because of space charge effects in the insulation as described in Volume I, Section 2.6.

For submarine cables, their weight, which is typically about 20 kg/m, becomes a limiting factor as far as depth of laying is concerned.[169] This weight, which is often due to the watertight lead sheath which protects the impregnated-paper insulation, corresponds to a laying tension of the order 8 ton for a depth of 500 m. Much higher tensions are anticipated during recovery operations for repair work, which is not permitted in deep water. In this respect, skin divers and specially adapted repair ships can be helpful.

1.7.2. Designs and Performances

The design of a cable for operation at high DC stress is entirely different from that of a conventional cable used for AC. As described in Volume I, Section 2.6., this is because insulation behaves differently when subjected to DC stress. The DC condition is characterized by two distinctive features: the resistively distributed electric field which is significantly sensitive to temperature, and space charge formation in the insulation which distorts the electric field distribution.

A. Designs of Cables Already Installed

Figure 1.7.1. shows the cross section of the DC cables used in the France-England inter tie.[172] Unlike submarine links for feeding an island from the mainland networks, this link is capable of supplying power in either direction. Interest lies in pooling the emergency power available and supplying the excess loads of both countries. The link comprises 2 single-core cables capable of transmitting 160 MV at ± 100 kV DC which lie on the seabed to a depth of 55 m. The cables comprise a copper conductor of 344 mm² cross section, impregnated paper with a wall thickness of 7.5 mm, and a 2.6-mm thick lead sheath.

Compared to the France-England interconnection, the Sardinia-Corsica-Italy link shows

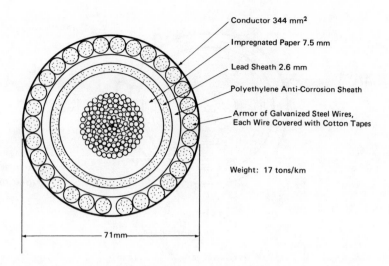

FIGURE 1.7.1. Cross section of DC ± 100-kV, 160-MW submarine cable for France-England interconnection.[172]

Table 1.7.1.
DESIGN PARAMETERS OF DC 200-kV 200-MW SOLID TYPE CABLE INSTALLED AS SARDINIA-CORSICA-ITALY LINK

Copper conductor 420 mm²
  (oval shape)

Impregnated paper insulation	11.8 mm
One lead sheath	2.5 mm
One polyethylene sheath	3.1 mm
A textile taping	
2 layers of steel tapes 0.3 × 34 mm (anti-twist effect)	
6.4 mm steel wire armor whipped to	7.4 mm diam.
Overall diam.	80 mm
Weight	21 t/km

From **Anon.**, *Submarine Power Cables*, Cables de Lyon Ad Release 7111-1051, (Lyon, year unspecified), 16. With permission.

important performance extrapolations in all areas: from 100 to 200 kV in voltage, 160 to 200 MW in power rating, 55 to 450 m in depth, 27 to 118 km in length, and 3 to 45 km between joints. The design parameters of the cables are shown in Table 1.7.1.[172] The conductor consists of oval, compacted copper strands, a 420-mm² section, having diameters of 22 mm and 26.8 mm. They thereby accommodate hydrostatic pressures in excess of 10 kg/cm² (100-m depth) during thermal cycles. The oval shape, which becomes more circular as the impregnating material expands during heating cycles, allows the insulation and the lead to return to their original shape under the combined action of hydrostatic pressure and the high-strength metal tape which is applied over the lead and interposed polyethylene sheath. The conductor is made of 60 wires distributed in 3 layers, the outer layer being laid up in the opposite direction to the inner 2. This assures almost perfect anti-twist characteristics. The conductor is screened by semiconductive carbon black paper tapes as a standard

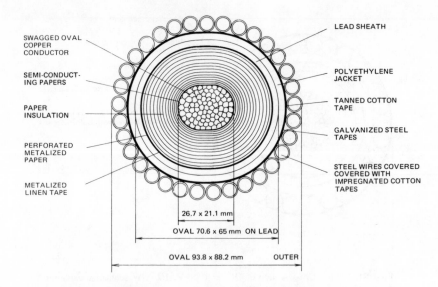

FIGURE 1.7.2.   Design of DC 300-kV cable (1 × 4000 mm, 600 A). (From Eyraud, I., Horne, L. R., and Oudin, J. M., *The 300 kV Direct Current Submarine Cables Transmission between British Colombia Mainland and Vancouver Island,* CIGRE Paper 21-07, CIGRE, Paris, 1970, 8. With permission.)

procedure. The insulation being 11.8-mm thick is subjected to a stress of 25 kv/mm at the conductor. The metal sheath which is made of rather hard lead alloy (alloy E: lead + 0.4% tin + 0.2% antimony), is 2.5-mm thick. This material is believed to exhibit better mechanical characteristics than the soft alloys usually employed in underground cables. The cable is covered with a polyethylene sheath, 3.2-mm thick with a melt index of 0.2. This is applied for mechanical reinforcement of the lead sheath and the bedding as well as for anticorrosion protection and water-tightness. Metal reinforcement is applied over the polyethylene sheath by the interposition of a fabric tape. It consists of galvanized steel tapes (24-mm wide, 0.3-mm thick) applied with triple start in 1 or 2 layers. This provides required antitwist action besides the simple reinforcing action. Bedding, composed of 2 compacted layers of chemically treated jute, is applied between reinforcement and armoring to a thickness of 2.5 mm. The armoring comprises 30 galvanized steel wires having a diameter of 6.4 mm, each individually wrapped with textile tape which builds up their external diameter to 7.6 mm.[173]

The novel feature of the Vancouver cable, compared with the Sardinia-Italy cable, is the rated voltage, 300 kV rather than 200 kV.[174] The current rating, on the other hand, is reduced from 750 to 600 A. Figure 1.7.2.[174] and Table 1.7.2.[172] show details of the dc 300-kV, 300-MW solid-type cable system which feeds Vancouver Island. Table 1.7.3. outlines DC cables installed throughout the world.

Table 1.7.4. lists performance data for DC ± 500-kV OF cables and DC ± 250-kV XLPE cables which have been developed, manufactured, and tested in Japan[171] as part of the research and development work initiated in 1971 for the Hokkaido link. This is a DC cable installation between Hokkaido and the mainland. Performance of selected papers are summarized in Table 1.7.5.[6] Deionized water-washed paper is preferred because it has high and stable electrical resistivity. High-density and high-air-impermeability paper is used to increase breakdown voltage and mechanical strength. Low-dissipation factor is a matter of lesser consequence.

Joints for DC cables are designed in a manner different from those for AC cables. Examples of a normal joint and flexible joint are shown in Figure 1.7.3. The normal joint is designed with a diameter as large as that for 275-kV AC class cables but is 40% longer. Humidity

Table 1.7.2.
DESIGN PARAMETERS OF DC 300-kV 300-MW SOLID TYPE CABLE INSTALLED FOR VANCOUVER ISLAND FEEDING

Deep seas:
    Copper conductor 400 mm² (oval shape)

Impregnated paper insulation	18.5 mm
One lead sheath	2.9 mm
One polyethylene sheath	3 mm
A textile taping	
A layer of 2 galvanized steel tapes for anti-twist effect	
Armor of galvanized and whipped 6.4 mm steel	
Overall diam.	91 mm
Weight	24.5 t/km

For nonpermanently submerged cables (about 3200 m)
    Copper conductor 650 mm² (oval shape)

Impregnated paper insulation	20.5 mm
One lead sheath	3 mm
One polyethylene sheath	3 mm
A textile tape	
Armor nonwhipped galvanized 5 mm steel wires	
2 Layers of jute servings	
Overall diam.	104 mm
Weight	30.5 t/km

Note: Total length of the various submarine links: 100 km. Transmitted power capacity: 300 MW at 300 kV DC through 3 single-core cables of which 1 is a spare cable.

From **Anon.**, *Submarine Power Cables,* Cables de Lyon Ad Release 7111-1051, (Lyon, year unspecified), 16. With permission.

and contamination are the chief concerns; they can cause thermal breakdown of joints. Even more care should be taken for flexible joints.

As far as potheads are concerned, the condenser cone type is preferred because of its excellent properties under transient voltage conditions (switching and lightning surges). The stress cone type is another candidate being favored because the strength of the internal insulation can be secured without exactly adjusting the potential distribution. This is assured by the length of the porcelain insulator, especially where a contamination-resistant insulator is used. The insulation length of normal splice joints is long in comparison to their outer diameter. Great care should be taken to minimize moisture absorption during splicing operations to avoid acute unbalance in the radial insulation resistivity distribution.

B. Maximum Operating Voltage Gradient

High-density paper is preferred for DC cable insulation because dissipation factor is not a consideration. This is distinctly different from AC cable insulation where lower density and higher air impermeability are important criteria. Values for design stress at the conductor are shown in Table 1.7.4.

Long-term DC strength of solid-type cables, particularly under thermal cycling, limits the DC working stress to about 25 kV/mm, with maximum cable temperature not exceeding 50 to 60 °C, in order to avoid excessive displacement of the insulating compound. The maximum working voltage ranges from 250 to 300 kV, depending on the test voltage level. Cables so

Table 1.7.3.
OUTLINE OF DC CABLES INSTALLED

System	Power (MW)	Voltage (kV)	Current (A)	Cable Kind	Cross section (mm²)	Insulation thickness (mm)	Emax (kV/mm)	Cable length (km) Total	Cable length (km) Sea	Sea bed depth (m)	Year	Remark
Sweden-Gotland	20	−100	200	Solid	90 Cu	7	25	96	96	140	1954	Sea return
Kashira-Moscow	30	200	150	Solid	150 Al	11.2	31.2	112	0	—	1955	
England-France	160	±100	800	Solid	340 Cu	9.3 / 7.5	13.5 / 18.5	2 Cables / 64	64	60	1961	
Volgograd-Donbass	750	±400	900	Solid	1000 Al	18	31	—	—	—	1964	
Volgograd-Donbass	750	±400	900	OF	550 Cu	16	43	—	—	—	1964	
Konti-Skan	250	250	1000	Solid	625 Cu	16	25	22	22	86	1965	Divided into two sections
Konti-Skan	250	250	1000	OF	2 × 310 Cu	12.4	33	58	58	86	1965	
Cook Strait (New Zealand)	600	±250	1200 (600 × 2)	GF	2 × 500 Cu	14.5	25	42	42	256	1965	Gas pressure 30 kg/cm²
Sardinia-Italy	200	200	1000	Solid	420 Cu	11.8	25	2 Cables / 116	116	450	1967	Sea return
Vancouver the Third	624	±260	1200	Solid	2 × 400 Cu	18.5	25	28	23	—	1969	
Gotland Power-Up	10	50	200 (600 × 2)	—	—	—	—	3 Cables / 96	96	—	1970	Sea return
London-Kingsnorth	500	±250	1000	OF	800 Cu	10.3	32	60	0	—	1971	
Juneau (Alaska)	70	70 ~ 125	—	—	—M	—	—	45	—	—	1972	
Churchill Fall (Canada)	4500	±500	4500	—	—	—	—	150	—	—	—	Plan
Spain-Morocco	200	220	910	—	—	—	—	32	—	—	—	
Italy-Yugoslavia	720	300	2400	—	—	—	—	120	—	—	—	
Hokkaido Link (Japan)	600	±250	1200	OF	600 Cu	14.5	—	43	42	300	1978	
Hokkaido Link Neutral Line	—	—	1200	XLPE	500 Cu	7.0	—	43	42	300	1978	

Table 1.7.4.
DESIGNS OF DC, OF, AND XLPE CABLES FOR FIELD EVALUATION

Item		Unit	±500 kV OF Cable		±250kV XLPE Insulated Cable
			Submarine cable	Land Cable	
Oil duct	Inner dia.	mmφ	16.0	22.0	-
	Spiral tape	mm	0.8	0.85	-
	Outer dia.	mmφ	17.6	23.7	
Conductor	Size	mm^2	1,000	1,000	1,000
	Shape	---	Hollow concentric strand		Concentric strand
	Component	Numbers/mmφ	160/2.74	144/3.0	127/3.2
	Outer dia.	mmφ	45.0	47.6	41.6
Insulation thickness *		mm	21.0	21.0	25.0
Screen		mm	0.5	0.5	0.5
Lead Sheath	Thickness	mm	4.0	4.0	-
	Outer dia.	mmφ	96.0	98.6	-
Reinforcement layer		mm	0.5	1.0	-
Protective covering		mm	5.0	5.0	4.5
Cable outer dia.		mm	107.0	110.6	101.6

* including inner and outer semi-conductive layer

designed will withstand internally generated overvoltages because of their important AC component.

Experience of long-term strength of gas-filled cables is limited, but it seems likely that a DC working stress higher than 25 kV/mm (perhaps 30 kV/mm) could be accepted. Like the solid type, this cable may be technically and economically interesting only for long submarine links.

The long-term DC strength of OF cables, with thermal cycling, involving temperatures up to 85 °C, is certainly sufficient to tolerate a DC working stress of 35 kV/mm. With required impulse withstand voltages of 2.5 to 3 times the working voltage, its nominal stress can be 35 to 40 kV/mm. Overvoltages of internal origin constitute no normal design problem. This cable can be used on land as well as for submarine links up to a maximum length of approximately 60 km. With a maximum service voltage of ± 750 kV, a power rating of 2000 MVA can be achieved and 3000 MVA seems feasible.[170]

C. Jointless Submarine Cables
The cost of repairing submarine cable is enormous in comparison to the value of the cable itself, therefore the risk of any repair work should be reduced to an absolute minimum.[175] For this reason, joints should be avoided to whatever extent this is possible, since they are always a potential source of trouble.

D. Challenge of DC XLPE Cables
DC cables with extruded insulation have not as yet acquired any service experience,

Table 1.7.5.
CHARACTERISTICS OF DC INSULATING PAPER

Classification	Thickness (μm)	Density (g/cm³)	Impermeability (s/1000 cc)	Tensile strength (kg/15 mm)	Breakdown voltage (kV/mm)				Dielectric constant	Dielectric loss tan δ 100°C (%)	Insulation resistivity		
					DC (room temp.)	Impulse (room temp.)	AC (room temp.)				ρ_0 (Ω·cm)	α(1/°C)	β(mm/kV)
AC low loss paper	125	0.72	1,300	14.0	138	127	45	3.4	0.21	19.5×10^{18}	0.10	0.043
	80	1.04	9,000	14.5	160	150	55	4.2	0.32	6.6×10^{18}	0.10	0.044
Newly developed	100	1.00	4,500	16.5	145	135	50	4.1	0.33	6.3×10^{18}	0.10	0.043
DC cable insulating	125	1.02	4,000	16.5	145	130	45	4.0	0.33	6.0×10^{18}	0.10	0.045
paper	150	1.06	3,000	17.5	140	130	40	4.0	0.32	5.0×10^{18}	0.10	0.045

Note: Columns to the right of "Breakdown voltage" give characteristics of insulating paper impregnated with synthetic oil. This paper has a higher density, better permeability, and greater breakdown voltage than AC cable insulting paper.

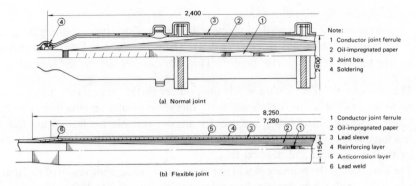

(a) Normal joint

(b) Flexible joint

(a) The normal jonit has the same outside diameter as and about 40% larger length than 275kV AC cable joints. (b) The flexible joint has approximately the same outside diameter as cables to give flexibility. The length is made longer to ensure the required insulation strength.

FIGURE 1.7.3. Structure of normal and flexible joints for 500-kV DC OF cable. (From Ando, N., Nanano, Y., Abe, H., Hayashi, K., and Numajiri, F., *Hitachi Rev.*, 23(12), 472, 1974. With permission.)

although they are considered to be a good replacement for solid and OF cables for the same reasons that apply to AC cables. They would be particularly advantageous for submarine use if the lead sheath could be eliminated. As noted in Volume I, Section 2.6., space charge effects appear to be much more serious contributors to insulation failure in extruded insulation than in taped insulation. Details of 1 design for a ± 250-kV experimental XLPE cable are indicated in Table 1.7.4.[171] The conductor size is 1000 mm^2 for a transmission capacity of 500 MW and a permissible conductor temperature of 80°C (set lower than the temperature 90°C for AC XLPE cables). Many of the failures experienced during testing occurred in splices and terminations. They appeared to be associated with the interface between the cable insulation and the molded insulation, and in the cable sections where the temperature distribution was most distorted.

E. DC POF Cables

Pipe-type cables, though already economically competitive with OF cables, are handicapped in AC applications by eddy current losses in the metallic pipes (see Section 1.2.). These losses increase very rapidly with current density and limit the power capacity of such cables to about 300 MW for voltages up to 220 kV. This handicap disappears with DC, a fact which should amply balance the current density limitations, making them a particularly attractive option for DC. The third cable which is normally provided in the pipe can act as an earth cable with reduced insulation. Using this cable in this way reduces the risk of soil drying.

1.8. RADIATION RESISTANT CABLES

1.8.1. Design Requirements for Power Cables in Nuclear Power Generating Stations

The environment of cables in nuclear power stations is significantly different from environment around conventional cables. It includes radiation as well as heat and water vapor. Effects of these triple factors are often interdependent, which emphasizes the need to investigate the combined effects.[180,181] Besides the requirement of a 40-year life under normal operating conditions, these cables must operate at the time of a loss-of-coolant accident (LOCA). A LOCA is assumed to take place once during the lifetime. By mid-1970, design acceptance in the U.S. called for qualification testing of preaged specimens by gamma radiation to 3.5 × 10^7 rad, followed by a simulated LOCA.[182] Further requirements concern flame-retardant properties in case of fire.

A. The Nuclear Radiation Environment

Our interest is in light water nuclear reactors which are of two types, boiling water reactors (BWR), and pressurized water reactors (PWR). There are three radiation environments to consider irrespective of the reactor type.[183] They are associated with the following areas:

1. The area near the reactor core within the primary shield. In this area, where, exposures up to 10^{12} rad/hr occur, only essentially inorganic insulation structures can function. Elastomer-based insulations and jackets are not suitable for use within the primary reactor shield because the covalent bonds of the organic materials are easily disrupted by the high gamma and neutron flux.
2. The area outside the primary shield but within the containment vessel of thermal reactor (drywell). In this area gamma dose rates ranging from 0.5 to 160 rad/hr, temperatures up to 70°C, and relative humidity as high as 90% are to be anticipated during operation. If abnormal bursts of energy should develop as a result of a nuclear or primary coolant incident, radiation levels may increase to 10^6 rad/hr, while the temperature in the area may rise rapidly to 150°C with steam pressure building to 50 psig (3.5 kg/cm²G, 441 KPa). Table 1.8.1. indicates the typical values of the dose rate corresponding to the zone specified for BWR reactors.
3. The area outside the drywell. This area suffers almost no radiation dose; temperature and humidity are normal.

B. LOCA Tests

The radiation environment of interest to us is in the second location, i.e., inside the drywell. Type tests for cables and their accessories intended for nuclear power stations should include all the prospective conditions that they will encounter during their installed life, in both normal and emergency conditions (referred to as design basis events, DBEs).

IEEE Std. 323-1974 and Std. 383-1974 specify the type tests for IE class cables and accessories.[184,185] The normal operating conditions considered indicate that the cable, as installed, should be suitable for operation at maximum ambient temperature, radiation, and atmospheric conditions and normal electrical and physical stresses for its installed life, as specified.

DBEs correspond to a LOCA (for cables in containment only), fire, and other severe hazards to cable operation. Throughout their normal lives the cables, field splices, and connections should be capable of operating through postulated environment conditions resulting from a LOCA. In addition, the cables, as installed, should be fire-retardant.

The DBE environment conditions to be simulated for a PWR or BWR, resulting from a postulated LOCA, generally consist of exposure to hot gases or vapors (e.g., steam), and a spray or jet of water, chemical solution, or other fluids. The test profile to simulate the environmental conditions anticipated within the primary containment in current PWR and BWR plants are presented in Table 1.8.2. and Table 1.8.3.[184] Figure 1.8.1. shows a representative test chamber profile for a combined PWR/BWR test.[184] If the actual conditions are different from these curves, the parameters may be adjusted accordingly. Simulation sequences suggested include:[184]

1. Exposing cable samples to simulated aging, radiation, and vibration
2. Conducting a stabilizing operation in normal environment to establish reference conditions
3. Injecting steam and chemical sprays at rates simulating service conditions, raising the temperature and pressure to test profile levels required
4. Maintaining these conditions for 3 hr. De-energizing for static readings including measurement of insulation resistance to ground. Re-energizing for one additional hour.
5. Reducing the environmental conditions to the normal operating conditions within 2 hr

Table 1.8.1.
RADIATION DOSE RATES IN PRIMARY CONTAINMENT ZONE

| Equipment or area | Radiation type | Operating dose rate | Design basis accident | | Integrated dose | |
			Type	Dose rate	Normal	Accident
Drywell,	γ	6.5×10^4	LOCA	1.3×10^6	2.3×10^{10}	2.6×10^7
no vessel shield	Neutron	6.3×10^7			7.9×19^{16}	
With vessel shield	γ	25.0	LOCA	1.3×10^6	8.8×10^6	2.6×10^7
Zone 1	Neutron	5×10^4			6.3×10^{13}	
With vessle shield	γ	50.0	LOCA	1.3×10^6	1.8×10^7	2.6×10^7
Zone 2	Neutron	1.4×10^5			1.8×10^{14}	
With vessel shield	γ	7.2	LOCA	1.3×10^6	2.5×10^6	2.6×10^7
Zone 3	Neutron	<1			$<1.3 \times 10^9$	
With vessel shield	γ	25.0	LOCA	1.3×10^6	8.8×10^6	2.6×10^7
Zone 4	Neutron	2×10^3			2.5×10^{12}	
With vessel shield	γ	4.0	LOCA	1.3×10^6	1.4×10^6	2.6×10^7
Zone 5	Neutron	2×10^3			2.5×10^{12}	
With vessel shield	γ	0.1	LOCA	1.3×10^6	3.5×10^4	2.6×10^7
Zone 6	Neutron	2×10^2				

Note: Dose rate γ-ray $=$ rad/ hr; neutron $=$ 1/cm^2·sec.

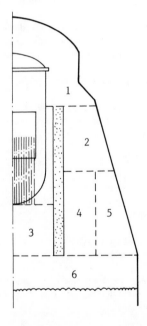

6.   Repeating the first cycle, but at the end of 3 hr, lowering pressure in steps to simulate postevent profiles for which the cables are to be qualified

B. Type Test Procedures

Type tests for control cables comprise quality acceptance tests, LOCA tests, and burning tests as shown in the flow chart of Figure 1.8.2.[187] The simulation of DBE conditions in qualification tests sometimes requires that a moisture-saturated environment be maintained at temperatures at which commercially available relative humidity and dewpoint sensors are capable of functioning.

Flame tests include the following:[186]

Table 1.8.2.
TEST CONDITIONS FOR PWRS

Typical In-Containment Design Basis Event Test Conditions

1. Exposure to nuclear radiation[a]
   4 Mrad after 1 hr
   20 Mrad after 12 hr
   24 Mrad after 1 day
   40 Mrad after 10 days
   55 Mrad after 1 month
   110 Mrad after 6 months
   150 Mrad after 1 year
2. Exposure to steam and chemicals
   a. Steam exposure

Time	Temperature		Pressure	
	°F	°C	(lbf/in.², gauge	kPa
0 to 10 sec	120 to 300	48.9 to 148.9	0 to 70	0 to 482.6
10 sec to 10 hr	300	148.9	70	482.6
10 hr to 4 days	210	98.9	40	275.8
4 days to 1 year	167	75.0	5	34.5

b. Spray exposure. Continuously spray vertically downward for first 24 hr with a solution of the following composition at a rate of 0.15 (gal/min)/ft² (6.1 (mℓ/ min)/ m²) of area of the test chamber projected on to a horizontal plane.

0.28 M H_3BO_3 (3000 parts per million boron)
0.064 M $Na_2S_2O_3$ (%)
NaOH to make a pH of 10.5 at 77°F (about 0.59%)
Dissolve chemicals on a 1-ℓ basis in the following order:
   1. 600 mℓ potable water
   2. H_3BO_3
   3. aOH
   4. $a_2S_2O_3$
   5. Add remainder of water to volume of 1 ℓ
   6. Add NaOH to make a pH of 10.5 at 77°F as required for the initial spray solution

Note: The values given in this table may vary from plant to plant and may or may not contain an adequate margin.

[a]   Conservative calculation of radiation dose to containment atmosphere resulting from beta and gamma radiation emitters released from the primary system and at a location within the primary containment.

1.   A vertical tray flame test on grouped cables. This is typically performed with a cable tray of the vertical ladder type which might be 6-in. (15-cm) wide, 3-in. (7.6-cm) deep, and 8-ft (2.44-m) high, for example. The flame source may be a gas burner with oil and burlap.
2.   A vertical flame test for individual wires. Test procedures in this case are described in ASTM D-2220-68, Section 5; IPCEA S-19-81, Section 6-19-6; and U/L 44, paragraph 242.

1.8.2. Designs and Performance

Two basic designs are shown in Figure 1.8.3. The first consists of a radiation-resistant,

Table 1.8.3.
TEST CONDITIONS FOR BWRS

Typical In-Containment Design Basis Event Test Conditions

1. Exposure to nuclear radiations[a]
   26 Mrad integrated over the accident
2. Exposure to steam and spray
   a. Steam exposure

Time	Temperature		Pressure	
	°F	°C	lb$f$/in^2, gauge	kPa
0 to 20 sec	135 to 280	57.2 to 137.8	0 to 62	0 to 427.5
20 sec to 5 min	280 to 340	137.8 to 171.1	62	427.5
5 min to 3 hr	340	171.1	40	275.8
3 hr to 6 hr	320	160.0	40	275.8
6 hr to 4 days	250	121.1	25	172.4
4 to 100 days	200	93.3	10	68.9

If it is not practical to reproduce the specified pressure and temperature profiles combined; it is acceptable during the first 4 days to follow the temperature profile and allow the pressure to conform to saturated conditions (100% relative humidity). This procedure is justified by the fact that temperature is the most important parameter and increasing the pressure (to maintain saturated conditions) will increase the severity of the test, if anything.

   b. Spray exposure. Continuously spray vertically downward with demineralized water at a rate of 0.15 (gal/min)/ft^2 (6.1 (mℓ/min)/m^2) of area of the test chamber projected onto a horizontal plane.

Note: The values in this table may vary from plant to plant and may or may not contain an adequate margin.

[a] See footnote to Table 1.8.2.

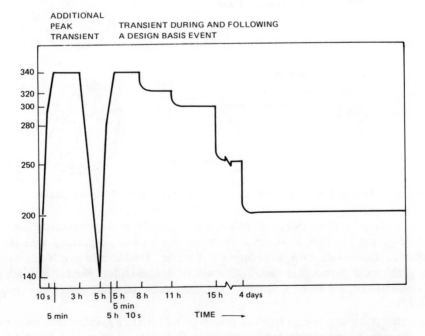

FIGURE 1.8.1.   Test chamber temperature profile for environment simulation (combined PWR/BWR). (From *IEEE Standards for Qualifying Class IE Equipment for Nuclear Power Generating Station*, IEEE Std. 323-1974, IEEE, Piscataway, N.J., 1974, 18. With permission.)

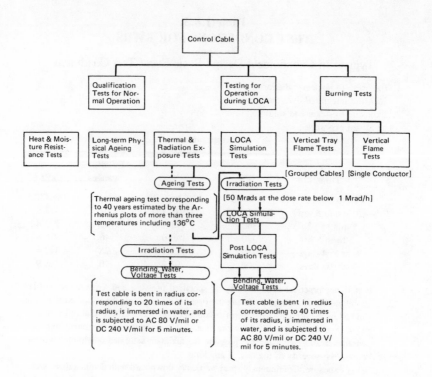

FIGURE 1.8.2. Type test flow based on IEEE Std. 323 and 383.

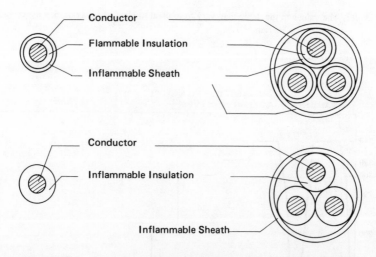

FIGURE 1.8.3. Insulation structures of cables in use for nuclear power stations.

plastic insulated conductor covered with a nonflammable or flame-retardant sheath. This design does not sacrifice its original insulation performance, especially with respect to radiation, but allows the cable to be larger in diameter. The increase in cable diameter may bring problems of permissible current. The second design utilizes insulation which is both radiation-resistant and flame-retardant to make the cable structure reasonably simple and slender.

Saturated polymers such as ethylene propylene rubber (EPR), chlorosulfonated polyethylene (Hypalon®) and chlorinated polyethylene (Chloroprene) seem best suited for antiradiation and flame-retardant purposes. EPR also has good insulating properties. There is

Table 1.8.4.
EXAMPLES OF NUCLEAR-USE CABLES FOR EXPERIMENTAL PURPOSE

Conductor	Insulation	Jacket or cover
Used for Nuclear Simulation Tests		
1/C 14 AWG coated copper	0.060 in. EPR base insulation	
1/C 14 AWG coated copper	0.030 in. Hypalon base insulation	
1/C 14 AWG coated copper	0.030 in. EPR base insultaion	0.015 in. Neoprene base cover
7/C 14 AWG coated copper	0.030 in. EPR base insulation cabled plus tape	0.060 in. Neoprene jacket
	Hand-wrapped EPR tape splice	
1/C 4/0 5 kV	EPR base + Hypalon with EPR/Neoprene splice	
Used for Burning Tests		
7/C 12 AWG Cu	0.030 in/0.015 EPR base plus Neoprene cabled	0.060 in. thermoset jacket
7/C 16 AWG Cu	0.030 in. silicone plus glass braid cabled plus tape	glass braid, OD = 0.45 in.
7/C 16 AWG Cu	0.030 in. flame resistant XLPE cabled plus tape	Neoprene jacket, OD = 0.50 in.
7/C 12 AWG Cu	0.020 in/0.010 in. PE plus PVC cabled plus tape	0.060 in. PVC jacket, OD = 0.58 in.
19/C 14 AWG Cu	0.030 in/0.015 in. EPR/Neoprene cabled plus tape	0.060 in. general purpose thermoset plus 0.020 in. steel interlock armor
19/C 14 AWG-Cu	0.030 in/0.015 in. EPR/Neoprene cabled plus tape	

belief in some quarters that the most attractive combination of cable materials for nuclear generating plants comprises chlorosulfonated PE- or chlorinated PE-jacketed insulations based on nonfilled XLPE, carbon black (or clayfilled XLPE), 90 °C oil-base, EPDM or EPM.[183] Adding antiflame agents and antiradiation agents to a specified insulation appears to be a standard practice. Sheath materials, even sulfochlorinated PE and chloroprene are sometimes endowed with the higher degree of fire resistance. Similar care is taken for fillings and tapes.

Design examples are as follows:[186]

1. Low voltage cable:
    a. No. 12 AWG, 7/W, tinned copper conductor, 30-mil (0.76 mm) ethylene propylene rubber insulation, with a 15-mil (0.38 mm) chorosulfonated polyethylene jacket
    b. No. 2 AWG, 7/W, tinned copper conductor, 45-mil (1.14 mm) ethylene propylene rubber insulation, with a 15-mil (0.38 mm) chorosulfonated polyethylene jacket
2. Medium voltage cable:
    a. No. 2 AWG, 7/W, copper conductor, extruded semiconducting strand shield, 90-mil (2.3 mm) ethylene propylene rubber insulation, with six corrugated copper wires embedded in a 75-mil (1.9 mm) semiconducting chlorinated polyethylene jacket. Other designs are shown in Table 1.8.4.[182]

Simulation test methods are still controversial because of complicated environments and their complicated combined effects on insulating materials. Consequently no definite design criteria have yet been fixed. Attention has been directed recently to the effect of radiation dose rate. When exposed to air during irradiation, it has been observed that insulation degradation depends on the dose rate as well as the total dose, as is described in Volume I, Section 2.4.6.

REFERENCES

1. **Bahder, G. and Eager, G. S.,** *Review of Present and Future Underground Transmission Systems,* APPA Engineering and Operations Workshop, Am. Publ. Power Assoc., Washington, D.C., 1975.
2. **Rosato, D. V.,** Electrical wire and cable coatings, *Wire Wire Prod.,* March, 1970, p. 49.
3. **Wanser, G. and Wiznerowicz, F.,** Cigre 1972 — Aktuelle Kabelfragen auf der Internationalen Hochspannungs Konferenz, *Elektrizitatswirtschaft,* 71(26), 771, 1972.
4. **Erb, J., Heinz, W., Hoffman, A., Köfler, H. J., Konarek, P., Maurer, W., Nahar, A., and Heller, I.,** Comparison of Advanced High Power Underground Cable Designs, Gesselschaft fur Kernforshung, M. B. H., Karlsruhe, W. Germany, Sept. 1975.
5. Arther D. Little, Inc., *Underground Power Transmission,* ERC Publ. No. 1-72, Electric Research Council, New York, 1971.
6. **Iizuka, K., Ed.,** *Handbook of Power Cable Technology,* (in Japanese), Denki-shoin, Tokyo, 1974.
7. **Ohtani, K.,** Cross-linked polyethylene and cross-linking processes, (in Japanese), *Jpn. Plast.,* 25(5), 31, 1974.
8. U.S. Patent, 3,054,142.
9. Japanese Patent, 48-13473, 48-14768.
10. Japanese Patent 48-8346.
11. **McKean, A. L., Oliver, F. S., and Trill, S. W.,** Cross-linked polyethylene for higher voltages, *IEEE Trans. Power Appar. Syst.,* 86(1), 1, 1967.
12. **Kreuger, F. H.,** *Endurance Tests with Polyethylene Insulated Cables,* CIGRE Paper 21-02, CIGRE, Paris, 1968.
13. **Jocteur, R. and Osty, M.,** *R & D in France in the Field of Extruded Polyethylene Insulated HV Cables,* CIGRE Paper 21-07, CIGRE, Paris, 1972.
14. **Bahder, G., Eager, G. S., Silver, D. A., and Lukae, R. G.,** Criteria for determining performance in service of XLPE insulated power cables, *IEEE Trans. Power Appar. Syst.,* 95(5), 1552, 1976.
15. **McKean, A. L.,** Breakdown mechanism studies in crosslinked polyethylene cables, *IEEE Trans. Power Appar. Syst.,* 95(1), 253, 1976.
16. **Patsch, R., Wagner, H., and Hermann, H.,** *Inhomogenities and Their Significance in Single-Layer Extruded Polyolefine Insulations for Cables,* CIGRE Paper 15-11, CIGRE, Paris, 1976.
17. **Doepkin, H. C. and Carroll, J. C.,** Recent Advances in Solid Dielectric Cables, personal communication, 1976.
18. **Yamauchi, H., Okada, M., and Fuwa, M.,** *Void Distribution in Crosslinked Polyethylene Insulated Cables Made by Conventional and New Processes,* IEEE/PES Meeting C72 504-9, IEEE, Piscataway, New Jersey, 1972, 1.
18a. **Yamauchi, H., Kaneko, R., Okada, M., Otsuji, M., Masui, M., and Sugiyama, K.,** New High Voltage XLPE Cable Curing Process Development and Electrical Characteristics, UT & D Conf. Rec., 1976, 337.
19. **Goosens, R. H.,** CIGRE Paper 209, CIGRE, Paris, 1966.
20. **Eager, G. S. and Bahder, G.,** Discharge detection in extruded polyethylene insulated power cables, *IEEE Trans. Power Appar. Syst.,* 86(1), 10, 1967.
21. IEEE Committee Report, Guide for calibration of test equipment for measurement of corona pulses, *IEEE Trans. Power Appar. Syst.,* 86(1), 1185, 1967.
22. **Blodgett, R. B. and Eigen, D.,** Cable corona signals — their origin & detection, *IEEE Trans. Power Appar. Syst.,* 87(6), 1492, 1968.
23. **Eager, G. S., Bahder, G., and Silver, D. A.,** Corona detection experience in commercial production of power cables with extruded insulation, *IEEE Trans. Power Appar. Syst.,* 88(4), 342, 1969.
24. **Eager, G. S., Bahder, G., Heinrich, O. X., and Suarez, R.,** Identification and control of electrical noise in routine-reel corona detenction of power cables, unspecified.
25. **Beneke, V.,** Messung von Teilentladungen in langen Kabeln und deren Bewertung, *Elektrizitatswirtshaft,* 72(18), 636, 1973.
26. **Lukaschewitsch, A. and Puff, E.,** Messung von Eilentlagungen an langen Kabeln, *Energie,* 2, 32, 1976.
27. **Kreuger, F. H.,** *Câble Isolé au Polyéthylène à Imprégnation Gazeus,* CIGRE Paper 21-02, CIGRE, Paris, 1970.
28. **Anon.,** HV Test WG for Rubber and Plastic Insulated Power Cables, *Electric Test Recommendation for Cross-linked Polyethylene Insulated Power Cables Rated 11,000 to 77,000 Volts,* Tech. Rep., (in Japanese), No. 112, Institute of Electrical Engineers of Japan, Tokyo, 1975, 1.
29. **Kojima, K., Hosokawa, K., Tsumoto, M., Yoda, B., Kanazawa, K., and Kaneko, R.,** Technical Progress of HV and EHV Cross-linked Polyethylene Insulated Cables in Japan, *CIGRE Paper G-21,* CIGRE, Paris, 1975.
30. **Matsuoka, S.,** Hypothesis of voids in semicrystalline polymers, *J. Appl. Phys.,* 32, 2334, 1961.

31. **Nagasaki, S., Miyauchi, H., Yoneyama, O., and Sanjo, K.,** *High Voltage Cross-Linked Polyethylene Power Cables and Their Accessories,* IEEE Underground T & D Conference Trans. Paper, IEEE, Piscataway, New Jersey, 1974, 482.

32. **Fujisawa, Y., Yasui, T., Kawasaki, Y., and Matsumura, H.,** Performance of 66-77 kV Cross-Linked Polyethylene Insulated Cable and New Developments, *IEEE Trans. Power Appar. Syst.,* 87(11), 1899, 1968.

33. **Itoh, Y., Hayashi, K., and Kawasaki, Y.,** *Effects of Additives and Contaminants on Electrical Properties of Polyethylene,* IEEJ-EIM Study Meeting, IM-74-43, (in Japanese), Institute of Electrical Engineers of Japan, Tokyo, 1974.

34. **Bahder, G., Katz, C., Lawson, J., and Vahlstrom, W.,** Electrical and electrochemical treeing effect in polyethylene and cross-linked polyethylene cables, *IEEE Trans. Power Appar. Syst.,* 93(3), 977, 1974.

35. **Vahlstrom, W.,** *Investigation of Treeing in 15 and 22 kV Polyethylene Cables Removed from Service,* Record of CEIDP Session III, National Academy of Sciences, Washington, D.C., 1972, 91.

36. **Vahlstrom, W., Jr.,** Investigation of insulation deterioration in 15 kV and 22 kV polyethylene cables removed from service, *IEEE Power Appar. Syst.,* 91(3), 1023, 1972.

37. **Lawson, J. H. and Vahlstrom, W., Jr.,** Investigation of insulation deterioration in 15 kV and 22 kV polyethylene cables removed from service. II, *IEEE Power Appar. Syst.,* 92(2), 824, 1973.

38. **Tanaka, T., Fukuda, T., and Suzuki, S.,** Water tree formation and lifetime estimation in 3.3 kV and 6.6 kV XLPE and PE power cables, *IEEE Trans. Power Appar. Syst.,* 95(6), 1892, 1976.

39. **Tanaka, T., Fukuda, T., Suzuki, S., Nitta, Y., Goto, H., and Kubota, K.,** Water tree in cross-linked polyethylene power cables, *IEEE Trans. Power Appar. Syst.,* 93(2), 693, 1974.

40. **Tabata, T., Nagai, H., Fukuda, T., and Iwata, Z.,** Sulfide attack and treeing of polyethylene insulated cables — cause and prevention, *IEEE Trans. Power Appar. Syst.,* 91(4), 1354, 1972.

41. **Fukuda, T., Hisatsune, T., Nagai, H., and Hasebe, M.,** Sulfide Attack to Polyethylene Insulated Control Cable and Development of Sulfide Capture Sheath, Proc. 21st Internal Wire Cable Symp., Atlantic City, N.J., December, 1972, 75.

42. U.S. Patent 3,841,810, Apparatus for Detection of Shield Imperfections in Electrical Conductors, Robinson, P. E. and Schmidt, R. A., General Cable Corp., October 15, 1974.

43. **Hayami, T.,** Development of liquid-filled type cross-linked polyethylene cable, *IEEE Trans. Power Appar. Syst.,* 88(6), 897, 1964.

44. **Fujiki, S., Furusawa, H., Kuhara, T., and Matsuba, H.,** *The Research in Discharge Suppression of High Voltage Crosslinked Polyethylene Insulated Power Cable,* IEEE PES Meeting 71 TP 195-PWR, IEEE, Piscataway, New Jersey, 1971.

45. **Kato, H., Maekawa, N., Inoue, T., and Fujita, H.,** *Favorable Effects of Additives on XLPE Cable Insulation,* IEEJ/EIM Study Meeting IM-74-45, in Japanese, Institute of Electrical Engineers of Japan, Tokyo, 1974.

46. **Lever, R. C., Mackenzie, B. T., and Singh, N.,** Influence of inorganic fillers on the voltage endurance of solid dielectric power cables, *IEEE Trans. Power Appar. Syst.,* 92(4), 1169, 1973.

47. **Charoy, M. A. and Jocteur, R. F.,** Very high tension cables with extruded polyethylene insulation, *IEEE Trans. Power Appar. Syst.,* 90(2), 777, 1971.

48. **Bahder, G., McKean, A. L., and Carrol, J. C.,** Development and installation of 138 kV cable system for tests at EEI Waltz Mill Station: cable 21, *IEEE Trans. Power Appar. Syst.,* 91(4), 1434, 1972.

49. **Endacott, J. D.,** EHV Underground Cable Technology — Present and Future, *Electron. & Power,* January 16, 1975, p. 20.

50. **Fujita, H., Ishitobi, M., and Itoh, H.,** *Insulation for Extra-High-Voltage Cable,* (in Japanese), Dai-nichi-Nippon Wire & Cable Review No. 47, Amagasaki-shi, 1971, 69.

51. **Miranda, F. J. and Gazzana-Priaroggia, P.,** Self-contained oil-filled cables — a review of progress, *Proc. IEE,* 123(3), 229, 1976.

52. **Fukagawa, H.,** *Research on Transmission Capacity Enlargement of EHV Cables,* (in Japanese), Review of CRIEPI No. 28, Central Research Inst. Electric Power Industry, Tokyo, 1975, 1.

53. **Arkell, C. A., Johnson, D. F., and Ray, J. J.,** 525-kV Self-contained oil-filled cable systems for Grand Coulee third power plant design proving tests, *IEEE Trans. Power Appar. Syst.,* 93(2), 468, 1974.

54. **Ray, J. J., Arkell, C. A., and Flack, H. W.,** 525-kV Self-contained oil-filled cable systems for Grand Coulee third power plant design and development, *IEEE Trans. Power Appar. Syst.,* 93(2), 630, 1974.

55. **Suetsugu, T., Ishizu, K., Oka, K., and Ishii, K.,** *Test Results of 500 kV of Cables for the Fukushima Nuclear Power Station,* Rec. National Convention of Institute of Electrical Engineers of Japan, No. 8-1108, 1445, Tokyo, 1975.

56. **Sato, N., Miyahara, A., Murakami, T., Shirado, R., Tsumoto, M., Yoshida, N., Murakami, M., Nakamoto, S., and Karasawa, S.,** *Installation of 500 kV OF Cables,* Fujikura Densen Review No. 55-12, Tokyo, 1976, 19.

57. **Izumi, Y., Hattori, M., Yokoyama, H., Gomi, T., and Nishino, I.,** *550 kV OF Cables for the Okutataragi Pumped Storage Power Station,* Rec. National Convention of Institute of Electrical Engineers of Japan, No. 7-886, Tokyo, 1976, 1193.

58. **Sato, N., Miyahara, A., Matsuo, J., and Sudo, K.,** *500 kV OF Cables for the Sodegaura Thermal Power Station,* Rec. National Convention of Institute of Electrical Engineers of Japan, No. 8-1106, Tokyo, 1975, 1443.

59. **Watanabe, T., Sato, N., Watanabe, Y., Miyahara, A., Kato, S., Hikino, K., and Urabe, Y.,** 500 kV Aluminum sheathed oil-filled cable for Sodegaura Power Station of Tokyo Electric Power Co., *Hitachi-Hyoron,* 56(10), 1013, 1974.

60. **Bahder, G., Corry, A. F., Blodgett, R. B., McIleen, E. E., and McKean, A. L.,** *550 kV High Pressure Oil-Filled Pipe Cable Development in the U.S.A.,* CIGRE 21-11, CIGRE, Paris, 1976, 1.

61. **Anon.,** Pipe type systems still dominate, *Electr. World,* October 1, 1976, p. 49.

62. **Anon.,** Field research on 345 kV underground cable system, *IEEE Trans. Power Appar. Syst.,* 85(4), 316, 1966.

63. **Burrel, R. W. and Young, F. S.,** EEI-manufacturers 550/550 kV cable research project Waltz Mill testing facility, *IEEE Trans. Power Appar. Syst.,* 90(1), 180, 1971.

64. **McAvoy, F. M. and Waldron, R. C.,** EEI-manufacturers 500/550 kV cable research project Waltz Mill testing facility — cable A, *IEEE Trans. Power Appar. Syst.,* 90(1), 191, 1971.

65. **Eich, E. D.,** EEI-manufacturers 500/550 kV cable research project Waltz Mill testing facility — cable B, *IEEE Trans. Power Appar. Syst.,* 90(1), 212, 1971.

66. **McKean, A. L., Merrell, E. J., and Moran, J. A., Jr.,** EEI-manufacturers 550/550 kV cable research project Waltz Mill testing facility — cable C, *IEEE Trans. Power Appar. Syst.,* 90(1), 224, 1971.

67. **Eager, G. S., Jr., Cortelyou, W. H., Bahder, G., and Turner, S. E.,** EEI-manufacturers 500/550 kV cable research project Waltz Mill testing facility — cable D, *IEEE Trans. Power Appar. Syst.,* 90(1), 240, 1971.

68. **McKean, A. L. and Garcia, F. G.,** EPRI-Manufacturers 500/550 kV cable research project- Cable C — summary of field test performance at Waltz Mill, *IEEE Trans. Power Appar. Syst.,* 95(1), 261, 1976.

69. **McIleen, E. E., Waldron, R. C., and Garrison, V. L.,** EPRI-Manufacturers 500/550 kV cable research project — Cable A — effect of increased conductor temperature, *IEEE Trans. Power Appar. Syst.,* 95(1), 279, 1976.

70. **Bahder, G., Eager, G. S., Jr., Silver, D. A., and Turner, S. E.,** 550 kV and 765 kV High pressure oil-filled pipe cable system, *IEEE Trans. Power Appar. Syst.,* 95(2), 478, 1976.

71. **Gubinsky, A. I., Khanukov, M. G., Kusnetsov, L. A., Masold, V. I., Peshkov, I. B., and Sorochkin, N. K.,** *500 kV Cable Liners for Hydro Power Stations,* CIGRE 21, CIGRE, Paris, 1976.

72. **Cavalli, M. and Mosca, W.,** A New Plant for Life-Tests on UHV Power Cables, Proc. Int. High Voltage Symp., Zurich, 1975, 674.

73. **Lawson, W. G., Simmons, M. A., and Gale, P. S.,** *Thermal Ageing of Cellulose Paper Insulation,* Conf. Rec. 1976 IEEE Int. Symp. EI, IEEE, Piscataway, New Jersey, 1976, 14.

74. **Arkell, C. A., Arnaud, U. C., and Skipper, D. J.,** *The Thermo-Mechanical Design of High Power Self-Contained Cable Systems,* CIGRE Paper 21-05, CIGRE, Paris, 1974, 1.

75. **Fujita, H., Itoh, H., Okamoto, T., and Nagatsuna, F.,** *Extra High Voltage Cable with Synthetic Polymer Paper Insulation,* (in Japanese), Dai-nichi-Nippon Cable Review No. 53, Amagasaki-shi, 1973 49-68.

76. **Fujita, H.,** Cable insulation and oils in the next generation, (in Japanese), *J. Jpn. Pet. Inst.,* 17, 588, 1974.

77. **Iwata, Z., Shii, H., Kinoshita, M., Hirukawa, H., and Kawai, E.,** *New Modified Polyethylene Paper Proposed for UHV Cable Insulation,* IEEE PES Winter Meeting F77 182-9, IEEE, Piscataway, New Jersey, 1977, 1.

78. **Reed, C. W.,** The Application of Polymer Films and Synthetic Papers for Use as Electrical Insulation, Proc. 7th Symp. Elec. Insulating Mater., Tokyo, 1974, 15.

79. **Edwards, D. R. and Reynolds, E. H.,** Paper/Plastic Laminate — A New Possibility for EHV Cable Insulation, unpublished, Central Research & Eng. Div., BICC Ltd., United Kingdom Atomic Energy Assoc., Harwell, Buckinghamshire, England.

80. **Yamamoto, T., Isshiki, S., and Nakayama, S.,** Synthetic paper for extra high voltage cable, *IEEE Trans. Power Appar. Syst.,* 91(6), 2415, 1972.

81. **Matsuura, K., Kubo, H., and Yamazaki, T.,** Development of Polypropylene Laminated Paper Insulated EHV Power Cables, Rec. 1976 Underground T & D Conf., Atlantic City, N.J., 1976, 322.

82. **Fujita, H. and Itoh, H.,** Synthetic polymer papers suitable for use in EHV underground cable insulation, *IEEE Trans. Power Appar. Syst.,* 95(1), 130, 1976.

83. **Numajiri, F., Oka, K., and Hanano, Y.,** FEP/C laminated paper of oil-filled cable for 1,000 kV and above, *Hitachi Rev.,* 25(2), 91, 1976.

84. **Edwards, E. R., Counsell, J. A. H., Gibbons, J. A. M., and Scarisbrick, R. M.,** *Polymer and Polymer/Paper Laminated Tapes for EHV Oil-Filled Cables,* CIGRE Paper 15-05, CIGRE, Paris, 1972, 1.

85. **Fujita, H., Itoh, H., and Matsuda, S.,** *A Novel Type of Synthetic Paper for Use in EHV Underground Cable Insulation,* paper to CEIDP C-11, National Academy of Sciences, Washington, D.C., 1976, 1.

86. **Itoh, H. and Fujita, H.,** Effect of impurities on dissipation factor of oil-impregnated synthetic paper, *IEEE Trans. Electr. Insul.,* 12(2), 125, 1977.

87. **Takahashi, S.,** *Recent Technical Progress of High Voltage Power Cables in Japan,* CIGRE Paper 21-04, CIGRE, Paris, 1970, 1.

88. **Roughley, T. H., Corbett, J. T., Winkler, G. L., Eager, G. S., Jr., and Turner, S. E.,** *Design and Installation of a 138 kV High-Pressure, Gas-Filled Pipe Cable Utilizing Segmental Aluminum Conductors,* IEEE PES Summer Meeting T73 491-8, IEEE, Piscataway, New Jersey, 1973, 1.

89. **Bahder, G., Walker, J. J., Eager, G. S., Lawson, J. H., and Wheton, L. B.,** Procedure to Increase Voltage Rating of Field Installed 15 kV Cable by Replacement of Nitrogen with Sulfur Hexafluoride, *IEEE PES Summer Meeting F76 357-4, IEEE,* Piscataway, New Jersey, 1976, 1.

90. **Gibbons, J. A. M., Howard, P. R., Skipper, D. J., Body, R. S., and Hawley, W. G.,** Gas-Pressurized Polythene Tape Dielectric for Extra-High Voltage Cables, *CIGRE Paper 201,* CIGRE, Paris, 1964, 1.

91. **Gibbons, J. A. M., Howard, P. R., and Skipper, P. J.,** Gas-pressurized lapped-polythene dielectric for extra-high-voltage power-cable systems, *Proc. Inst. Elec. Eng.,* 112(1), 89, 1965.

92. **Rhodes, R. G., Wootton, R. E., and Nugent, H.,** Assessment of the Possible Use of Polythene/Gas Dielectrics in H. V. Cables, *Proc. Inst. Electr. Eng.,* 112(8), 1617, 1965.

93. **Nakagawa, M., Saito, M., and Ohtaka, H.,** *Some Basic Characteristics of Gas-Pressurized Lapped Polyethylene Tape Dielectric for Extra-High Voltage Cables,* (in Japanese), Dai-nichi Nippon Cable Review No. 34, Amagasaki-shi, 1966, 29.

94. **Nakagawa, M. and Ohtaka, H.,** Some Basic Characteristics of Gas-Pressurized Lapped Polyethylene Tape Dielectric for Extra-High Voltage Cables, (in Japanese), Part 2, Dai-nichi Nippon Cable Review No. 35, Amagasaki-shi, 1967, 37.

95. **Nakagawa, M. and Ohtaka, H.,** *Some Basic Characteristics of Gas-Pressurized Lapped Polyethylene Tape Dielectric for Extra-High Voltage Cables,* (in Japanese), Part 3, Dai-nichi Nippon Cable Review No. 36, Amagasaki-shi, 1967, 21.

96. **Short, H. D.,** A theoretical and practical approach to the design of high-voltage cable joints, *Trans. Am. Inst. Electr. Eng.,* 68, 1275, 1949.

97. **Fisher, R. G.,** *138 kV Splice for Extruded Dielectric Cables,* EPRI EL-354, Electrical Power Res. Inst., Palo Alto, Calif., 1977, 1.

98. **Rittman, G. W. and Heyer, S. V.,** Water contamination in a cross-linked polyethylene cable joint, *IEEE Trans. Power Appar. Syst.,* 95(1), 302, 1976.

99. **Eager, G. S., Jr. and Silver, D. A.,** Development and installation of 138 kV cable for tests at EEI Waltz Mill Station: cable 22, *IEEE Trans. Power Appar. Syst.,* 91(4), 1434, 1972.

100. **Anderson, H. C., Rutherford, M. A., and Cox, E. B.,** Development and installation of 138 kV cable systems for tests at EEI Waltz Mill Station: cable 23, *IEEE Trans. Power Appar. Syst.,* 92(4), 1443, 1972.

101. **Balaska, T. A., Blais, L. D., and McBeath, R. H.,** 69 kV Underground neutral tee joints for solid dielectric cable system, *IEEE Trans. Power Appar. Syst.,* 93(3), 950, 1974.

102. **Chroy, M. A. and Jocteur, R. F.,** Very high tension cables with extruded polyethylene insulation, *IEEE Trans. Power Appar. Syst.,* 90(2), 777, 1971.

103. **Matsumura, H.,** *Practical Design of Prefabricated Cable Accessories for Cross-Linked Polyethylene Insulated Power Cable,* (in Japanese), Sumitomo Elec. Tech. Rev. No. 110, Samitomo Electric Ind., Tokyo, 1975, 109.

104. **Nagasaki, S., Matsumura, H., Sanjo, K., Hayami, H., Yoneyama, O., and Ogura, I.,** *Development of 154 kV Cross-linked Polyethylene Insulated Power Cable,* (in Japanese), Sumitomo Elec. Tech. Rev. No. 110, Sumitomo Electric Ind., Tokyo, 1975, 45.

105. **Sanjo, K. and Shiraoka, K.,** *General Report about 66—77 kV Cable,* (in Japanese), Sumitomo Elec. Tech. Rev. No. 110, Sumitomo Electric Ind., Tokyo, 1975, 31.

106. **Brealey, R. H., Arone, N. F., Nakata, R., and Fischer, F. E.,** *Injection Molded Cable Splice for Connecting Oil-Paper Cable to Polyethylene Cable,* IEEE PES Winter Meeting Rec. C75 205-0, IEEE, Piscataway, New Jersey, 1975, 1.

107. **Walker, J. J. and Juhlin, E. O.,** Laboratory evaluation of 345 kV cable 4 low-pressure oil-filled self-contained type after field testing at Cornell University, *IEEE Trans. Power Appar. Syst.,* 85(4), 337, 1966.

108. **Watanabe, T., Kikuchi, K., Matsuura, K., Yoshida, N., Haga, K., and Numajiri,** *Higashi-Tokyo EHV Cable Test Project in Japan,* IEEE PES Winter Meeting Rec. C75 208-4, IEEE, Piscataway, New Jersey, 1975, 1.

109. **Allam, E. M. and McKean, A. L.,** 138 kV Prefab capacitive graded joint for oil-filled cable systems, *IEEE Trans. Power Appar. Syst.,* 96(1), 20, 1977.

110. **Bianchi, G., Durse, M. A., and Occhini, E.,** *Design and Test of A Flexible EHV Cable Connection for A Floating Plant,* IEEE PES Summer Meeting Rec. F76 313-7, IEEE, Piscataway, New Jersey, 1976, 1.

111. **Gear, R. B., Heppner, D. R., Lusk, G. E., and Nicholas, J. H.,** EEI-manufacturer 500/550 kV cable research project. Pothead No. 2 — high pressure oil paper pipe type, *IEEE Trans. Power Appar. Syst.,* 90(1), 199, 1971.

112. **Lanfranconi G. M., Maschio, G., and Occhini, E.,** Self-contained oil-filled cables for high power transmission, *IEEE Trans. Power Appar. Syst.,* 93(5), 1535, 1974.

113. **Maeno, T., Gomi, Y., Ando, N., and Ide, S.,** Development of new type normal joint for 275 kV of cables, (in Japanese), *Hitachi Hyoron,* 57(8), 77, 1975.

114. **Iizuka, K., Uchiyama, T., Kojima, K., Hayashi, T., and Mitsui, T.,** *Transmissible Current Capacity of High Pressure Oil Pipe-Type Cable by Forced Cooling,* (in Japanese), Sumitomo Review No. 99, Sumitomo Electric Ind., Tokyo, 1969, 22.

115. **Muto, K. and Tsumoto, M.,** *Installations of Forced-Air-Cooling-Type Station Cables,* (in Japanese), Fujikura Cable Review No. 17, Fujikura Cable Works, Ltd., Tokyo, 1959, 8.

116. **Okada, N., Hirose, K., Nagahama, I., Memida, N., Shiroya, M., Takada, M., Hayashi, T., and Kimura, Y.,** *Artificial Cooling Method of Underground Power Cables,* (in Japanese), Sumitomo Review No. 79, Sumitomo Electric Ind., Tokyo, 1962, 23.

117. **Numajiri, F. and Amino, H.,** Forced water cooling of the duct installed power cables, (in Japanese), *Hitachi Hyoron,* 45(11), 110, 1963.

118. **Numajiri, F. and Kato, Y.,** Forced cooling water of the duct installed power cable, (in Japanese), *Hitachi Hyoron,* 46(11), 80, 1964.

119. **Yamamoto, T. and Yokose, K.,** Direct cooling of power cables and its verification, (in Japanese), *Denki Hyoron,* No. 11, 1266, 1971.

120. **Iizuka, K. and Kagaya, S.,** *Field Test on Forced Cooling of High Pressure Oil Filled Pipe Cable,* (in Japanese), Fujikura Cable Review No. 34, Fujikura Cable Works, Ltd., Tokyo, 1967, 9.

121. **Takada, M.,** *On Cooled Cable of A System Utilizing Conductor Oil Passage,* (in Japanese), Sumitomo Review No. 60, Sumitomo Electric Ind., Tokyo, 1956, 22.

122. **Numajiri, F.,** Forced water cooling of conductor for power cables, (in Japanese), *Hitachi Hyoron,* 47(6), 1113, 1965.

123. **Cox, H. N.,** Assisted cooling of transmission cables, *Electr. Rev.,* December 1967, p. 2.

124. **Lanfranconi, G. M., Maschio, G., and Occhini, E.,** Self-contained oil-filled cables for high power transmission in the 750—1,200 kV range, *IEEE Trans. Power Appar. Syst.,* 93(5), 1535, 1974.

125. **Watson, E. P. C., Brooks, E. J., and Gosling, C. H.,** Canals as cable routes, *Proc. Inst. Electr. Eng.,* 114(4), 510, 1967.

126. **Sakamoto, I.,** *Development of 500 kV Bulk Capacity Underground Power Cable in Japan,* CIGRE Paper 21-01, CIGRE, Paris, 1974, 1.

127. **Watanabe, T., Ninomiya, K., Arai, T., Hiyama, S., Sugiyama, N., and Kitani, K.,** *Higashi-Tokyo EHV Cable Test Project in Japan. II. Forced Cooling Tests on EHV Underground Transmission Line,* IEEE PES Winter Meeting Paper C75 209-2, IEEE, Piscataway, New Jersey, 1974, 1.

128. **Müller, U., Peschke, E. F., and Hahn, W.,** *The First 380 kV Cable Bulk Power Transmission in Germany,* CIGRE Paper 21-08, CIGRE, Paris, 1976, 1.

129. **Beale, H. K., Hughes, K. E. L., Endacott, J. D., Miranda, F. J., Flack, H. W., and Nicholson, J. A.,** *The Application of Intensive Cooling Techniques to Oil-Filled Cables and Accessories for Heavy Duty Transmission Circuits,* CIGRE Paper 21-09, CIGRE, Paris, 1968, 1.

130. **Birnbreier, H., Fischer, W., Rasquin, W., Grosse-Plankermann, G., and Schuppe, W. D.,** *High Power Cable with Internal and External Water Cooling,* CIGRE Paper 21-09, CIGRE, Paris, 1974, 1.

131. **Hayashi, H., Kamiyama, T., Torigoe, Y., Ichino, T., and Tanaka, N.,** *22 kV Internally Water-Cooled 15 kV Capacity XLPE Cable,* IEEE PES Winter Meeting Paper A76 167-7, IEEE, Piscataway, New Jersey, 1976, 1.

132. **Falke, H. and Schlang, P.,** Ausbaufähige 110 kV-Kunststoff-Kabelanlagen, *Elektrotech. Z. Ausg. B,* 25(20), 554, 1973.

133. **Lacoste, A., Royere, A., Lepers, J., and Benard, P.,** *Experimental Construction Prospects for the Use of 225 kV—600 MVA Links Using Polyethylene Insulated Cables with Forced External Water Cooling,* CIGRE Paper 21-12, CIGRE, Paris, 1974, 1.

134. **Williams, J. L., Shimshock, J. F., Bankoske, J. W., and Purnhagen, D. W.,** *Comprehensive Forced-Cooled Tests on Pipe-Type Cables at Waltz Mill,* CIGRE Paper 21-07, CIGRE, Paris, 1976, 1.

135. **Buckweitz, M. D. and Pennell, D. B.,** Forced cooling of UG lines, *Transmission Distribution,* April 1976, p. 51.

136. **Burrell, R. W.,** Application of oil-cooling in high-pressure oil-filled pipe-cable circuits, *IEEE Trans. Power Appar. Syst.,* 84(9), 795, 1965.

137. **Anon.,** Oil-flow and pressure calculations for pipe type systems, AIEE Committee Report, *Trans. Am. Inst. Electr. Eng.,* Part III, 74, 251, 1955.

138. **Williams, J. A., Eich, E. D., and Aabo, T.,** *Forced Cooling Tests on 230 kV and 345 kV HPOF Cable Systems,* IEEE PES Winter Meeting Paper A76 201-4, IEEE, Piscataway, New Jersey, 1976, 1.

139. **Abdulhadi, R. S. and Chato, J. C.,** *Combined Natural and Forced Convective Cooling of Underground Electric Cables,* IEEE PES Summer Meeting Paper F76 312-9, IEEE, Piscataway, New Jersey, 1976, 1.
140. **Nicholas, J. M. and Selsing, J.,** *High Ampacity Potheads,* NTIS PB-248 342, Natl. Tech. Inf. Syst., Springfield, Va., 1975, 1.
141. **Trump, J. G. and Cooke, C. M.,** *Final Report on ERC Research Project RP78-2 (Phase I) at the MIT on Gas Dielectric Insulation,* Massachusetts Institute of Technology, Cambridge, 1972, 1.
142. **Doepkin, H. C., Jr.,** Compressed-gas insulation in large coaxial systems, *IEEE Trans. Power Appar. Syst.,* 88(4), 364, 1969.
143. **Diessner, A. and Trump, J. G.,** Free conducting particles in a coaxial compressed-gas-insulated system, *IEEE Trans. Power Appar. Syst.,* 89(8), p. 1970, 1970.
144. **Bortnik, I. M. and Cooke, C. M.,** Electrical breakdown and the similarity law in SF_6 at extra-high voltages, *IEEE Trans. Power Appar. Syst.,* 91(5), 2196, 1972.
145. **Cooke, C. M. and Trump, J. G.,** Post-type support spacers for compressed gas-insulated cables, *IEEE Trans. Power Appar. Syst.,* 92(5), 1441, 1973.
146. **Cooke, C. M.,** Ionization, electrode surfaces and discharges in SF_6 at extra-high-voltages, *IEEE Trans. Power Appar. Syst.,* 94(5), 1518, 1975.
147. **Cooke, C. M., Wotton, R. E., and Cookson, A. H.,** *Influence of Particles on AC and DC electrical Performance of Gas Insulated Systems at Extra-High-Voltage,* IEEE PES Summer Meeting Paper F76 323-6, IEEE, Piscataway, New Jersey, 1976, 1.
148. **Cronin, J. C. and Dethlefsen, R.,** *Guide to the Use of Gas Cable Systems,* NTIS PB-247 898, Natl. Tech. Int. Serv., Springfield, Va., 1975, 1.
149. **Graybill, H. W., Cronin, J. C., and Field, E. J.,** Testing of gas insulated substations and transmission systems, *IEEE Trans. Power Appar. Syst.,* 93, 404, 1974.
150. **Cookson, A. H., Bolin, P. C., Doepkin, H. C., Jr., Wotton, R. E., Cooke, C. M., and Trump, J. G.,** *Recent Research in the United States on the Effect of Particle Contamination Reducing the Breakdown Voltage in Compressed Gas-Insulated Systems,* CIGRE Paper 15-09, CIGRE, Paris, 1976, 1.
151. **Johnson, B. L., Doepkin, H. C., Jr., and Trump, J. G.,** Operating parameters of compressed-gas-insulated transmission lines, *IEEE Trans. Power Appar. Syst.,* 88(4), 369, 1969.
152. **Pederson, B. O., Doepkin, H. C., Jr., and Bolin, P. C.,** Development of a compressed-gas-insulated transmission line, *IEEE Trans. Power Appar. Syst.,* 90(6), 2631, 1971.
153. **Cookson, A. H., Farish, O., and Sommerman, G. M. L.,** Effect of conducting particles on AC corona and breakdown in compressed SF_6, *IEEE Power Appar. Syst.,* 91(4), 1329, 1972.
154. **Cookson, A. H. and Wotton, R. E.,** Movement of Filamentary Conducting Particles under AC Voltages in High Pressure Gases, Int. High Voltage Symp., Zurich, 1975, 416.
155. **Emanuel, A. E., Doepkin, H. C., Jr., and Bolin, P. C.,** Design and test of a sliding plug-in conductor connector for compressed gas-insulated cables, *IEEE Trans. Power Appar. Syst.,* 95(2), 570, 1976.
156. **Westinghouse Corp.,** Descriptive Bulletin 33-650 D WE A, *Type CGIR Compressed Gas Insulated Bus,* Pittsburgh, 1.
157. **Nakata, R.,** *Controlled Particle Scavenging Technique For Use in HVDC SF_6 Gas Bus,* IEEE PES Summer Meeting Paper A76 410-1, IEEE, Piscataway, New Jersey, 1976, 1.
158. **Ferici, N. J.,** Forces et Charged de Petits Objets en Contact avec Une Électrode Affectée dún Chap Électrique, *Rev. Gen. Electr.,* 75(10), p. 1145, 1966.
159. **Takuma, T., Kita, K., and Watanabe, T.,** *Calculation of the Electric Field in Three-Phase Gas Insulated Cables,* IEEE PES Summer Meeting Paper C72 500-7, IEEE, Piscataway, New Jersey, 1972, 1.
160. **Mashikian, M. S., Whitney, B. F., and Freeman, J. J.,** *Optimal Design and Laboratory Testing of Post-Type Spacers for Three-Phase SF_6 Insulated Cables. Part I,* IEEE PES Summer Meeting Paper A76 396-2, IEEE, Piscataway, New Jersey, 1976, 1.
161. **Whitney, B. F. and Mashikian, M. S.,** *Optimal Design and Laboratory Testing of Post-Type Spacers for Three-Phase, SF_6 Insulated Cables — Part II,* IEEE PES Summer Meeting Paper A76 397-0, IEEE, Piscataway, New Jersey, 1976, 1.
162. **DeMaris, S.,** 500 kV CGI System is installed in the cascades, *Transmission Distribution,* April 1975, p. 4.
163. **Takagi, T., Hayashi, H., Higashino, T., Nishihara, S., and Itaka, K.,** Dielectric Strength of SF_6 Gas and 3-Core Type CGI Cables under Inter-Phase Switching Impulse Tests, *IEEE Trans. Power Appar. Syst.,* 93, p. 354, 1974.
164. **Furukawa Electr. Ind.,** *Outline of CGITLs,* Ichihara-shi, Japan, 1975.
165. **Eidinger, A. and Dobsa, J.,** *Efficiency of Direct and Indirect Cooling of Underground Transmission Systems Having Solid or Gaseous Insulation,* CIGRE Paper 21-04, CIGRE, Paris, 1976, 1.
166. **Cookson, A. H., Lapen, R. J., and Kothmann, R. E.,** *Analysis of Forced Cooling of Compressed Gas Insulated Transmission Lines,* EPRI EL-228, Project 7840-1 Final Report, Electric Power Research Inst., Palo Alto, Calif., Dec. 1976, 1.

167. **EPRI TD-2 Progress Report,** *Underground Transmission,* Electric Power Research Inst., Palo Alto, Calif., March 1975, p. 21.
168. **Anon.,** *Design & Test of Three-Conductor Gas-Insulated Cable,* NTIS PB-244 392, Natl. Tech. Inf. Serv., Springfield, Va., 1975, 1.
169. **Oudin, J. M.,** D.C. for submarine cables, *Energy Int.,* September 1967, p. 22.
170. **Maschio, G. and Occhini, E.,** *High Voltage Direct Current Cables: The State of the Art,* CIGRE Paper 21-10, CIGRE, Paris, 1974, 1.
171. **Yoshida, K., Sakamoto, Y., Tabata, T., Numajiri, F., Tsumoto, M., and Kojima, K.,** Research and Development of HVDC Cables in Japan, *CIGRE Paper 21-03,* CIGRE, Paris, 1974, 1.
172. **Anon.,** Submarine Power Cables, *Cables de Lyon Ad Release 7111-1051,* Lyon, (year unspecified), p. 1.
173. **Gazzana-Priaroggia, P. and Palandri, G. L.,** 200 kV D.C. Submarine Cable Interconnection between Sardinia and Corsica and between Corsica and Italy, *CIGRE Paper 21-05,* CIGRE, Paris, 1968, 1.
174. **Eyraud, I., Horne, L. R., and Oudin, J. M.,** *The 300 kV direct Current Submarine Cables Transmission between British Colombia Mainland and Vancouver Island,* CIGRE Paper 21-07, CIGRE, Paris, 1970, 1.
175. **Gazzana-Priaroggia, P. and Maschio, G.,** *Continuous Long Length AC and DC Submarine HV Power Cables — The Present State of the Art,* IEEE PES Winter Meeting Paper T73 127-8, IEEE, Piscataway, New Jersey, 1972, 1.
176. **Last, F. H., Gazzana-Priaroggia, P., and Miranda, F. J.,** The underground HV DC link for the transmission of bulk power from the thames estuary to the centre of London, *IEEE Trans. Power Appar. Syst.,* 90(4), 1893, 1971.
177. **Oudin, J. M., Eyraud, I., and Constantin, L.,** *Some Mechanical Problems of Submarine Cables,* CIGRE Paper 21-08, CIGRE, Paris, 1972, 1.
178. **Nelson, R. A.,** *High Voltage DC Field Testing Solid Dielectric Cables,* IEEE Annual Pulp. Pap. Ind. Tech. Conf. Rec., IEEE, Piscataway, N.J., 79, 1976.
179. **Ando, N., Nanano, Y., Abe, H., Hayashi, K., and Numajiri, F.,** Development of 500 kV DC oil-filled cable, *Hitachi Rev.,* 23(12), 469, 1974.
180. **Cambell, F. J.,** Combined environments versus consecutive exposures for insulation life studies, *IEEE Trans. Nucl. Sci.,* 11, 123, 1964.
181. **Cambell, F. J.,** Radiation Effect Standards for Electrical Insulation, Proc. 9th Elec. Insul. Conf., Boston, September 1969, 68.
182. **McIleen, E. E., Garrison, V. L., and Dobrowolski, G. T.,** Class IE cables for nuclear power generating stations, *IEEE Trans. Power Appar. Syst.,* 93, 1121, 1974.
183. **Blodgett, R. B.,** Insulations and jackets for control and power cables in thermal reactor nuclear generating stations, *IEEE Trans. Power Appar. Syst.,* 88(5), 529, 1969.
184. **Anon.,** *IEEE Standards for Qualifying Class IE Equipment for Nuclear Power Generating Station,* IEEE, Piscataway, New Jersey, Std. 323-1974.
185. **Anon.,** *IEEE Standards for Type Test of Class IE Electric Cables, Field Splices, and Connections for Nuclear Power Generating Stations,* IEEE, Piscataway, New Jersey, Std. 383-1974.
186. **Ling, T. H. and Morrison, W. F.,** *Qualification of Power and Control Cable for Class IE Applications,* IEEE PES Winter Meeting Paper C74 045-1, IEEE, Piscataway, New Jersey, 1974, 1.
187. **Kuriyama, I., Hasegawa, T., Ogura, J., Onishi, T., and Kimura, H.,** Development of cables for nuclear power generating stations, *Hitachi Hyoron,* 58(3), 247, 1976.

Chapter 2

NEW TYPES OF CABLES

2.1. CRYOGENIC RESISTIVE CABLES

2.1.1. Introduction[1-9]

Recognizing the future need for underground power transmission in single-circuit capacities that exceed extrapolation of established cable technology[1] as explained in connection with forced cooling of oil-filled (OF) type cables and compressed gas insulated cables in the previous chapter, the electric utility industry is concentrating considerable research and development effort on bulk underground transmission. Cryogenic cable systems, capable of transmitting single-circuit capacities of 2000 MVA and above have evolved as the most promising candidates.

With respect to the conductor through which the current will flow, cryogenic cables are of two types: resistive and superconducting. Cryogenic resistive cables utilize the reduced electrical resistivity of their conductors at cryogenic temperatures; typically at liquid nitrogen temperatures (equal to and lower than 77 K). Superconducting cables make use of "zero electrical resistivity" of some special metals which occurs below a certain low temperature (usually near liquid helium temperature, 4.2 K). This will be described in the next section.

A cryogenic resistive cable (CR cable) is an insulated conductor which has been cooled to cryogenic temperatures to reduce the electrical resistivity of the conductor and thereby its ohmic loss. The result is a dramatic increase in current capacity. By nature it is a forced cooled cable because the refrigerant extracts the heat produced in the conductor. The electrical resistivity of two electrical conductor materials, copper and aluminum, is shown in Figure 2.1.1. At liquid nitrogen temperatures, the factor-of-ten increase in the electrical conductivity of electrical wire grade aluminum or copper permits operation of the cable conductors at high current density with small associated electrical losses. At liquid hydrogen temperatures, a conductivity of approximately 500 times the room temperature value is achieved with materials which are about 99.99% pure.[2]

There are two types of CR cables with respect to electrical insulation. One is a liquid nitrogen-impregnated tape-insulated cable: the concept is similar to the oil-filled type cable in which liquid nitrogen replaces oil and is expected to act as both coolant and part of the electrical insulation. The other is a vacuum-insulated liquid nitrogen-cooled cable: this concept is similar to CGI cables. It uses vacuum in place of SF_6 gas, and is internally cooled by passing liquid nitrogen through the center of the conductor. In this case the liquid nitrogen acts as the coolant only.

Unlike conventional cables, cryogenic resistive cables naturally require thermal insulation, because the cables, or at least their conductors, are cooled down to liquid nitrogen temperature. Vacuum is used in part or for the entire thermal insulation in the second cable system. Foams, powders, fibers, and multilayer insulations are all candidates for the thermal insulation of CR cables. Powders and fibers can be evacuated to increase thermal resistivity. Multilayer insulation is evacuated, in which condition it is called superinsulation.

Since CR cables are forced-cooled, they need a refrigeration system to cool and transfer liquid nitrogen under pressurized conditions. The flow rate depends on the allowable temperature rise in the fluid, which is bounded on the lower end by the freezing point of the coolant and on the upper end by a practical working pressure to suppress boiling, or just reach the boiling point, depending on the design. Boiling is undesirable; it affects electrical insulation, as described in Volume I, Section 2.3.7. and creates flow instabilities.

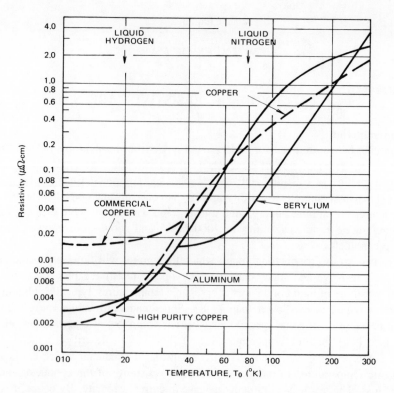

FIGURE 2.1.1. Electrical resistivity of Cu, Al, and Be at cryogenic temperatures. (From Arthur D. Little Inc., *Underground Power Transmission*, Electric Research Council, New York, 1971, p. 10.4. With permission.)

2.1.2. Electrical and Thermal Insulation

A. Electrical Insulation

Conceivable electrical insulation systems for cryogenic resistive cables are categorized in Volume I, Section 2.2.2. and repeated below for convenience:

1. Vacuum and spacers
2. Refrigerant (liquid or gas) and spacers
3. Refrigerant-impregnated paper (cellulose or synthetic)
4. Refrigerant-impregnated plastic tape
5. Extruded plastic cooled by refrigerant

Of these, the first and third are considered promising. The second appears to have insufficient withstand-voltage capability for cable insulation. With the third type of insulation, it is extremely difficult to impregnate any refrigerant other than helium through taping interstices. As regards the fifth configuration, it may be possible to immerse low-voltage XLPE cables (3 kV or 6 kV class), in liquid nitrogen without any trouble, because the insulation is thin, but it does not seem likely that the same procedure can be applied to HV and EHV cables with heavy insulation. The insulation is susceptible to cracking during cool down and when subjected to a small mechanical force or shock, because of its glassy state at liquid nitrogen temperature.

B. Vacuum Insulation[7-25]

The dielectric strength of vacuum is greater than that of SF_6 gas for practical gap lengths. The main reason for this is that breakdown in vacuum is initiated by field emission of

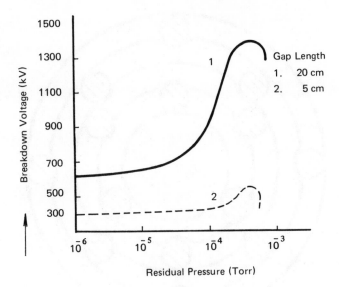

FIGURE 2.1.2. Effects of residual pressure on breakdown strength of vacuum.

electrons from the metal of the electrode. One might expect an extremely high value for the dielectric strength, perhaps as high as 1000 kV for a 1-cm gap. However, it is found in practice that vacuum exhibits much smaller values and obeys a power law $V_B = Ad^n$, where $n = 0.5$ and $A \simeq 300kV/(cm)^{0.5}$, as compared with $n \simeq 0.8$ for impulse breakdown of SF_6 gas at a pressure of 10 kg/cm²G.[8]

The above experimental results established conclusively that "pure vacuum" can perform well enough for HV or EHV cable insulation. In point of fact, vacuum has some inherent unstable characteristics. In the first place, it is liable to degrade to low vacuum as a consequence of real or virtual leaks. These latter include the liberation of occluded gas in cable materials such as conductors and dielectric spacers, and the vaporization of cable materials when subjected to electrical failure. Figure 2.1.2. indicates the effects of residual pressure on breakdown strength of "vacuum" or quasi-vacuum. According to the figure, the pressure effect becomes more marked as the gap length increases. A maximum is observed at a pressure of about 4×10^{-4} torr for a 20-cm gap. It is dangerous to use a pressure close to 4×10^{-4} torr for cable design, and furthermore, 4×10^{-4} torr is considered too high to operate as high-quality thermal insulation. Adsorbed gases could be removed from metals by baking, but synthetic resins that might be used for spacers could not be baked at high temperature.

Secondly, although the precise details of electrical breakdown in vacuum remain obscure, field emission of electrons from the conductor surfaces seems to be a major determining factor. Another important fact is that field emission increases the dielectric loss tangent of a vacuum insulation system, while the contribution from the dielectric spacers is expected to be negligible because they occupy only a very small fraction of the insulation space. Early work indicated that field emission currents from bare aluminum conductors would be appreciable for EHV cables, but that anodizing of the aluminum surfaces offered a potential solution to this problem.[19] The magnitudes of field emission currents depend to a considerable extent on the amount of water, gases, carbon, and organic matter adsorbed on the metal surfaces.[13] Nevertheless, it is claimed by some people that bare aluminum with clean surfaces may be acceptable for practical applications.

The third problem arises as a consequence of a phenomenon which can be observed after an initial discharge takes place. Electrons and ions in vacuum readily gain energy from the

FIGURE 2.1.3. Spacer multi-shields for coaxial vacuum insulation system. (From Graneau, P. and Jeanmonod, J., *IEEE Trans. Electr. Insul.*, 6(1), 41, 1971. With permission.)

electric field because of their long mean free paths. Once a discharge occurs, it can form cathode spots preferentially on the conductor surface adjacent to the insulating spacers, without producing sufficient ions to sustain a high current vacuum arc. Subsequently, the cathode spots feed the discharge with metal ions. The insulators may suffer overheating and metal vapor deposition from these adjacent discharges.

To eliminate the hazards just described, electrostatically floating metallic ion shields have been devised. They suppress the energy gain of electrons and ions. A conceptual design for such shielding is illustrated in Figure 2.1.3.[19] Anodized aluminum or titanium is favored for the shielding material.

The conductor support insulators, or spacers, constitute the weakest points in a vacuum insulation system. When a solid dielectric object bridges the vacuum gap between a pair of high-voltage electrodes, the breakdown strength of the gap is reduced, for most of the ionic activity preceding and during breakdown takes place at the triple junctions between metal, dielectric, and vacuum, and along the dielectric surfaces. Firm bonding of the dielectric material to the metal might help, but such bonds would interfere with differential thermal contraction and expansion which occur during cool-down and warm-up of a cryogenic cable. Bonding of components might better be omitted so as to avoid problems from this cause.

One should also pay attention to Penning-type crossed-field discharges in designing this type of cable. The interaction of ions with the current-induced circumferential magnetic field and the radial electric field might have several consequences:

1. The breakdown strength of the gap may be lowered
2. The accompanying energy loss mechanism provides a further heat source
3. Both liberation and pumping gas occur so that inevitably the vacuum pressure is influenced

Some anticipated electron trajectories in a coaxial system are depicted in Figure 2.1.4.[16]

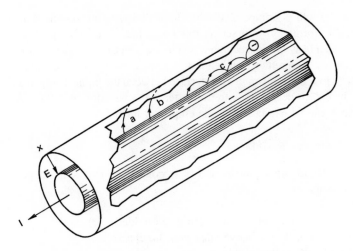

FIGURE 2.1.4. Possible electron trajectories in the coaxial cable. (From Graneau, P., *Magnetic Fields in Vacuum Insulated High Voltage Power Apparatus,* NSF-RANN Project, National Science Foundation, Washington, D.C., 1974, 2. With permission.)

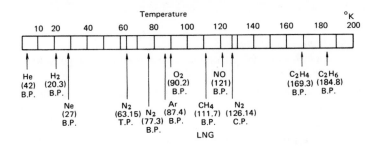

FIGURE 2.1.5. Temperature of cryogenic liquids.

The Lorentz force should drive an accidentally developed arc down the coaxial line away from the current source. Rapid arc motion would hopefully distribute heat over a larger area, thereby reducing the danger of a permanent damage. It is estimated that the velocity of an arc in a vacuum-insulated cryocable is 10^4 m/sec, which is surprisingly high.

For both electrical and thermal insulation, it is necessary to evacuate the long-length vacuum system to around 10^{-6} torr. (1.30×10^{-4} Pa). For this purpose, it is imperative to pump the system continuously and reliably throughout the service life of the vacuum-insulated cryocable.

C. Electrical Insulation by Refrigerants [26-43]

Several refrigerants are available as shown in Figure 2.1.5.[26] (see also Volume I, Section 2.2.2. and Section 2.3.5.). Among them are liquid nitrogen and liquid hydrogen. Even liquid neon is interesting as both insulant and coolant. Liquefied natural gas is attractive too because of its availability, but unfortunately its liquifaction temperature is too high to accord with a worthwhile reduction in electrical resistivity of cable conductors. Helium is clearly a strong candidate for superconducting cables. Liquid neon has an advantage over other competitive liquids with respect to its cooling capability but is less available. Liquid oxygen and hydrogen are both explosive by nature. Liquid hydrogen was once investigated as a coolant and insulant,[38] but with little success.

Liquid nitrogen is promising. As seen in Volume I, Figure 2.3.11. and Figure 2.3.12., its breakdown strength is comparable with those of insulating liquids, but it may be susceptible to bubble formation in practice; this could lead to instability of the insulation. Its dielectric loss tangent is quite acceptable under low electric stress condition, as one would expect from theory. Regrettably, it tends to increase as the applied voltage is increased, possibly due to a loss mechanism in the interface between the liquid nitrogen and the electrode (or cable conductor). It decreases with time after the application of voltage and under steady state condition it can be expressed by the empirical formula:

$$\tan\delta = A \exp (bE) \tag{2.1.1.}$$

where E = the electric stress and A and b = constants. This stress dependence may be interpreted in terms of:

1.   A surface layer on the electrode (such as an oxide layer)
2.   Electron injection from ''active sites'' on the electrode
3.   Migration of ions in the liquid

The value of $\tan\delta$ is appreciable at cable operating stress, e.g., 0.001 at 400 V/mil.

In any event, no one has attempted to construct an electrical insulation system based on liquid nitrogen or some other refrigerants alone. These materials in their gaseous state are quite unacceptable from the viewpoint of electrical strength.

D. Liquid Nitrogen-Impregnated Paper [26-43]

This insulation system is obviously analogous to that used in OF-type cables, so one might therefore expect it to be fairly stable in its properties. Cellulose paper, polyethylene paper, and polypropylene-cellulose laminate are now regarded as promising candidates for EHV class cryogenic resistive cables. The first two in particular have been the subjects of intensive study for the past 15 years.

Recent comparative investigations using model cables with 0.04-in. (1-mm) insulation thickness, have revealed the following.[1] Postcalendered polyethylene paper provides a breakdown strength of the order of 1350 V/mil (53.2 kV/mm). The density of polyethylene paper is identified as an important parameter in this respect. Although postcalendered paper has exhibited low losses and reasonably high breakdown strength, from a manufacturing standpoint it is presently not as well characterized and not as uniform as a high quality cellulose paper. Furthermore, polyethylene paper has poorer taping characteristics than cellulose papers, which necessitates a reduction of taping tensions in the taping operation.

High-quality EHV cellulose paper has the highest breakdown strength (1750 V/mil or 68.9 kV/mm) of the three candidate materials, but unfortunately has a relatively high dissipation factor (800 μrad or 0.08%). However, the characterization, uniformity, and good taping characteristics of cellulose paper cannot be overlooked.

The third material, a laminate of polypropylene film and cellulose paper, is expected to combine the better properties of both materials. Readily available forms of this cellulose-polypropylene paper laminate are presently quite limited. In general, the parameters of this material have been optimized for oil-paper cable or capacitor applications. A small cylindrical sample made of this material exhibited a breakdown strength of 1350 V/mil (53.2 kV/mm) and relatively low losses. The material has good taping properties.

Carbon black paper, carbon black polymer tape, and metallized plastic tape are available for use as screening materials. In general, the presence of a carbon black screen will increase the measured breakdown strength. It appears to have the greatest effect on polyethylene paper and a smaller and similar effect for both cellulose paper and cellulose-polypropylene laminate.

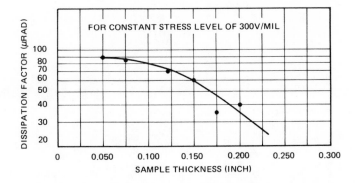

FIGURE 2.1.6. Polyethylene paper dissipation factor vs. insulation thickness.

The presence of carbon black screens increases the dissipation factor as might be predicted from the results of liquid nitrogen itself, described earlier. The measured effect on cellulose paper and cellulose-polypropylene laminate samples is small. For these two materials, the higher intrinsic losses of the materials tend to swamp the influence of the electrode interface losses. For polyethylene paper, the effect is pronounced and is inadmissible in normal cable design. The dissipation factor of this insulation system is dominated by the electrode interface loss.

It has been established that screens of polyethylene-based carbon black are superior to those of cellulose-based carbon black in that they improve the breakdown strength and reduce the dissipation factor of polyethylene paper insulation. Fortunately, the latter is not as dependent on the electric stress as carbon black paper. The loss depends on the resistivity of the carbon black materials, which should be as small as possible, but high enough to still function as a screen.

Metalized tape polymer is also a promising candidate as a screening material.[26] It does not appear to increase significantly the dissipation factor in polyethylene paper insulation systems. AC breakdown strength of 60 kV for test samples 1-mm thick are obtained. This is 20% greater than the values for carbon black paper screened polyethylene paper insulation with the same sample configuration. The dissipation factor of this insulation system can be less than 10 μrad or 0.001%.[26] The electric stress effect on the dissipation factor diminishes as the insulation thickness increases because it is an interfacial phenomenon, as evidenced by Figure 2.1.6.[1]

Neither cellulose paper nor polyethylene paper appears to be a fully satisfactory material for cryogenic cable applications. One might expect cellulose-polypropylene laminate to be better. A special grade should be developed for the purpose of this cable insulation. Table 2.1.1.[1] summarizes the material parameters for cable design and compares the three candidate insulating materials.

E. Thermal Insulation

Thermal insulation systems for use with cryogenic fluids are in general of three types: those that are gas-filled, those that are evacuated, and vacuum itself. Thermal conductivity ranges of typical insulating materials in the form of[44-47] foams, powders, fibers, and multilayers, both evacuated and unevacuated, are compared in Figure 2.1.7.[44] Since gas in an insulation space provides good thermal contact between the components of the insulation, its removal greatly increases the effectiveness of the insulation, as shown in Figure 2.1.8[43]

For cryogenic resistive cables, it is imperative to suppress the heat leak as much as possible because the refrigeration efficiency is not high, a power input of 7 to 12 kW being required to remove the heat of 1 kw generated at 77 K. With the exception of the potheads, the heat

Table 2.1.1.

INSULATION MATERIAL SELECTION FOR CRYOGENIC RESISTIVE CABLES

Material	Breakdown strength	Dielectric losses	Taping properties	Manufacturing process control	Availability
Cellulose paper	High	High	Excellent	Well established	Special grade paper, may require custom manufacture
Polyethylene paper	Acceptable	Low	Acceptable	Not well established for electrical insulation applications	Available
Cellulose-polypropylene laminate	Acceptable	Acceptable	Good	Established	Special grade material, will require custom manufacture

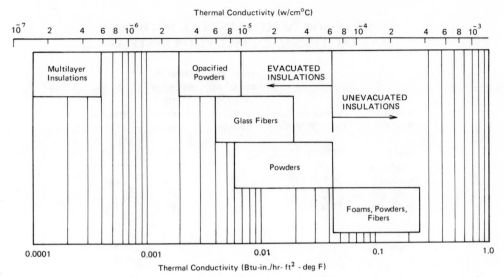

FIGURE 2.1.7.   Comparison of thermal conductivities of typical cryogenic insulation systems. (From Glaser, P. E., *Machine Design,* August 1967, p. 147. With permission.)

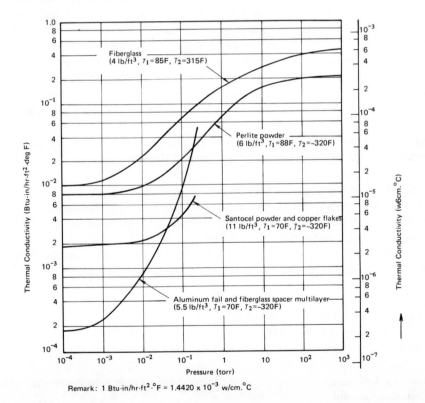

FIGURE 2.1.8.   Effect of gas pressure on thermal conductivities of powder, fiber, and multilayer insulations. (From Glaser, P. E., *Machine Design,* August 1967, p. 147. With permission.)

Table 2.1.2.
COMPARISON OF HEAT IN-LEAK BETWEEN
SEVERAL THERMAL INSULATION SYSTEMS

Kinds of thermal insulator	Thermal conductivity (W/m°C)	Inner diameter of thermal insulator (mmφ)	Outer diameter of thermal insulator (mmφ)	Heat leak (W/m)
Super insulation	5×10^{-5}	380	480	0.3
Vacuum (10^{-6} torr)	6.5×10^{-3}	380	580	21
Perlite evacuation (10^{-2} torr)	4×10^{-3}	380	580	13
Foam	2×10^{-2}	380	680	48

leak takes place through the thermal insulation which surrounds the pipe containing the liquid nitrogen and the cables. As one would expect from Figures 2.1.7. and 2.1.8., the heat leak depends on the kind of thermal insulation. Table 2.1.2. compares the heat leak of several thermal insulation systems. The choice of a specific system is made after considering coordination with other heat losses and, of course, the economics of the situation.

Multilayer insulations are so devised as to interleave highly reflective radiation shields with low-conductivity spacers, the gas pressure within the layers being reduced to less than 10^{-4} torr. Typical radiation shields are aluminum foil and metalized polyester films. Theoretically, the performance of thermal insulation improves with the number of radiation shields. Too dense an application, however, adversely affects performance because heat transfer tends to be dominated by the thermal conduction of the spacers between the multilayers and the reflecting metal surfaces. Typical multilayer insulations are composed of 20 to 30 layers per centimeter. They have a loss of about 1 W/m² between 77 K and 300 K, and about 50 mW/m² between 4 K and 77 K. The performance may degrade by one or two orders of magnitude depending on the weight of the insulation.

2.1.3. Cable System Design Considerations

Liquid nitrogen-impregnated paper (cellulose or synthetic) insulation appears to be stable and therefore promising. In this section, the application of this electrical insulation system to cryogenic resistive cables will be considered.

A cryogenic resistive cable system is similar in structure to a pipe-type, oil-filled, paper-insulated cable system; it comprises the following subsystems:

1. Cable cores
2. Cryogenic envelope with thermal insulation
3. Cable splices and terminations
4. Refrigeration system

There are marked differences in cryogenic envelope and refrigeration systems as compared with conventional cables.

A. Cable Conductor

Annealed copper and aluminum with 99.9% or 99.99% purity are acceptable for conductors used at liquid nitrogen temperatures. The further purification of those materials would not

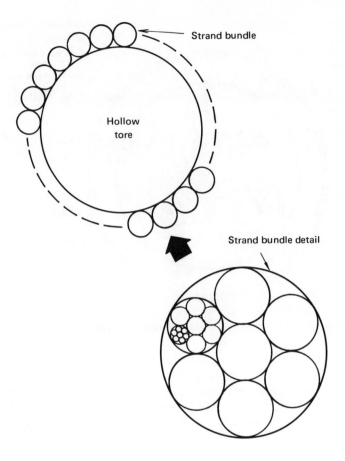

FIGURE 2.1.9. Stranded conductor concept for cryogenic resistive cables. (From Minnich, S. H. and Fox, G. R., *Cryogenics,* June 1967, p. 168. With permission.)

be profitable. Other metals, such as berylium, may give much lower electrical resistivities at liquid nitrogen temperatures as indicated in Figure 2.1.1. but the availability and experience are both limited, and their technical feasibility and economic viability are uncertain.

For copper and aluminum, the DC resistivities at liquid nitrogen temperatures are approximately one tenth of their values at room temperature. Their AC resistivities, however, are appreciably higher due to skin effect, thus, much more care should be taken as regards the conductor configurations. In principle, a cylindrical conductor may be used, but more effective conductor utilization may be obtained by stranding, provided the strands are properly insulated and transposed.[40] Transposition means the geometrical arrangement of the strands so that the inductances of the strands are equal. This assures that they carry equal currents. A simple hexagonal stranding sequence is the most elementary form of transposition. Six elementary strands are wound around a central core (not used as a conductor): six of these bundles are wound around a larger core; six of the resulting bundles are wound around a third core, and so on. This configuration is called "litz" wire; its space utilization is poor if many strands are required. A conductor model composed of litz wires is shown in Figure 2.1.9. In this figure, a hollow former is wrapped with bundles of transposed strands (hexagonal stranding shown in this example). For this conductor configuration, the AC resistance, including the contribution from eddy currents, can be computed from the following expression:

$$R_{eff} = \frac{\rho_c}{S} \left[1 + \frac{(2\pi fdS)^2 \times 10^{-18}}{12(\rho_c r)^2} \right] \; [\Omega/cm] \qquad (2.1.2.)$$

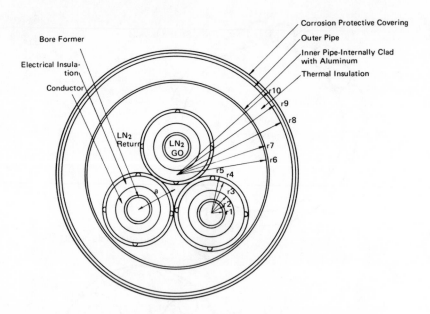

FIGURE 2.1.10. Cross section of cryogenic resistive cable for design calculation.

where ρ_c = the DC resistivity of a given conductor [Ω-cm], S = the conductor cross section [cm²], r = the conductor radius [cm], d = the strand diameter [cm], and f = the power frequency [sec⁻¹]. It is obvious from the above formula that there is an optimum conductor cross section or radius for a given strand diameter. Furthermore the magneto-resistance effect contributes to an increase in the AC resistance for high-purity metals. The effect is smaller for aluminum than for copper.

B. Cable Insulation

As explained in Section 2.1.2., cellulose paper, polyethylene paper, or polypropylene cellulose laminate can be used for a prototype cable. Carbon black (cellulose or polyethylene) paper, metalized polymer tape should function satisfactorily as cable screening. The insulated cable core with screens and other necessary electrical and mechanical reinforcements is pulled into a pipe and impregnated with liquid nitrogen. Thereafter the pipe system is evacuated. In favorable circumstances the insulation can operate at a field strength around 15 kV/mm at the conductor surface for EHV cryogenic resistive cables.

C. Cable Design Methods

One design shall be described by way of illustration. The particular model of liquid nitrogen-impregnanted, synthetic paper-insulated cable is shown in Figure 2.1.10. As explained in the previous section, this type of cryogenic cable is forced cooled and thereby designed to optimize the heat balance between generation and removal. The various parameters in the calculation are as follows:

Skin depth of conductors

$$\delta_c = \frac{1}{2\pi} \frac{\sqrt{\rho_c \times 10^9}}{f} \ [cm] \qquad (2.1.3.)$$

where ρ_c = the electrical resistivity (Ω-cm) and f = the power frequency (Hz).

The coefficient of eddy current loss

$$\lambda = \left[\frac{k \cdot t \cdot s}{\delta_c^2} \right]^2 \times \frac{1}{3} \qquad (2.1.4.)$$

where k = the ratio of the real to the apparent conductor cross section, t = the conductor thickness (t = $r_3 - r_2$) (cm), and s = the diameter of a single segmental wire (cm).

AC resistance of conductors per unit length

$$R_{ac} = \frac{\rho_c}{S} (1 + \lambda) \ (\Omega/cm) \tag{2.1.5.}$$

where S = the conductor cross section (cm^2).

Pipe loss

$$W_p = 1.4 \left\{ \frac{I \cdot a}{(1 - a^2/r_6^2)} \right\}^2 \cdot \frac{\rho_p}{\delta_p \cdot r_6^3} \ [w/cm] \tag{2.1.6.}$$

where I = the current through a cable conductor [A], a = $r_5/\cos 30°$[cm], ρ_p = the electrical resistivity of metal clad to the inner surface of the pipe [Ω-cm], and δ_p = the skin depth of the backing metal. Only the trefoil configuration of three cables is being considered.

Dielectric loss

$$W_d = 2\pi \ fC \ \frac{V_0^2}{3} \ \tan \delta \times 10^5 \ [w/cm]$$

$$\text{with } C = \frac{\epsilon^*}{18 \ \ell n \ (r_4/r_3)} \ [\mu F/km] \tag{2.1.7.}$$

where ϵ^* = the specific dielectric constant of the cable insulation, V_0 = the line voltage (kV), and tanδ = the loss tangent of the cable insulation.

Maximum electric stress

$$E_{max} = \frac{V_0}{\sqrt{3} \ r_3 \ \ell n(r_4/r_3)} \ [kV/cm] \tag{2.1.8.}$$

Heat-leak

$$W_{th} = k_1 \frac{2\pi(T_h - T_i)}{\ell n(r_8/r_7)} \ [W/cm] \tag{2.1.9.}$$

where k_1 = the thermal conductivity of the thermal insulation [w/cm ·K] and T_h = the ambient temperature in the cryogenic cable system.

Friction loss of a pipe

$$W_f = \frac{A\rho_m f^3}{r} \times 10^{-7} \ [W/cm]$$

$$\text{with } f_r = 0.0014 + 0.125 \ Re^{-0.32} \tag{2.1.10.}$$

where A = the cross section of refrigerant flow path [cm^2], p_m = the average density of the refrigerant [g/cm^3], v = the flow velocity of the refrigerant [cm/sec], f = the friction coefficient, and r = the effective radius of the flow path.

Pressure drop

$$\Delta p = 4f_r \ \frac{\rho_m v^2}{2d} \times 9.87 \times 10^{-7} \ [atm/cm]$$

with

$$d = \begin{cases} 2r_1 \text{ for the "go" path} \\\\ 2\left(\dfrac{r_6^2 - 3r_4^2}{r_6 - 3r_4}\right) \text{ for the "return" path} \end{cases} \qquad (2.1.11.)$$

Flow rate of refrigerant

$$Q = \frac{L\left\{3(W_c + W_d + W_f) + W_p + W_{th} + W_{f'}\right\}}{\rho_m C_p \Delta T} \qquad (2.1.12.)$$

where L = the cable length in one cooling system [cm], W_c = the conductor loss [W/cm/ϕ], W_f = the friction loss in the "go" path [W/cm/ϕ], $W_{f'}$ = friction loss in the return path, [W/cm/3ϕ], C_p = the specific heat of the refrigerant [J/g·K], and ΔT = the permissible temperature rise [K].

Temperature rise in the go and return paths

$$\Delta T_{GO} = \frac{3(W_c + 1/2 W_d + W_f)}{\theta C_p \rho_m} \text{ [K]} \qquad (2.1.13.)$$

$$\Delta T_{RE} = \frac{W_p + 3/2 W_d + W_{th} + W_f}{\theta C_p \rho_m} \text{ [K]} \qquad (2.1.14.)$$

where ΔT_{GO} = the temperature rise in the go path and ΔT_{RE} = the temperature rise in the return path.

Jam ratio

$$RJ = \frac{2r_6}{2r_4 + 1.5h_S} \qquad (2.1.15.)$$

where h_S = the height of the skid wires.

The optimum condition, derived from the temperature and flow rate of the refrigerant, occurs when the friction loss is half the sum of all the other losses. However, it is almost impossible to satisfy this condition in designing cryogenic resistive cables because conductor loss would be too high for a practical permissible design. High flow rate to remove conductor heat is imperative. As a consequence of this, the length of a single cable section is determined by the pressure head rather than the temperature rise.

A flow chart for making these calculations is presented in Figure 2.1.11. In the cable configuration shown in Figure 2.1.10., the cryogenic system requirements are most likely to be determined by the "go" paths. Table 2.1.3. shows some calculated results. The ratio of (W_t - W_f) to $2W_f$, which should be unity under the above optimum flow condition, is around 25. Computations reveal that for a cable design such as shown in Figure 2.1.10., the interrefrigeration station distance are too short if one intends to transmit a power great as 10 GVA by a cryogenic resistive cable, within the present framework of manufacturing and transporting technologies, which limits the maximum cable diameter to 150 mmϕ.

D. Cryogenic Envelope

To select a suitable thermal insulation for a cryogenic resistive cable out of the several kinds developed, bearing in mind the available facilities for storage and transportation of cryogenic fluids, one must consider thermal coordination with the heat generated in the cable conductors which may exceed 100 W/m for a reasonably designed several thousand MVA

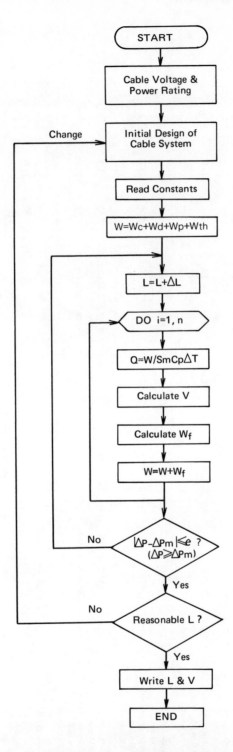

ΔT and ΔPm are to be predetermined in this program

FIGURE 2.1.11. Flow chart to calculate the flow rates or velocities and inter-refrigeration-station distances under the constant temperature and pressure change conditions (only ''Go'' path).

Table 2.1.3.
HEAT LOSSES, COOLANT FLOW RATES, AND INTER-REFRIGERATION-STATION DISTANCES FOR SOME CRYORESISTIVE CABLE DESIGNS

500 kV, 10 GVAM 20-10 kg/cm² G

Multilayer thermal insulation	Circuit	Current (A)	Conductor loss	Dielectric loss	Pipe loss	Heat in-leak	Friction loss	Total	Flow rate (cm/sec)	L^a (km)	RC^b (kW)	$(w_r-W_f)/2Wf$
			Generated heat (W/cm)									
Temperature 95—65 K Conductor 1200 mm²	1 cct Emergency	11547.0	8.5857	0.0015	3.1327	0.0172	0.2472	11.9843	328.25(Go) 56.01(Re)	2.72	1630.8	23.74
Insulation thickness 30 mm LN₂ bore 66.0 mmφ Emax = 12.75 kV/mm	2 cct Normal	5773.5	2.1465	0.0015	0.7832	0.0172	0.0618	3.0102	201.27(Go) 34.34(Re)	6.64	999.9	24.35
Temperature 95—65 K Conductor 1950 mm	1 cct	11547.0	6.06	0.0014	2.59	0.0172	0.194	8.86	320(Go) 36(Re)	2	900	22.3
Insulation thickness 30 mm LN₂ bore 50.0 mmφ Emax = 13 kV/mm	2 cct	5773.5	1.51	0.00 14	0.646	0.0172	0.048	2.22	200(Go) 22(Re)	5	560	23.2
Temperature 95—65 K Conductor 3000 mm²	1 cct	11547.0	4.2576	0.0018	3.1327	0.0172	0.1569	7.5662	290.80(Go) 37.95(Re)	2.92	1105.0	24.112
Insulation thickness 25 mm LN bore 57.7 mmφ Emax = 14.4 kV/mm	2 cct	5773.5	1.0644	0.0018	0.7832	0.0172	0.0396	1.9062	178.64(Go) 23.31(Re)	7.12	678.8	24.068
Temperature 95—77 K	2 cct 3cct									3.50 5.80	425.4 320.9	

500 kV 3.5 GVA

Multilayer thermal insulation	Circuit	Current (A)	Conductor loss	Dielectric loss	Pipe loss	Heat in-leak	Friction loss	Total	Flow rate (cm/sec)	L^a (km)	RC^b (kW)	$(w_r-W_f)/2Wf$
Temperature 87—65 K Conductor 2600 mm² Emax = 16.0 kV/mm	35 GVA	4041.45	0.641	Negligible	0.359	0.059 (Go) 0.039 (Re)	0.021 Pumping loss	1.119		16	872 kW/km ÷ 7.8 112 kW/km	26.14

a Inter-refrigeration-station distance.
b Refrigeration capacity — required electrical input is about 10 times.

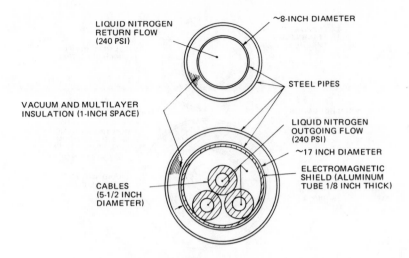

FIGURE 2.1.12.   Evacuated multilayer insulated pipes.

cable. Superinsulation is excellent, but evacuated powder insulation or even a foam insulation system would be acceptable judging from the values listed in Table 2.1.2.

A thorough investigation has been made of two types of coolant flow systems with respect to the maximum cable length of one cooling section. The systems considered were

1.   A single-pipe system with counterflow cooling, as illustrated in Figure 2.1.10. employs the so-called interstitial return concept. The coolant flows down the cable bores and returns in the interstices between the cables and the cable pipe. The heat leakage is less than for any other configurations because of the minimal perimeter. As explained previously, the length of a cooling section is likely to be limited by the pressure head, this being the case, a larger cable bore is required, which in turn makes for a stiffer cable.
2.   A two-pipe system: one pipe is similar to that just described: it carries unidirectional flow of coolant in three cable bores as well as in the pipe itself. A second, separate pipe, without cables carries the return flow.

The second concept is illustrated in Figure 2.1.12., Figure 2.1.13., and Figure 2.1.14. In this flow structure, the coolant in the bores can be exchanged with that outside the cable cores (but still inside the pipe) at points such as cable joints. In this way the cable length that can be handled by one refrigeration station can be extended.

Figure 2.1.12. shows a cross section of such a structure. It comprises three flexible cables with taped electrical insulation, pulled through a cryogenic envelope which is insulated with evacuated multilayer insulation. It has a stainless steel pressure pipe with an internal electromagnetic shield. A separate vacuum-insulated line returns the liquid nitrogen to the refrigerator. This is an expensive system. It is estimated that the cable and return pipe envelopes probably account for roughly half of the total installed system cost, thus, significant savings could be achieved if less costly envelopes could be found.[1]

Some people consider that foam insulation may be an acceptable alternative to superinsulation for the cryogenic cable pipe. The first embodiment of such a foam-insulated design is shown in Figure 2.1.13[1] where it will be seen that foam insulation of an appropriate thickness has been directly substituted for the superinsulation. The second foam-insulated structure shown in Figure 2.1.14[1] is radically different from a thermal standpoint, in that the electromagnetic shield is located outside the thermal insulation. As a consequence of

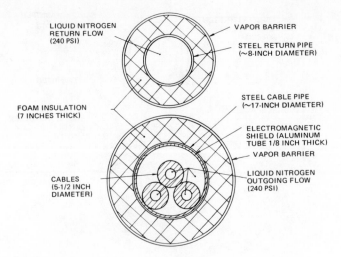

FIGURE 2.1.13. Foam insulated pipes — Concept A.

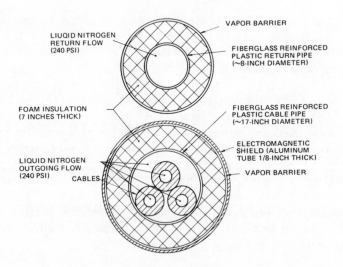

FIGURE 2.1.14. Foam insulated pipes — Concept B.

this arrangement, the shield losses are primarily dissipated to the soil, thereby significantly reducing the refrigeration load and offsetting the effects of the increased heat leak through the foam insulation. This concept requires that the inner steel pipe be replaced by a fiber reinforced plastic pipe to eliminate electrical losses in the low-temperature region.

Some other designs for cryogenic envelopes with foam insulation are shown in Figure 2.1.15.[1] The double pipe structure (Figure 2.1.15. [A]) increases heat transfer between the outgoing and returning coolant streams. This increase is potentially undesirable because it can cause one or both streams to develop a temperature peak between the cable endpoints. The effect is less than one might suppose, especially if the pipes are separated by sufficient distance (1/10 in. for example). A triple pipe structure (Figure 2.1.15. [B]) may offer the advantage of ease in pulling cables during installation and the use of smaller diameter pipes. The heat leakage for this arrangement is somewhat greater than for the other foam envelopes, because of the larger perimeter in contact with the soil.

Table 2.1.4.[1] compares five foam insulation schemes with the super insulation scheme with respect to refrigeration load. The most promising of these five appears to be the

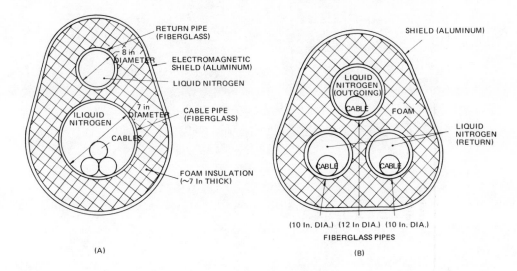

FIGURE 2.1.15. (A) Cryogenic envelopes comprising double pipes and (B) triple pipes.

arrangement with separately insulated pipes — Concept B. It seems likely that this configuration will be easier and less expensive to manufacture than either of those with multiple pipes. Moreover, it has smaller thermal losses than Concept A.

E. Cable Splice Design

Cryogenic cable can be spliced by the conventional stress cone splice approach described in Volume II, Section 1.4. Alternatively, capacitively graded splices can be used. The latter offers no substantial improvements from the standpoint of reduced size, but does have the potential for reducing preparation and assembly time.

F. Cryogenic Pothead Design

Potheads for cryogenic resistive cables can be of two types, ambient-temperature potheads and low-temperature potheads. No thermal insulation is needed inside the bushing for the first of these structures. Thermal grading is done on the cable side, below the bushing. High-ampacity potheads under development are of this type.

Potheads of the second type, which can be called the cryogenic potheads, have thermal insulation within the bushing. Conventional potheads, such as those described in Volume II, Section 1.4. cannot be applied directly to the termination of a cryogenic cable. The following additional functions must be provided by the termination:[1]

1. Means for removing the cryogenic fluid from the bore of the cable
2. Means for thermally insulating the bushing from the low-temperature region of the cable
3. Means for thermally grading the conductor from the low-temperature region to ambient temperature

G. Refrigeration Systems

The cooling power needed for cryoresistive cables is more than 100 W/m. Efficiency of existing liquid nitrogen refrigerators is around 10%, a value of 7.8% being sometimes quoted. Because of the economy of size in refrigerators, the distance between the adjacent refrigerators should be as long as can be tolerated by flow impedance considerations of the coolant in the cable.[5] This implies refrigerator distances of 10 to 20 km and cooling powers of

Table 2.1.4.
COMPARISON OF REFRIGERATION LOADS

Source of heat	Vacuum[30] insulation	Foam insulation[a]				
		Separate pipes[b] concept A	Separate pipes[b] concept B	Double pipe	Interstitial return	Triple pipe
Conductor, eddy current, and dielectric losses (kW/mi)	117.0	117.0	117.0	117.0	117.0	117.0
Pump losses (kW/mi)	3.3	3.3	3.3	3. 3	2300.0	3.3
Shield losses (kW/mi)	59.0	59.0	—	—	—	—
Heat leak[c] (kW/mi)	15.8	94.8	97.8	80.2	62.0	112.0
Total (kW/mi)	195.1	274.1	218.1	200.5	2479.0	232.3
Increase over vacuum insulation (%)	—	40	12	3	1200	19

a   Foam thickness of 7 in. has been assumed in all configurations.

b   Heat leakage reported for separate pipes differs by a small amount from that recorded in the quarterly reports.

c   Heat leak computed on the basis of 75 K fluid temperature in the cryogenic pipes.

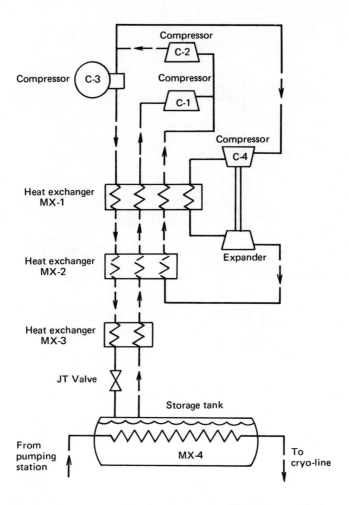

FIGURE 2.1.16. Simplified flow diagram of an LN$_2$ refrigerator (Claude Cycle).[5]

several megawatts at liquid nitrogen temperature. This translates into several tens of megawatts as electric power imput for conversion.

Reliable refrigerators of this size can be provided by existing technology in the fields of air separation and natural gas liquefaction. The Claude cycle with N$_2$-refrigerant and the Bell-Coleman or Brayton cycle deserve serious consideration. The Claude cycle as illustrated in Figure 2.1.16. has the advantages of a low cost refrigerant (nitrogen) and the direct application of the well-developed technology of air separation plants.[5]

Figure 2.1.17. shows the Brayton cycle with a neon refrigeration system.[5] This is preferably equipped with centrifugal or turbine compressors and expanders. Its capacity can be adjusted economically to accommodate a growing demand by simply installing additional compressors and expanders. Use of lower temperature refrigerants such as He, H$_2$, and Ne make liquid nitrogen at about 65 K available to feed the cable.

Heat exchangers, transfer-by-pressure pumps, and vacuum pumps are needed for cryogenic cable refrigeration systems in addition to the refrigerators themselves.

2.1.4. Several CR Cable Designs and Systems

In spite of much effort, no cryogenic resistive cable has as yet been placed in service. The several designs for cryogenic cables and their accessories to be briefly described in this

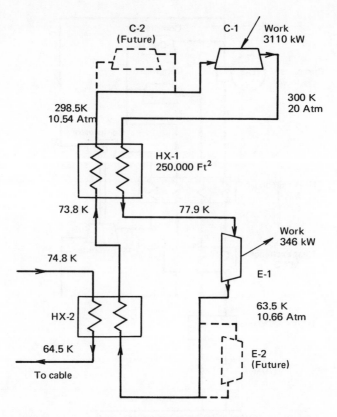

FIGURE 2.1.17.   Neon cycle refrigerator Bell-Coleman or Brayton Cycle.[5]

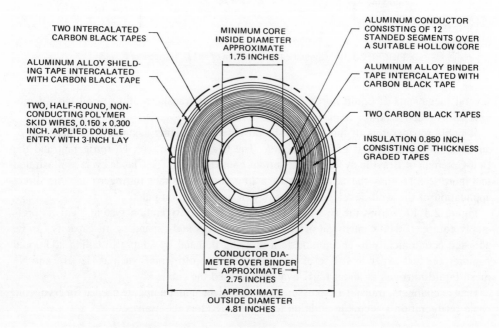

FIGURE 2.1.18.   Cross section of proposed cryogenic resistive cable.

section are therefore either test prototypes or simply conceptual ideas. Most of the activity has been directed towards the development of AC cryoresistive three-conductor cables.

Table 2.1.5.
SUMMARY OF PROTOTYPE AND PROPOSED CR CABLE DESIGNS

	G.E. U.S.	G.E. U.S.	Furukawa Japan	Hitachi Japan	Fujikura Japan	Showa Japan	CRIEPI Japan	U.P.C. U.S.	CGE France
Design values									
AC-voltage (kV)	345/500	500	154	275	154	154	500	138	DC ± 125
Power (MV·A)	3000	3500	3000	3000	(2400 A)	(5000 A)	10000	1000	2000
Conductor									
Material	1 Conductor Al	1 Conductor	3 Conductors Al (99.99)	Al	1 Conductor Al	1 Conductor Cu(99.99%)	3 Conductors Al	1 Conductor Al	1 Conductor Al(99.997%)
Cross section (mm²)	1800	2750	2400	2040	600	450	2000		2000
Performance	12 Segments wound on a spiral (45 mm I.D.) 37 strands (0.25-0.3 mm φ) per segment	24 Segments wound on a spiral former (76.2 mm I.D., 82.6 mm O.D.)	7 Segments wound on a spiral (45 mm I.D.), each wire (2.0 mmφ) formal coated	12 Stranded segments wound on a corrugated tube	ca. 200 wires wound in 4 layers on a hollow core	126 Wires wound in 2 layers on a spiral former of 40 mm diameter	Segmental		4 Layers of 0.55 mm rectangular strands
Diameter (mm)	70	102	75	75	30		79	50	70
Electrical insulation									
Material	Synthetic polyethylene paper (Tyvek®)	PE Paper, cellulose paper, PP-CP, laminate	PE Paper (Tyvek®), polycarbonate film and PE film	PE Paper		Cellulose paper	PE Paper	Vacuum	Polyethylene telephthalate films 25 μm thick

Table 2.1.5. (continued)
SUMMARY OF PROTOTYPE AND PROPOSED CR CABLE DESIGNS

Insulation thickness (mm)	21.6 (Limited by the manufacturing capability)	12	≈ 25	12.5	13.5	30		3
LN₂-Pressure a. temp.	5.6 bar at 80 K	5—10 bar at 77 K	16 bar 70—85 K		5 bar	10—20 bar 65—95 K		Gas helium 20 bar 20—25 K
LN₂-pipe Diameter (mm)	122	Stainless steel 146	380		82	145, 150		
Thermal insulation								
Material	Superinsulation Foam	Superinsulation	Superinsulation Polyurethane foam	Superinsulation	Evacuated powder (perlite)	SI * Powder** Foam***	SI or foam	SI 10⁻⁴ torr
Thickness (mm)			≈150 2.0 mm × 40					
Overall diameter (mm)	ca. 700	650	718			160* 600** 700***		
Remarks	12 m	20 m	20 m	15 m	30 m	Conceptual	—	20 m

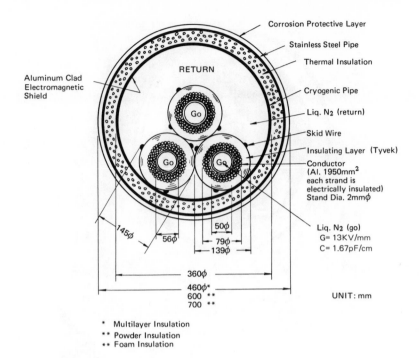

FIGURE 2.1.19. Cross section of CR cable line designed for 500 kV 10 GVA/Rout with 2 or 3 circuits. (From *Cryogenic Power Transmission*, (in Japanese), CRIEPI Advanced Power Transmission Series No. 17, 1973, 33. With permission.)

A. Liquid Nitrogen-Impregnated Paper-Insulated Cables

Figure 2.1.18. shows the cross section of a cryoresistive cable proposed by the General Electric Co. after a 10-year investigation.[1] The selection of insulating materials for this cable is still uncertain. In order to increase refrigeration distances, a modified design is under consideration as indicated in Table 2.1.5. A 3-phase cryoresistive cable, with 2 or 3 circuits, operating at 500 kV, is shown in Figure 2.1.19. Such a cable should be rated at 10 GVA or thereabouts. This corresponds to a CRIEPI design as shown in Table 2.1.3. and Table 2.1.5. It will be noted from Table 2.1.3. that with this design a single refrigeration distance is relatively short — several kilometers. A photograph of a cable of this type is shown in Figure 2.1.20. Table 2.1.5. summarizes several designs of prototype and proposed cryoresistive cables, some of which have been manufactured for test.

B. Splices

A capacitively graded splice, designed for a 500-kV, 3500-MVA cryogenic cable, is shown in Figure 2.1.21.[1] The maximum stress occurs near the high-voltage end of the capacitor system, at the midpoint of the splice, being 365 V/mil (14.6 kV/mm) and 18.2 V/mil (720 V/mm) in radial and axial directions, respectively. The point of lowest stress is located at 80% of the half-splice length as measured from the midpoint of the splice. Here the radial and axial stress are 250 V/mil (10 kV/mm) and 12.5 V/mil (500 V/mm) in their respective directions. The splice is designed to provide a stress uniformity factor, defined as the ratio between maximum and minimum stresses, of 1.45. Coolant stop joints have not yet been developed, but one conceptual design for such a joint is shown in Figure 2.1.22.

C. Cryogenic Potheads

Figure 2.1.23. illustrates a cryogenic pothead termination designed for 500 kV, 4000 A.[1] In this design, voltage grading is achieved by appropriate shielding and by using a stress

FIGURE 2.1.20.   Three-phase 154 kV CR cables. (Courtesy of the Fu-rukawa Electric Co.)

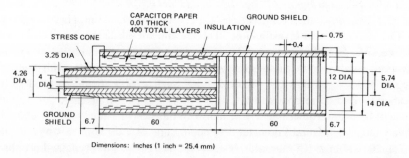

Dimensions: inches (1 inch = 25.4 mm)

FIGURE 2.1.21.   Capacitively graded splice concept for 500-kV, 3500-MVA cryogenic cable.

cone made with a wrapped dielectric. This can be replaced by capacitively graded insulation in which the voltage is uniformly graded over the insulation by using a stress cone and a number of series-connected capacitors distributed over the length of the pothead.

The transition from low temperature to ambient temperature for the current lead is made in the vacuum space at the top of the pothead. Copper bellows or copper braid may be used ahead of the lead-out. In the vacuum space, a conductor with a cross-sectional area of 1/4 in.^{2} (160 mm^2) and a length of 6 in. (150 mm) will carry a current of 4000 A, with a heat leak into the pothead of about 160 W. Under no-load conditions, the heat leak is about 80 W. The bellows between the conductor support plate and the glass-epoxy cylinder accommodate the differential contraction between the copper conductors and the cylinder.

The glass-epoxy cylinder serves as return path for the liquid nitrogen which flows through the cable conductor bore to the top of the pothead. An evacuated space between the glass-epoxy cylinder and the outer porcelain shell provides an effective thermal and dielectric barrier and ensures that the porcelain operates at ambient temperature.

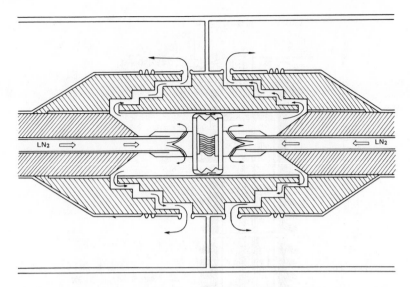

FIGURE 2.1.22. Stop joint concept for cryoresistive cables. (From *Cryogenic Power Transmission*, (in Japanese), CRIEPI Advanced Power Transmission Series No. 17, 1973, 42. With permission.)

Some other pothead designs which as yet are only conceptual are shown in Figure 2.1.24.

D. Vacuum-Insulated Cryoresistive Cables

The scaled cross section of a proposed 138-kV liquid nitrogen cryocable is shown in Figure 2.1.25.[13] It will be seen to comprise four extruded aluminum pipes, three-phase spacers, and telescopic contraction joints. Three hollow sector-shaped aluminum conductors contain flowing subcooled liquid nitrogen. The spacers, though not illustrated, locate the conductors inside an aluminum alloy pipe. The space between the enclosure and the conductors is evacuated to provide both thermal and electrical insulation. The spacers have to be designed to minimize heat flow from the warm vacuum enclosure and they must be made from inorganic materials so as to keep the out-gasing rate low.

A corresponding termination is designed by using bushings, 3- ft long, with vacuum seals. It has a tabular aluminum conductor surrounded by a conducting shell of titanium, on which the ion shield assembly is mounted. The series of titanium discs is held together with aluminum rods which are joined to the metal. This termination design should be protected by a high-voltage surge diverter which suppresses all voltage surges in excess of 200 kV.[25]

E. Research Project Plans

In 1976, a research group in Japan established plans for the complete development of a cryoresistive cable. Figure 2.1.26. presents the steps of a research program aimed at the installation of 500-kV, 10-GVA cryoresistive cable in the relatively near future. It would be equipped with a (1 + 1) or a (1 + 2) circuit system having no more than 1 refrigeration station every 10 km.

The work plan calls for three phases, which will be described below. Much has already been accomplished.

Phase I — Fundamental research and design

1. Reviewing of presently available technologies
2. Conceptual design of the cable systems
   a. Fundamental design of the cable structure, thermal insulation, and refrigeration methods

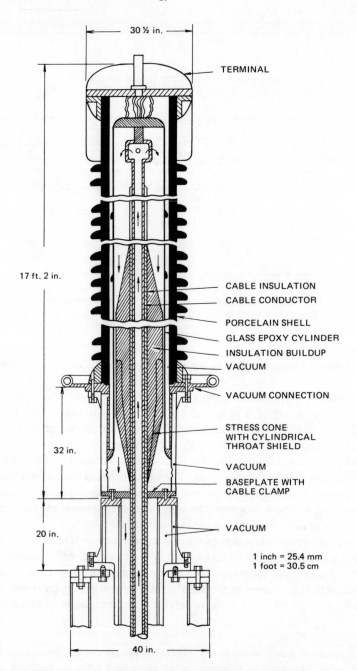

FIGURE 2.1.23.   Pothead termination for cryoresistive cable (500 kV, 4000 A).

    b.   Investigation of an optimum design

3.   Research and development of system components

    a.   Conductors

        ● Selection of conductor materials (purity and manufacturing technology)

        ● Investigation of conductor structures to reduce AC losses (size, stranding methods)

    b.   Electrical insulation

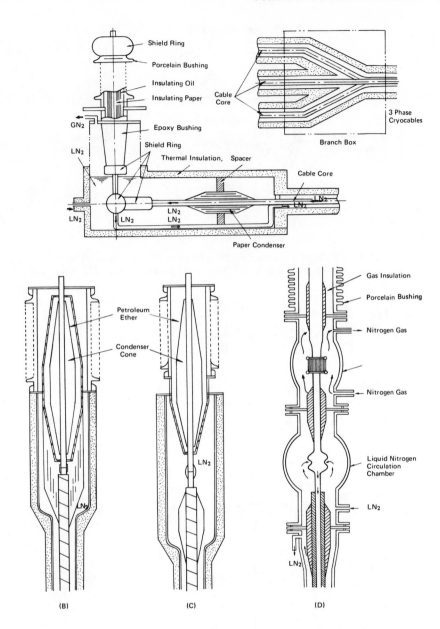

FIGURE 2.1.24. Several cryogenic pothead designs (conceptual).

- Electrical performance of liquid nitrogen (long gap and surface flashover characteristics)
- Selection of insulating tapes (electrical and mechanical characteristics of candidate tapes probably in paper form and their tapeability)
- Long-term stability of liquid nitrogen-impregnated insulation system
- Electrical tests on short samples

c. Pressure pipes (inner pipes)
- Investigation of pipe materials and structures (internally clad conductors, jointing methods, FRP pipes)
- Evaluation of pipe losses

d. Thermal insulation

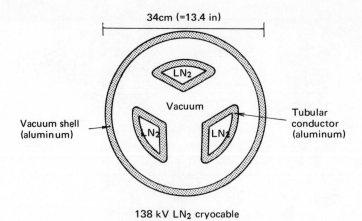

138 kV LN₂ cryocable

FIGURE 2.1.25.   Cross section of a 138-kV vacuum-insulated liquid nitrogen cooled cable. (From Graneau, P., *Cryogenics*, January 1975, p. 27. With permission.)

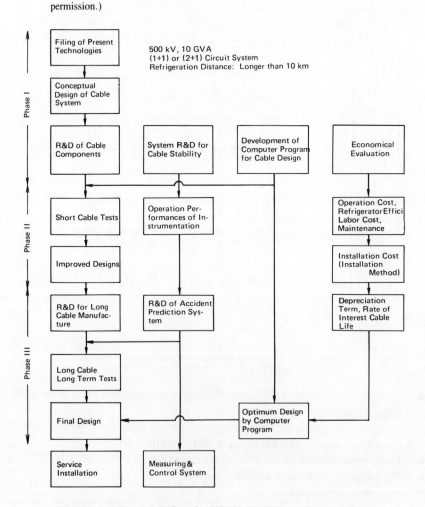

FIGURE 2.1.26.   R & D flow for 500-kV, 10-GVA cryoresistive cables.

- Investigation of insulation materials and structures (the degree of vacuum, jointing methods)

- Total thermal insulation performance of the cable system including spacers, joints, and terminations
- Long-term stability (effects of shock at time of shipping and installation, and persistant small vibrations)

e. Cable accessories
- Investigation of structures of joints and terminations
- Investigation of heat leak from terminations and cooling methods

f. Thermal contraction of cable systems
- Methods of offsetting the thermal contraction of the cable cores
- Thermal contraction in the pipes and in the thermal insulation (contraction joints)

g. Development of refrigeration systems
- Specification of refrigerators and evaluation of their reliability
- Investigation of efficiency, maintenance, and noise characteristics of existing refrigerators
- Design of refrigerant circulation systems (reliability evaluation of pressure heads)
- Methods of initial cooling (related to the research item (f)
- Long-term automatic operation

4. R & D of cable system stability
a. Circuit systems and economics
b. Refrigeration power reserve
c. Probable accidents and countermeasures
d. Investigation of accident prediction systems (instrumentation, data processing, safety devices)
e. Judicial problems

Phase II — Short model cable tests

1. Short model cable (3 phases, 500 kV) specifications
- Length: 30 to 50 m
- Voltage: 500 kV
- Current: 12,000 A
- Installation: aboveground
- Accessories: termination to air and gas, normal joints, feeding joints
- Refrigeration system: closed cycle or heat exchanger system

2. Investigation items in short-length-cable tests
- Electrical: withstand voltages, partial discharges, dielectric losses, conductor losses
- Cooling: temperature at selected locations, refrigerant temperatures, refrigerant pressures, pressure head, refrigeration system efficiency, automatic-control performance
- Others: initial-cooling characteristics, short-circuit characteristics (temperatures, partial discharges, mechanical forces)

3. R & D prior to field tests
a. Sufficient analysis of the test results from the short model cable evaluation, iteration to improve designs.
b. Fabrication methods
c. Maintenance and inspection methods
d. Countermeasures against probable accidents
e. Determination of refrigeration power reserve

Phase III — Field tests on long-length cables

1.   Long-length cable (3-phase, 500 kV) specifications
      ● Cable length: 750 m feeding joints, normal joints, air-terminations, miniclad
         terminations, trifurcating box, curved and uneven cable installation
      ● Voltage: 500 kV
      ● Current: 6000 A (or 12,000 A)
      ● Installation: direct buried and tunnel
      ● Refrigeration: closed cycle with an automated refrigeration station
2.   Test items
      ● The same items described for the short model cable tests
      ● Contraction: cooldown characteristics
      ● Records of refrigeration operation
      ● Expansion and contraction of cable system
      ● Records of installation (installation methods, installation time, initial cooling)
3.   Final design
      ● Sufficient analysis of the test results from the long-length cable evaluation,
         iteration to improve designs

2.2. SUPERCONDUCTING CABLES

2.2.1. Introduction

As an ultimate long-range solution to bulk power transmission, superconducting cables show considerable promise. Economic and technical feasibility studies have been made on cables capable of transmitting at least 2 GVA and preferably 5 GVA to 10 GVA. Superconducting cables make use of the phenomenon of the superconductivity which certain special metals exhibit below their respective critical temperatures. Superconducting materials are those which lose their electrical resistance entirely when cooled to within a few degrees of absolute zero. The idea of using such materials for loss-free power transmission dates back almost as far as the discovery of superconductivity itself. In 1913, only 2 years after Kamerlingh Onnes discovered the effect at Leiden, the editor of the *Electrician* speculated on the possibility of a power transmission line of zero potential drop.[49] Serious consideration was not given to the subject until McFee's first proposal for Pb-conductor cables in 1961 and 1962.[50,51]

Superconducting cables appear to represent the "ultimate" in bulk power transmission systems which may be realized in the future. Such cable may well constitute a major or central part of power systems in the 21st century, recognizing that their present economic break-even capacity is well in excess of 2 GVA (the highest power rating that conventional cables might be expected to attain even by intensive modification). The real potential is for future cables to handle extra-large blocks of power, but at the present time they are under investigation as a competing substitute for conventional bulk power transmission cables. Most investigations are concentrating on rather small-scale superconducting cables.

The chief advantage of superconducting cables lies in their relative compactness which arises as a consequence of their higher power densities compared to conventional cables. Contrary to early expectations, their zero electrical resistivity does not constitute a major saving of transmission losses since this is partly offset by the need for refrigeration power.

Both DC and AC superconducting cables have received attention as R & D projects. The former is the more efficient method for the transmission of large blocks of electric power because no conductor loss is anticipated. With the latter, there will be some AC loss in the superconductors, giving rise to complicated problems in conductor design. As yet no DC power networks exist, though there are a number of individual lines. Present cost of conversion equipment at line terminals mandates relatively long distances for such installations if they are to be economically viable, unless they are asynchronous ties. The use of AC, on the other hand, is more readily accepted, because it requires no modification of present power systems and does not need a long transmission distance to be justified.

All superconducting cable designs so far proposed comprise at least two subsystems; the conductor system and the thermal envelope. In this respect they are similar to the cryoresistive cables described in Section 2.1. Superconductors must be cooled to around 4 K, thus double thermal insulation is attractive. This is accomplished by providing a heat (or radiation) shield, kept at liquid nitrogen temperature, between the liquid helium and the outside ambient air. Unlike the situation with cryoresistive cables, multilayer insulation is preferred for the thermal insulation between the liquid nitrogen and ambient temperatures.

The following three concepts for cable and envelope assembly are now under consideration:[5]

1.   The rigid or pipe type concept — Conductors and envelope are both rigid. This is a superconducting version of the rigid, compressed-gas insulated cable. In this case, helium serves the dual purposes of coolant and electrical insulation. Three concentric pairs of conductors form a three-phase circuit: the rigid pipes are contained within a thermally insulated envelope. Such a coaxial design may also be applied for DC superconducting cables. This approach allows fabrication of lengths of about 20 m only and, like rigid CGI cables, entails many joints. To accommodate cable contraction during cool-down it is necessary to install bellows or use materials with low thermal contraction coefficients (such as invar).

2.   The semiflexible concept — The thermal envelope consists of rigid tubes with thermal-contraction-compensating bellows. The conductor system is flexible and comprises either corrugated tubular conductors or flexible, hollow conductors made of wires or strips helically wound on a carrier. This structure is similar to that of pipe-type OF cables. It is fabricated in lengths of about 200 to 500 m.

3.   The completely flexible concept — Both the conductor system and the thermal envelope are flexible. The thermal envelope is made of corrugated tubes, while the conductor system is like the semiflexible design just described. Thus it is roughly analogous to the conventional oil-filled, flexible cable. Flexible cables can be reeled and installed in lengths up to 300 m (possible 1000 m), which greatly reduce the number of field-fabricated joints. Another advantage of flexible cable is that it can be designed to accommodate the contraction of the various parts as the cable is cooled.

2.2.2. Superconductor Materials and Conductor Configuration

A. Superconductivity

Superconductivity is a phenomenon that causes certain metals to suddenly lose all electrical resistance when their temperature is lowered to near absolute zero. An essentially perpetual current can be induced in a loop of such material. The phenomenon is accompanied by marked changes in the magnetic properties of the material.

Understanding superconductivity has taken time; it proceeded in several historical stages. A certain degree of understanding comes directly from thermodynamic considerations. Many attributes can be described by London's and Landau-Ginzburg's equations. A successful explanation in terms of quantum theory was given by Bardeen, Cooper, and Schrieffer for Type I superconductors. This is now called the BCS theory. Ginzburg, Landau, Abrikosov, and Gorkov, in what is called the GLAG theory, give an explanation for Type II superconductors.

The zero-resistivity property of superconductors is interpreted in terms of the behavior of electron pairs called Cooper pairs. One electron interacts attractively with another electron through their mutual interaction with phonons, thereby overcoming the inherent repulsive force between them. The resulting Cooper pairs are separated from each other by a small but measurable energy gap. The pairs obey Bose-Einstein statistics in contrast to electrons which obey Fermi-Dirac statistics. According to the BCS theory, the critical temperature for superconductivity above in which superconductivity disappears, depends on the electron density at the Fermi energy level E_F of the metal and the attractive inter-electron interaction U. When $UN(E_F) \ll 1$, this temperature is given by

$$Tc = 1.14 \; \theta \exp(-1/UN(E_F)) \qquad\qquad (2.2.1.)$$

where θ = the Debye temperature.

Besides the critical temperature, there is another limiting factor, the critical magnetic field. A sufficiently strong magnetic field will disrupt the superconductivity. Even without an external magnetic field, quenching of the superconductivity can take place due to the self magnetic field induced by current flowing through a superconductor, if the field is high enough. The current at which this occurs is termed the critical current, with a corresponding I_C, and critical current density, J_C. The critical magnetic field, in general, decreases with increasing temperature. In summary, superconductivity can exist only between certain limits of current, magnetic field, and temperature.

Superconducting materials are usually classified as Type I or Type II according to their magnetic behavior. Magnetization takes place in such a way as to expel magnetic flux from superconducting materials; this is called the Meissner effect. There are two modes as shown in Figure 2.2.1. The first exhibits the perfect Meisner effect, or the perfect diamagnetism, up to its critical magnetic field H_C. In the diamagnetic state, induced electric currents circulate in a very thin surface layer (a few hundredths of a micron thick) so that all internal fields are exactly canceled and the interior is electromagnetically screened from the outside world. Such materials are Type I superconductors, or soft superconductors; they are chiefly some pure metals and a few alloys. The second group is Type II superconductors, or hard superconductors. The magnetic behavior of Type II superconductors is different from Type I. They obey the Meissner effect within a relatively narrow range of magnetic field; up to H_{C1} as indicated in Figure 2.2.1. Above H_{C1}, they expel the magnetic field partially, but they are still superconductors. In other words, they allow currents to penetrate relatively deeply into the bulk of a sample once the critical surface-current density has been exceeded. The resulting internal magnetic flux is accommodated in an array of microscopic filaments of nonsuperconducting or ''normal'' materials, each of which is wrapped in a vortex of current. Much greater magnetic fields, represented by H_{C2} are generally required for the complete quenching of superconductivity in Type II materials than in Type I materials. High critical-current densities and low losses for alternating currents in the mixed state depend on the presence of compositional inhomogeneities or defects which can ''pin down'' the vortex array and counter its tendency to drift.

The essential distinction between the two kinds of superconductors is found in the length of their mean free paths for normal conduction electrons at low temperatures. Superconductors belonging to Type I have a coherence length larger than the magnetic flux intrusion length. This is characteristic of almost all pure metal superconductors. The reverse is true for Type II superconductors, which are often alloys.

Except in DC superconducting cables, AC losses are of prime importance. They arise from hysteresis characteristics of magnetization. If the magnetization curve is reversible, there will be no hysteretic effect and the losses in the superconductor will be zero. This condition is very nearly achievable with some materials provided conditions are proper and the frequency slow. The very low loss region is confined to the perfectly diamagnetic or Meissner part of the magnetization curve at low fields. At high fields, where only Type II

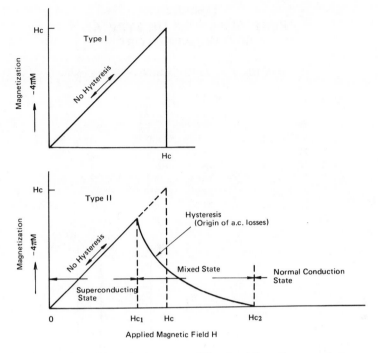

FIGURE 2.2.1. Magnetization curves of Type I and Type II superconductors.

materials remain superconducting, the hysteretic loss depends on the current carrying capabilities of the conductor and its physical dimensions.

B. Superconducting Materials

Superconductivity has been recognized in at least three types of material: (1) pure metals, (2) alloys (physical mixtures of metals), and (3) compounds (chemically bonded elements). The first group is Type I superconductors, while almost all within the other two are Type II. Properties of selected superconductors are shown in Table 2.2.1.

One of the most important parameters to be considered when choosing a superconductor for a cable application is the temperature to which it must be cooled, since a few additional degrees of cooling can easily double the costs of refrigeration. It is therefore desirable to keep the operating temperature as high as possible.[49] Compounds and alloys are generally the answer to this requirement, as is clearly seen in Table 2.2.1. These materials can also carry higher currents. For these reasons, Type II superconductors, such as NbTi, Nb_3Sn, and V_3Ga are usually selected for DC cable applications. Compounds Nb_3Sn and V_3Ga are commercially available in tape form and can be produced in filamentary form by solid-state diffusion. The alloy NbTi is available in a great variety of configurations (and compositions) ranging from rods, to foils, to fine filaments only a few microns in diameter.[54] The ternary alloys Nb-Ti-Zr and $Nb_3(AlGe)$ are presently under investigation.

The choice of superconducting materials for AC cables entails cognizance of the magnetic hysteretic loss. Compounds preferred for DC applications exhibit an unacceptable hysteretic loss; consequently attention has been given to Type I superconductors. Pure niobium is favored because of its high H_{C1} (0.126 Tesla at 5 K), high critical temperature Tc (9.2 K at H = 0) and low AC losses. The compound Nb_3Sn is another candidate for AC applications. It has higher AC losses than Nb, but due to the high Tc ($\sim$18 K), the operating temperature of the cable can be raised, thereby reducing the required cooling power.

C. System Requirements for Superconductors

It is important to define the critical parameters for superconducting materials as applied

Table 2.2.1.
PERFORMANCES OF TYPICAL
SUPERCONDUCTORS

Material	Tc(K)	Hc(kG)	Jc(A/cm²)
Sn	3.7	0.2	
Pb	7.2	0.8	
Nb	9.2	1.5	$\sim 10^4$
Nb Zr	11.0	100	$\sim 10^5$
Nb Ti	9.7	120	$\sim 10^5$
Nb₃Sn	18.2	220	$\sim 10^6$
V₃Ga	14.6	220	$\sim 10^6$
Nb₃Ge	23.2	>350	
Nb₃(Al₀.₂G₀.₈)	20.7	350	

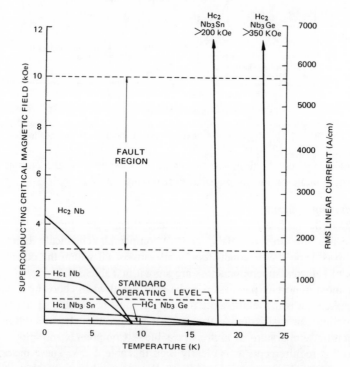

FIGURE 2.2.2.   Projected operating and fault levels for a superconducting power transmission line and their relationship to the properties of some conductors.[55]

to superconducting cables. These are reasonably clear for an AC cable.[55] For the superconductor or superconducting composite, AC losses must be below ($< 10\ \mu W/cm^2$) at the field level required. In this regard, system considerations dictate a linear current density of about 550 A/cm rms, which corresponds to a peak surface field of about 1000 Oe. At the same time it is necessary to provide adequate fault capacity, i.e., the ability to carry three to ten times the rated current for short periods of time and then recover successfully.

Figure 2.2.2. illustrates how well the properties of superconducting materials of practical interest match the above requirements.[55] It also indicates two possible approaches for conductor design. The first is to use Nb, which has the largest H_{C1} of any superconductor known, but a relatively low critical temperature of only 9 K, and to operate at temperatures below 6 K so that $H < H_{C1}$. H_{C2} exists in practical niobium, and is actually very low as is indicated in Figure 2.2.2. Therefore, the superconductor alone cannot handle fault currents, a parallel normal conducting path must be incorporated for this purpose. If the helium refrigerant is supercritical (> 5 K), the temperature range (5 K $< T < 6$ K) in which Nb superconducting cables must operate seems inconveniently narrow.

An alternate approach is to use one of the so-called A15-type superconductors such as Nb_3Sn or Nb_3Ge, which will hopefully allow higher and less constrained operating temperatures, and facilitate the handling of fault currents with superconductor alone. Problems may arise from their hysteretic losses and brittle nature.

Type II superconductors are advantageous for DC superconducting cables. They can carry about 10^5 A/cm^2 as their critical current under normal operating conditions, which requires a cross section of a few square centimeters (for currents of the order of 10^5 A and a safety factor of 2 or 3). DC superconducting transmission lines store magnetic energy in the order of 5 MJ/km. Disposing of such an amount of energy from a 1000-km line is a far from trivial problem. To minimize this energy, the inductance should be kept low. This may be achieved by causing the current to flow in opposite directions along flat closely spaced ribbons or strips, or in a coaxial configuration.[54]

D. Conductor Configurations

Superconductors are used in the form of thin surface layers, 25- to 50-μm thick, with a normal metal (Al or Cu) acting as the substrate. The arrangement which is shown in Figure 2.2.3.(A) is needed for structural reinforcement and electrical stabilization. It can be fabricated by vacuum vapor deposition or by plasma plating. Should the superconductor quench, the substrate must carry the total current. Furthermore, it must bypass a high-fault current if such an event occurs under fault conditions.

Superconducting cable core structures are generally coaxial, with the superconductor on the outside of the inner pipe and on the inside of the outer pipe, as illustrated in Figure 2.2.4.[49] External electromagnetic fields and forces can be eliminated for a superconducting cable core by arranging the phase and neutral conductors as a coaxial pair. Cylindrical symmetry also minimizes surface current densities in the superconducting layers, which are required on the outer surface of the inner conductor and on the inner surface of the outer conductor.

Coaxial conductor systems can be made out of rigid tubular conductors, corrugated tubular conductors, flexible hollow conductors, or flexible hollow conductors built up of wires and strips. Rigid versions have been under development for nearly a decade. One obvious limitation of such a design is conductor length which must be kept short (say 20 m) for reasons of transportation. It follows that such a design would need a lot of jointing work in the field at the time of installation. A second technical problem would be thermal contraction. Various kinds of discrete contraction elements such as bellows, "omega" bends, or even sliding joints could help, although as yet, none of these has in fact been tried. A different concept is being explored.[57] The conductors are bonded to a layer of a low-contraction alloy (invar) twice as thick as the copper, in a Nb/Cu/invar tricomposite form, in which the invar dominates the mechanical and thermal properties of the composite as shown in Figures 2.2.3.(C) and 2.2.5.(A). Another approach to the contraction problem is to make use of conducting tubes with continuous helical corrugations, which can absorb thermal contraction by local flexing of the metal. A corrugated design would not, of course, be able to use niobium tin for its superconducting layers.

A third and increasingly popular scheme is to form the tubular conductors from wires or strips laid in a helical fashion about the cable axis as shown in Figure 2.2.5.(B).[49] Thermal contraction now appears as a shrinking in diameter of the helix, while its length is easily held constant. Like corrugated tubes, helically wound conductors can provide cores which

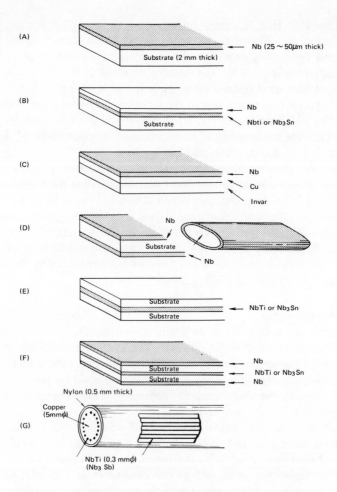

FIGURE 2.2.3. Various structures of conductors for superconducting cables.

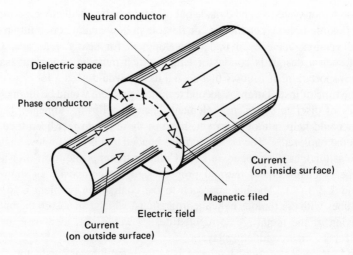

FIGURE 2.2.4. Basic structure of superconducting cable core.[49]

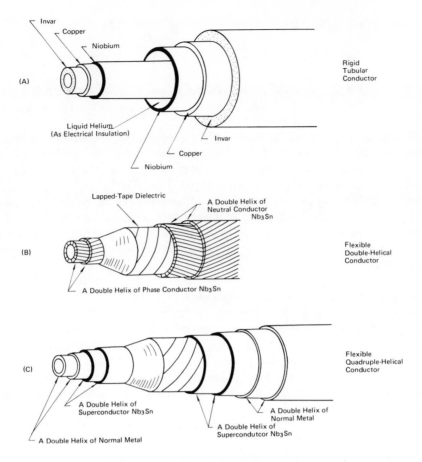

FIGURE 2.2.5. Several superconductor designs.

are sufficiently flexible to be manufactured in lengths of a few hundred meters and wound onto large drums for transportation. They would therefore need fewer jointing operations in the field, and in addition they would cope readily with bends along the cable route.

With AC superconducting cables, the current distribution on the conductors requires that the helical strips be coated with superconducting layers on both sides, as shown in Figure 2.2.3.(D) to preclude resistive losses in the copper substrate. This sandwich structure is acceptable for Nb but not for niobium tin because of its fracture strain. One solution might be to reverse the sandwich and put the niobium-tin layer in the middle as shown in Figure 2.2.3.(E), but this structure might produce considerable power dissipation in outer normal metal layers which would be exposed inevitably to the magnetic field of the current in the superconductor. This may possibly be avoided by using the type of composite shown in Figure 2.2.3.(F).

A simple helix generates an axial magnetic field proportional to the number of turns per unit length and the total current. An analysis of the behavior of the magnetic field in this configuration indicates the following three conditions should be satisfied simultaneously:[52]

1. Zero field in the cable interior, achieved when

$$P_1 = P_2 \qquad\qquad (2.2.1.)$$

2. Zero flux enclosed by the cable, achieved when

$$P_1/d_1^2 = P_2/d_2^2 \qquad\qquad (2.2.2.)$$

3. Equal radial contraction of both conductors, achieved when

$$P_1/d_1 = P_2/d_2 \qquad (2.2.3.)$$

where P_1 and P_2 = the pitches of the helices, and d_1 and d_2 = the diameters (the subscript 1 and 2 denote inner and outer conductors). In general, cable made from single-layer helical conductors will satisfy only one of these conditions.

With equal pitches or equal lay angles for the inner and outer conductors there will still be a net magnetic field inside the inner helix so that induced currents will flow on the inside of the inner conductor strips. Losses in the copper can be screened out for niobium-based composite by coating the strips as shown in Figure 2.2.3.(D).

A serious problem arises as a consequence of the imbalance in the total axial magnetic flux because $P_1/d_1^2 \neq P_2/d_2^2$. The nonzero axial magnetic flux will induce voltages and circulate currents in any enclosing metal work. In particular, there may be large eddy current losses in the helium ducting around each core. There are several ways in which this problem can be minimized:

1. By putting all three cores into a single duct
2. By reducing the lay angle of the outer conductors
3. By dividing each conductor into two separate helices

The last approach, the double helix, has received the most attention.

A double helix appears to provide a promising conductor configuration. The two layers would be of equal pitch but opposite sense. Each pair alone generates no axial magnetic field which makes it possible to satisfy thermal contraction conditions when choosing the pitch. However, a new problem may arise in this configuration: current flows on both surfaces of any one layer in each conductor making it necessary to have both faces superconducting. Recognizing the limited strain that Nb_3Sn can tolerate, an eight-layer configuration has been proposed[52] (see Figure 2.2.5.[C]). It consists of eight concentric helices arranged in pairs. Proceeding from the inside out, these would be a core comprising two helices of normal metal having opposite pitch; two helices of superconductor having opposite pitch, and finally, an outer conductor consisting of the same double pairs in reverse order.

The structure of DC superconducting cable cores is much simpler than that just described because in this case AC losses need not be considered. A simple coaxial construction should suffice. Care must be taken to avoid AC losses in the superconductors caused by AC ripple on the DC voltage. Very fine multicored wires of NbTi are available as shown in Figure 2.2.3.(G).[61] They can be stranded in such a manner as to provide alternate positive and negative polarities thereby cancelling magnetic field effects. A higher current density is achieved as a result.

E. Development of Low Loss and High Critical Current Materials of Type II

It was thought at one time that AC losses of Nb_3Sn would be unacceptably high for AC cable applications, but now it appears that they can be reduced to an acceptable level by suitable surface treatment such as chemical etching, electropolishing, and mechanical polishing. It is reported[57] that a significant reduction in the losses is attained through removal, by polishing of a porous outer layer of Nb_3Sn which forms at the Nb_3-Sn-liquid-Sn interface and extends into the material. The losses can be made smaller than 1 μW/cm^2 at 500 A/cm rms, so satisfying one of the system requirements for superconductors.

For $H_{C1} < H < H_{C2}$, the critical current density J_C is not an intrinsic property of a superconductor, but is governed by flux pinning. When large critical currents are obtained through flux pinning, flux penetration is minimized above H_{C1} and the hysteretic losses are

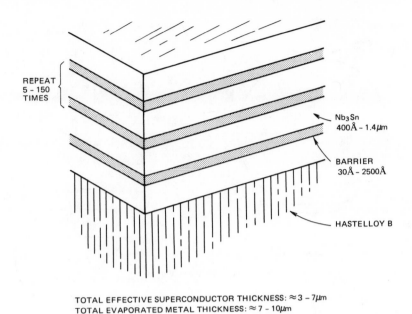

REPEAT
5 - 150
TIMES

Nb₃Sn
400Å – 1.4μm

BARRIER
30Å – 2500Å

HASTELLOY B

TOTAL EFFECTIVE SUPERCONDUCTOR THICKNESS: ≈ 3 – 7μm
TOTAL EVAPORATED METAL THICKNESS: ≈ 7 – 10μm

FIGURE 2.2.6.   Structure of a layered superconductor.[55]

likewise minimized. In order to achieve the desired strong flux pinning in Nb_3Sn, fine-scale layered composites are now proposed, which consist of layers of Nb_3Sn separated by flux pinning barrier layers of normal or weaker superconducting material as is schematically shown in Figure 2.2.6.[55] Because of the discontinuity in the superconducting properties that they provide, the barrier layers pin the vortices. The pinning effect is affected or determined by the deposition conditions, the type of barrier material, and the respective thicknesses of the superconducting and barrier layers. Successful results have been obtained with composite Nb_3Sn conductors with 90 Å yttrium barriers,[55] recognizing that the magnetic penetration depth and coherence length of the superconductor itself are 1580 Å and 50 Å, respectively. The DC critical current increases with increasing boundaries per unit length (10 to 50 boundaries per micrometer). At 60 Hz, instabilities arise in this structure which reduce the AC critical current, and so improved AC stability is required.

2.2.3. Electrical Insulation

As mentioned in Volume I, Section 2.3.7. and Section 2.1.2., electrical insulation at cryogenic temperatures can be vacuum, gas, liquid, or lapped polymer tape impregnated with liquid. For superconducting cable use, the preferred forms of electrical insulation are vacuum, liquid or supercritical helium, and lapped plastic insulation impregnated with helium.

For an AC superconducting cable, the prime consideration in selecting a dielectric material is dielectric loss, because other thermal losses are by design extremely low. A tanδ of < 10^{-5} is required at operating temperature and of course voltage. Dielectric strength is another important quantity. It should be as high as possible. Vacuum and helium have the advantage of an extremely low tanδ (less than 10^{-6}). Vacuum, as already discussed, can only be used with rigid or corrugated tubular conductors, and regrettably requires absolute leak-tightness of the system. Liquid helium, as pointed out in Volume I, Section 2.3.7., has a low dielectric strength, which leads to design problems. The most promising electrical insulation seems to be lapped plastic tapes impregnated with liquid helium, or even better, with supercritical helium. There is a variety of polymers available for which tanδ is around 10^{-5} at 4.2 K, but their mechanical and taping characteristics are a serious problem.

For DC superconducting cables, tanδ is of no consequence and therefore the choice of the dielectric material is dictated primarily by dielectric strength.

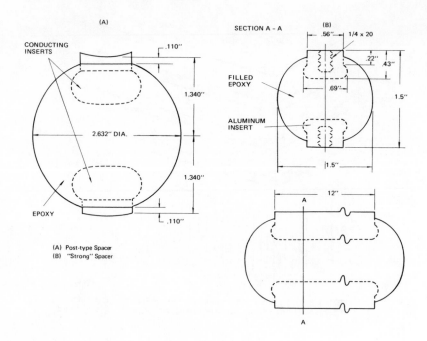

FIGURE 2.2.7.   Cross section of epoxy insulating spacers for rigid type superconducting cables.[58]

A. Liquid Helium

When liquid helium is used as electrical insulation, spacers are required to hold the cable cores in their correct positions in the cryogenic envelope. Pressurized liquid helium and supercritical helium are worthy of investigation.

Spacers mounted between each conductor and its respective shield must serve two functions. First, they must maintain the coaxial alignment of the conductors and shields (see Figure 2.2.3.[A]) and be capable of supporting the 120- (or 100-) Hz radial forces which result from any misalignment of the conductor within the shield. Second, they must be able to supply sufficient axial force to the niobium-copper-invar composite conductor to maintain a constant length during cycling between room temperature and the operating temperature of the line. Two types of spacers proposed to satisfy the above requirements are illustrated in Figure 2.2.7.[58,59] The high-voltage insulation is provided by heavily filled epoxy to match the thermal contraction of the electrically conducting inserts between room temperature and 4.2 K.

B. Lapped Plastic Insulation

The essential features of a lapped-tape insulation, suitable for a superconducting cable are shown in Figure 2.2.8.[49] At present, polyethylene is favored as a tape material because of its low tanδ, high dielectric strength, cost, and availability, although taping performances as it affects flexibility of a finished cable leave much to be desired. To give better flexibility, the tapes should be between 70- and 150-μm thick and be wound with small gaps between the adjacent turns in each layer. The mechanical properties can be tailored by orienting the tapes. Uniaxially or biaxially oriented polyethylene or polypropylene tapes should be considered.

Lapped insulation is impregnated with helium. Its insulating properties derive, therefore, in part from the high intrinsic electric strength of the polymer used and in part from the electrically weaker helium contained in the butt spacings. Electrical performance of this type of insulation with respect to AC and impulse discharge inception stresses is presented

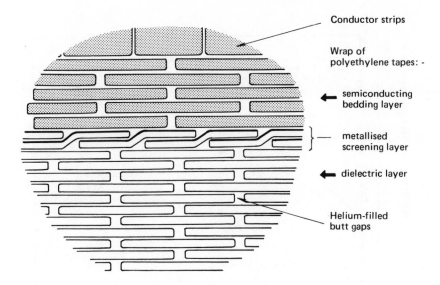

FIGURE 2.2.8.   Lapped plastic insulation for flexible superconducting cable.[49]

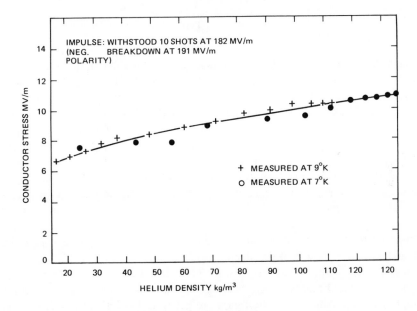

FIGURE 2.2.9.   Variation of discharge inception stress with helium density for 4 layers 101.6 μm valeron-type 4096.[60]

in Figure 2.2.9. and Figure 2.2.10.[60] Impulse voltages cause the inter-tape gaps to break down, transferring all the stress to the tapes themselves, which can withstand a short-duration impulse relatively easily. For normal operation, however, discharges in the gaps must be avoided, since over an extended period, such discharges could lead to serious deterioration of the polymer.[49]

Electrostatic screens are necessary to minimize the effects of enhanced stresses at the edges of the conductor strips. These screens consist of intercalated layers of one-face-metalized plastic tape. Direct contact between bare metal (cable conductor) and helium will reduce the breakdown strength. Thus, screens should ensure that in regions of high voltage there are no metal-helium interfaces. Double-depth butt gaps are likely to form between

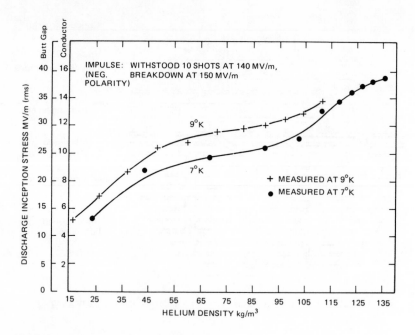

FIGURE 2.2.10.   Variation of discharge inception stress with helium density for 12 layers of 38-μm Mobil® Bicor No. 240-B.[60]

intercalated layers and the first layer of insulation, causing a deterioration of the electrical performance of the insulation system. It has been experimentally proven that use of thin tape without butt gaps adjacent to metal electrodes improves the performance.

It is claimed that by careful arrangement it is possible to avoid butt gaps adjacent to the metal.[49] A bedding layer of semiconducting tape, as shown in Figure 2.2.8., supports the screen across the gaps between conductor strips, thus allowing the screen to follow the conductor potential, while preventing a direct contact which could provide closed paths for induced currents with attendant additional heating within the core.

The breakdown strength of a supercritical helium-filled gap depends on the gap depth as well as gas density (see Figures 2.2.9. and 2.2.10.). The discharge inception stress increases with decreasing tape thickness down to 80 μm. For thinner tape (less than 50 μm), there is a tendency for the tapes to fall into the butt gap so giving rise to a nonuniform field region and an intermediate butt-gap depth.

Embossing of polymer tapes may be useful in facilitating impregnation by liquid or supercritical helium. Embossed tapes have a very low discharge inception stress coupled with high discharge magnitudes. Butt-gaps are not well defined and discharges can occur at tape interfaces where gaps vary from zero to twice the embossing depth. Nevertheless, it appears likely that as long as the surface defects are less than a quarter of the tape thickness and small in number, their effects will be negligible.

2.2.4. Superconducting Cable Designs

A. AC Superconducting Cable Designs

Table 2.2.2. lists design parameters of several prototype AC superconducting cables.[5]

Figure 2.2.11. illustrates the conceptual design of the rigid superconducting transmission system being developed by the Linde Division of the Union Carbide Corporation.[62] Three concentric pairs of conductors form a three-phase circuit; the rigid pipes are contained within a thermally insulated envelope. Table 2.2.3. indicates the dimensions of the cable core.[63] The conductor and shields are niobium-copper-invar tricomposites as will be seen from

Table 2.2.2.
PERFORMANCES OF PROPOSED AC SUPERCONDUCTING CABLES

Company or laboratory	CERL	Siemens	CGE/EdF	ATF	Krzhizhanovsky Institute	Furukawa	BNL	Linde Union Carbide (UCL)	BICC
Rated voltage (line-to-line) (kV)	132 275	120	140 180	110	35	154	132	138 230 345	33
Rated current (kA)	6.1 8.5	12	12.4 16	2.65	10	3	13.8	7.1 11.8 17.75	13
Rated power capacity (MVA)	(1400) 4000	2500	3000 5000	500	600	1000	3000	1690 (3400) 4710 10590	750
Principles of design	Semiflexible three-phases, triaxial arrangement, Rigid tubular conductors / Helically wound conductors	Semiflexible three phases, triaxial arrangement Helically wound hollow conductors	Semiflexible three phases	Totally flexible three phases triaxial arrangement, corrugated tubes		Rigid coaxial conductor pairs	Semiflexible coaxial conductors, helically wound	Rigid coaxial conductor pairs rigid tubes of Invar® Cu-Nb composite	Rigid concentric tubes, triaxial arrangement
Conductor Superconductor	Nb	Nb	Nb	Nb	Nb	Nb (foil)	Nb₃Sn (ribbon)	Nb	Nb (foil)
Stabilization material	Cu/Al (Cu)/Al	Al/Cu	Cu	Cu/Al		Cu	Al/Cu	Cu	Al
Linear current density on inner conductor (A/cm)	400 340	550	520 555			240	320	580 580	190
Electrical insulation	Wrapped plastic multilayer foil (PE-tape) + He	Wrapped plastic foil	Wrapped foil insulation (PE)	He (10 atm)		Wrapped plastic foil	He impregnated tape wrap	Supercritical helium solid dielectric spacers	Vacuum helium
Cryogenic envelope	LN₂-Shield superinsulation steel pipe	LN₂-Shield (Invar®) superinsulation	LN₂-Shield (Invar®) alumina powder, vacuum steel pipe	LN₂-Shield vacuum flexible Dewar		LN₂-Shield superinsulation	LN₂-Shield superinsulation	He-gas cooled Cu-shield superinsulation	LN₂-Shield superinsulation
Overall diameter of cable (cm)	46.5	~50	60.4 70.8	25			42	~34 (60) ~47 ~63	
Losses Cable (kW/km)	87	85 (100)					211 (6.2—8.2 K)	(99)	
Per terminal (kW)	~125	50—100 (150)						(75—150)	
Heat in-leak at 4.2 K: Cable (W/km)	288 410	200	300 400	300			1314 (6.2—8.2 k)		
Per terminal (W)	~300	~400					161		
Cryogenic performance coefficient (W/W)									
Comments	8 m-Long 1-phase loss measurements actually tested model	30 m-Model cable, one phase (120 kV, 12 kA)	18 m Full scale cryogenic envelope		12 m-Long one phase test model	7 m-Long test section, current tests	2 × 20 m Long flexible cryogenic envelope	7 m-Long test facility for AC measurements	Historic 3 m-test cable 2080 A 1967, 1969 end of program

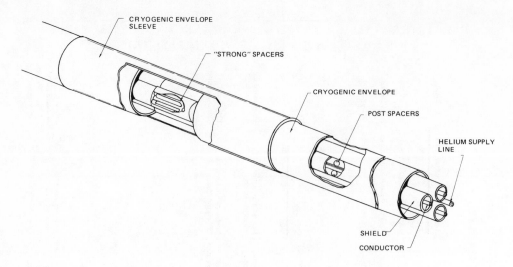

FIGURE 2.2.11.   Conceptual design of a rigid superconducting cable.[62]

Table 2.2.3.
CONDUCTOR AND SHIELD
DIMENSIONS FOR A 138-kV, 3400-
MVA SUPERCONDUCTING CABLE
OF RIGID TYPE[63]

	Conductor		Shield	
	(in.)	(cm)	(in.)	(cm)
O.D.	3.150	8.00	8.662	22.00
I.D.	2.954	7.50	8.562	21.75
Wall	0.098	0.249	0.050	0.127
Niobium	0.002	0.005	0.002	0.005
Copper	0.032	0.081	0.016	0.041
Invar®	0.064	0.163	0.032	0.081

Figure 2.2.3.(A). Insulating epoxy spacers are employed to maintain the coaxial conductor-shield configuration. A set of three small post spaces, typically located at 1-m intervals along the length of the line, withstand the radial forces. A single "strong" spacer, designed to support the axial load resulting from the thermal contraction of the conductors, is located in each section of about 20 m at most (see also Figure 2.2.7.). Helium at a temperature around 4.6 K and a pressure of about 5 atm maintains the desired operating temperature of the system and provides the electrical insulation between each conductor and its shield. The helium is supplied from closed cycle refrigeration stations located 8 km apart along the length of the cable system, with each refrigeration station serving a 4-km length on each side of the station. Some dimensions and electrical characteristics of the cable system under development are given in Table 2.2.4.[64]

Figure 2.2.12. shows a sketch of a CERL design[49] for a 5-GVA, 400-kV superconducting cable core of the flexible type, using niobium-clad copper-strip conductor with a maximum surface-current density of 40 A (rms) per millimeter at rated current. The dielectric is lapped polyethylene tape with a maximum electric stress at the inner conductor of 10 MV/m (10 kV/mm). The outer duct is a corrugated stainless-steel pipe. This core could remain operational after carrying the maximum anticipated through-fault current of 30 kA for a total of about 500 msec in any 5-hr period.

Table 2.2.4.
ELECTRICAL CHARACTERISTICS OF SUPERCONDUCTING CABLES OF RIGID TYPE WITH HELIUM DIELECTRICS[64]

Voltage (kV)	138		230		345		500	
BIL, kV	450	600	600	900	750	1050	900	1200
I, kA	10.6	14.1	14.1	21.2	17.7	24.8	21.2	28.3
Rating, maximum continuous (MVA)	2536	3380	5333	8449	10559	14792	18368	24491
Resistance under fault[a]	5×10^{-5}		4×10^{-5}		8×10^{-6}		$4 \times {}^{-6}$	
Resistance with rated current[a]	3×10^{-8}		8×10^{-9}		2×10^{-9}		1×10^{-9}	
Ind. reactance[a]	6.37×10^{-4}		2.29×10^{-4}		1.02×10^{-4}		4.85×10^{-5}	
Shunt capacitance (MVAR/mi)	0.6426		1.785		4.016		8.435	
Surge impedance loading (MVA)	318		882		1985		4169	

[a] P.U. mile on 100 MVA base.

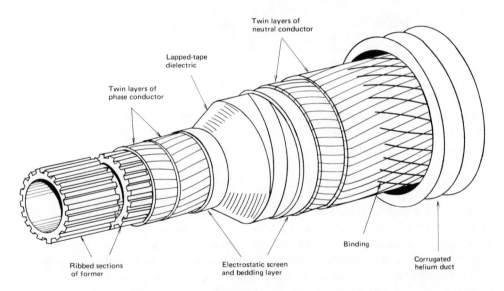

FIGURE 2.2.12.   Design of a 400-kV, 5-GVA superconducting cable core of flexible type.[49]

A superconducting cable design proposed by Brookhaven National Laboratory (BNL) is similar to this. A single-helical structure or a double-helical structure may be employed for the superconductor configuration. The inner conductor is supported by a steel or aluminum helix and the outer conductor is contained by flat steel or aluminum armor. The cable is cooled from inside the inner conductor and from outside the armor. Helium fills the interstices of the insulation.[53] To complete a single-phase superconducting cable, it can be installed in a cryogenic envelope developed by Kabelmetal, as shown in Figure 2.2.13.[65] This consists of four concentric corrugated pipes with multilayer insulation serving as thermal insulation. The middle annulus acts as an intermediate temperature shield. The inner and outer annuli must be continuously evacuated.

Three phases can be contained in one cryogenic envelope. An example of this in cross section is illustrated in Figure 2.2.14.[66] and Figure 2.2.15.[5,67] In the CGE/EdF design,

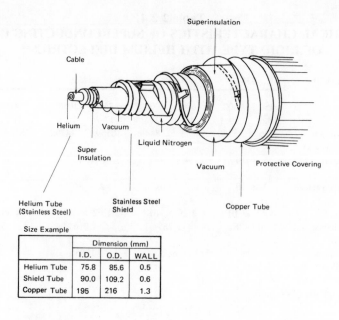

	Dimension (mm)		
	I.D.	O.D.	WALL
Helium Tube	75.8	85.6	0.5
Shield Tube	90.0	109.2	0.6
Copper Tube	195	216	1.3

FIGURE 2.2.13. A flexible superconducting cable design. (From Coitoru, Z., Dechamps, L., and Schwab, A. M., *Entropie,* N0.56, Mars-Avril 1974, p. 52. With permission.)

evacuated alumina powder is used for thermal insulation between ambient temperature on the steel pipe and liquid nitrogen temperature on the radiation shield. Vacuum serves as thermal insulation between liquid nitrogen and liquid helium temperatures. The three cable cores are cooled individually in the same manner, each core being refrigerated both internally and externally by the ''go'' flow. The helium returns through a separate pipe. In the CERL design, each phase of the cable is cooled by internal helium flow. The radiation shield is cooled by liquid nitrogen located in eight ducts, four ''go'' and four ''return'' ducts. The whole system is enclosed in a single steel pipe, 465 mm in outer diameter.

An AC cable with all three phase conductors arranged in coaxial form is possible. Such an assembly would require that the three tubular conductors be vacuum-insulated electrically, while the outer conductor be so-insulated both electrically and thermally. To have complete field compensation for a symmetrical load, one phase must be subdivided into two phases, so that the currents in succeeding coaxial conductor pairs are phase-shifted by 180°. This would require additional phase shifters. Slight axial misalignments of the conductors would generate vibratory forces which, during faults, would be several times the weight of the conductors.[5] The all-coaxial conductor arrangement is the most compact design, but it would be difficult to assemble.

A somewhat different design concept has been devised by ATF. It is a 110-kV, 500-MVA flexible AC cable, cooled by helium flow at around 4 bar. Electrical insulation is provided by maintaining the helium pressure at about 10 bar so as to take advantage of its higher dielectric strength at that pressure. Corrugated tubes are employed to give flexibility to the finished cable line.

B. DC Superconducting Cable Designs

As long as one is really making use of superconductivity, earlier discussions have pointed up a number of substantial advantages of using DC. First, no AC losses occur in steady state operation, and no superconducting screen is needed. These facts lead to much simpler conductor designs than in the AC case. A further advantage is that there is no dielectric

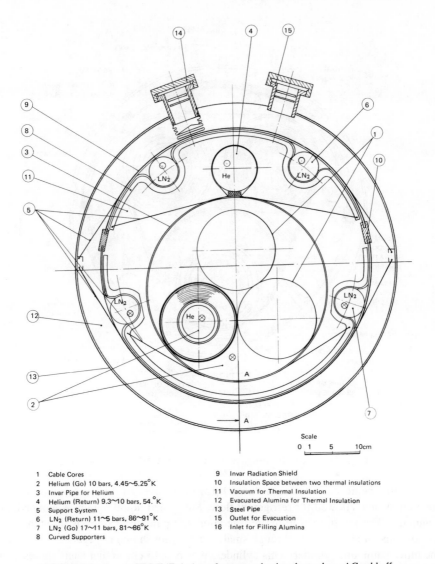

1	Cable Cores	9	Invar Radiation Shield
2	Helium (Go) 10 bars, 4.45~5.25°K	10	Insulation Space between two thermal insulations
3	Invar Pipe for Helium	11	Vacuum for Thermal Insulation
4	Helium (Return) 9.3~10 bars, 54.°K	12	Evacuated Alumina for Thermal Insulation
5	Support System	13	Steel Pipe
6	LN₂ (Return) 11~5 bars, 86~91°K	15	Outlet for Evacuation
7	LN₂ (Go) 17~11 bars, 81~86°K	16	Inlet for Filling Alumina
8	Curved Supporters		

FIGURE 2.2.14.   A CEG/EdF design of superconducting three-phase AC cable.[66]

loss. Consequently, for the same power transmission, the cable diameter of a DC cable is smaller and the thermal losses are lower. Indeed, overall losses of the cable are lower and there is no limitation as far as transmission distance is concerned. However, the potential application for DC superconducting cables seems to be less than for their AC counterparts because of the unavailability of systems where they can be installed and the overall system costs.

Table 2.2.5. lists performances of several proposed DC superconducting cables.[5] Most of them consist of two separate single conductor cables with voltage symmetrical to ground, in a common thermal envelope.

A DC superconducting cable design of coaxial conductor type proposed by Los Alamos Scientific Laboratory (LASL) is illustrated in Figure 2.2.16.[54] The design goal is a transmission capacity of 5 GW at 100 kV. The inner conductor is a nominally 10-cm I.D. aluminum tube of about 0.5-cm wall thickness. There is a 50-μm thick layer of NbTi on the outside surface of the aluminum. This superconducting shell is covered with about 3 mm of Mylar® or Kapton®, which is surrounded by another aluminum cylinder of 2-mm

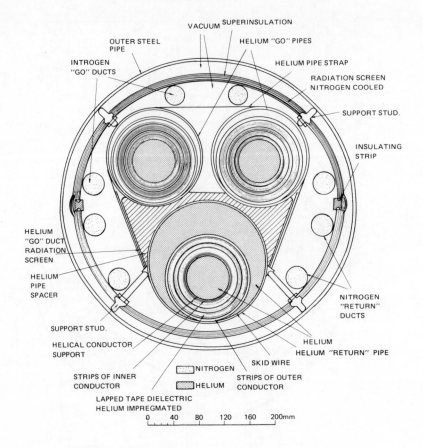

FIGURE 2.2.15. A CERL design of 275-kV, 4-GVA AC superconducting cable.[5]

thickness, this being coated with a 2nd 50-μm thick layer of NbTi. An open annulus of about 1 cm is then allowed for refrigerant flow. Subcooled liquid helium is pumped through this annulus. The return path to the refrigeration and pumping station is through either the innermost pipe or alternatively through smaller pipes (not shown in Figure 2.2.16.) attached to the third aluminum cylinder. This cylinder, which is 15 cm in diameter, serves as a heat shield. The shield itself is situated in a high vacuum; it is wrapped with a 5-cm layer of evacuated superinsulation. The entire assembly is enclosed in a steel pipe about 25 cm in diameter, which remains at ambient temperature.

Figure 2.2.17. shows a semiflexible design for a DC superconducting cable proposed by CERL.[5,67] The electric stress on the inner conductor is 20 kV/mm, and the linear current density is with a critical current density of 6×10^5 A/cm^2 at H = 1.2×10^3 A/cm ($\simeq$ 0.15 T). Carrying current would be 17.4 kA. Figure 2.2.18. shows a CGE/EdF design for a DC superconducting cable, which is obviously similar to the AC cable structure shown in Figure 2.2.14.[66]

Photographs of several designs for AC and DC superconducting cables are shown in Figure 2.2.19.

C. Joints and Terminations

Joints and terminations of superconducting cables do not appear to have received the same attention technically as the cable cores themselves. As mentioned previously, joints are critical in rigid-type cables. Some ideas for cryogenic enclosures currently under investigation are presented in Figures 2.2.20., 2.2.21., and 2.2.22.[59] Figure 2.2.20. is an example of the

Table 2.2.5.
CHARACTERISTICS OF PROPOSED DC SUPERCONDUCTING CABLES[5]

Company or laboratory	AEG-Kabelmetal-Linde	CGE/EdF	LASI	Furukawa	CERL	Siemens
Rated voltage (kV)	±200	±110 / ±140	100	110	230	230
Rated current (kA)	12.5	13.6 / 17.9	50	45.5	17.4	44
Rated power capacity (MVA)	5000	3000 / 5000	5000	5000	4000	10000
Principles of design	Totally flexible parallel single conductors, helically wound	Semiflexible coaxial conductors	Semiflexible parallel and coaxial conductors	Rigid coaxial conductor pairs	Rigid (pipe-type) coaxial conductors helically wound strips	Semiflexible helically wound hollow conductors
Conductor Superconductor	Nb_3Sn (ribbon)	Nb_3Sn / Cu	$Nb_3Sn/Nb_3(AlGe)$	NbTi	Nb-Ti-Zr	NbTi
Stabilization material	Cu		Al/Cu	Cu	Cu/(Al)	Cu
Linear current density on inner conductor (A/cm)		2070 / 2180				
Electrical insulation	Wrapped paper	Mylar®	Kapton® or Mylar® wrap	Wrapped plastic foil	Lapped polymer with He-gas	Wrapped plastic foil
Cryogenic envelope	LN_2-Shield superinsulation	LN_2-shield (Invar®) alumina powder, vacuum steel pipe	LN_2-Shield (Al) superinsulation steel pipe	LN_2-Shield superinsulation	LN_2-Shield superinsulation steel pipe	LN_2-Shield (Invar®) superinsulation
Overall diameter of cable (cm)		27.6 / 30	~25	~30		~45
Losses Cable (kW/km)	70				20	55
Per terminal (kW)	55				63	125—250
Heat inleak at 4.2 K Cable (W/km)	110				66	
Per terminal (W)	120	51 / 64	30		210	
Cryogenic performance coefficient (W/W)					~300	
Comments	Joints constructed, 16 m current tests, 20 m voltage tests					

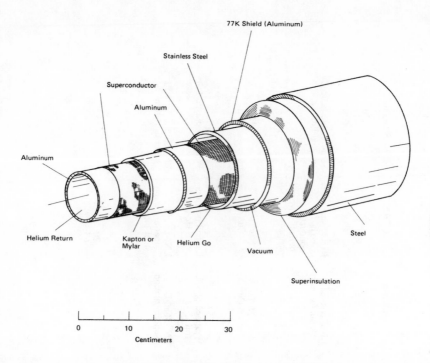

FIGURE 2.2.16.   Coaxial superconducting transmission line.[54]

bellows approach. The conductive heat leak is minimized because of the convolutions which create a long path for heat transfer from ambient temperature to helium temperature.

The ribbed-membrane concept is illustrated in Volume I, Figure 2.4.21. In this design, the bellows are replaced by a thin metal cylinder supported by ribs. Some of these ribs also act as the heat station rings.

Figure 2.4.22. in Volume I shows the supported membrane concept. Epoxy fiberglass reinforced cylinders replace the ribs as structural members, and a thin metal vapor barrier is placed over them. Some of the advantages and disadvantages of these three arrangements are indicated in Table 2.2.6.[59] It is thought that the heat leak can be reduced to 200 W/km of cable at 4.5 K.

A pothead design for a 138-kV superconducting cable termination developed by West-inghouse and Union Carbide[68] is shown in Figure 2.2.23. It uses cast epoxy as the primary electrical and thermal insulation. The vapor-cooled conductor is enclosed in a dewar tube in the center of the epoxy casting. The lower portion of the epoxy serves as the transition from liquid helium to ambient temperature. The upper portion is surrounded by SF_6 gas enclosed in a glass-reinforced epoxy tube.

2.2.5. Refrigeration Considerations

Generally speaking, superconducting cable systems require two refrigerants, helium and nitrogen, so refrigeration is clearly a major component of such systems. A reasonable refrigeration station spacing is 10 km or more. There is no choice other than helium for the principal refrigerant because of the temperature at which superconductors must operate. Since a Nb_3Sn cable could operate at a somewhat higher temperature than an Nb cable, the proposed refrigeration schemes for these two systems differ to some extent. There are essentially three refrigeration methods under consideration: "cascade", "single fluid", and "mixed fluid" systems.[59]

The cascade system is inherently the most efficient cycle since it approaches the ideal

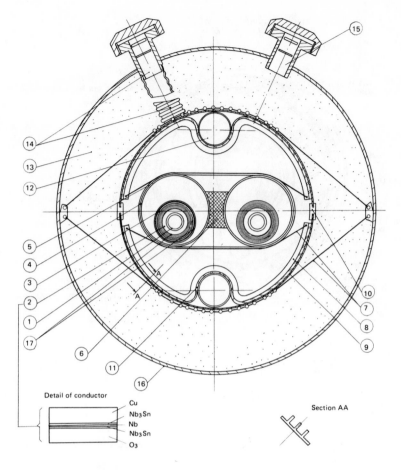

Detail of conductor

Cu
Nb_3Sn
Nb
Nb_3Sn
O_3

Section AA

1 Reinforced Plastic Tube, Perforated, 8mm O.D, 2mm thick	11 LN_2 "GO" Pipe (Invar), 28mm D.P. 1mm thick
2 Nb_3Sn Superconductor Backed by Copper, 3mm thick	12 LN_2 "Return" Pipe
3 Lapped Mylar Insulation 7mm thick	13 Alumina Thermal Insulation, 50mm thick
4 Invar Pipe, 58mm I.D. 1mm thick	14 Outlet for Evacuation
5 Iron Ring	15 Inlet for Filling Alumina Power
6 Support	16 Steel Pipe, 268mm I.D., 3.5mm thick
7 Steel Wires for Suspension	17 Aluminum Screen
8 Curved Supporters	
9 Invar Cover, 168mm O.D. 1mm thick	
10 Epoxy Resin Support & Polyamide Filter for Alumina	

FIGURE 2.2.17. A 230-kV, 4-GVA DC superconducting cable design.[5]

reversible system more closely than does adiabatic expansion using a single fluid, being more commonly used for a liquefier than a refrigerator. But in this system each loop must be completely leak-proof. Moreover, heat exchangers become more complex and a number of compressors are required to service the different refrigerators.

The mixed system involves the mixing of helium with a high molecular weight gas. The mixed refrigerant, with a higher average molecular weight is easy to compress with an efficient and economical air centrifugal machine. Its main advantages are in reducing equipment cost and improving leak-tightness and reliability. However, the heavy gas component added to the helium during the compression process needs to be separated to avoid contamination in the cold sections of the refrigerator.

A single-fluid system uses helium alone as a refrigerant. The corresponding refrigeration cycle is simple, allowing the use of standard heat exchangers and a "cold box" design. On the other hand, it may require some compromise as regards compressors: generally reliable

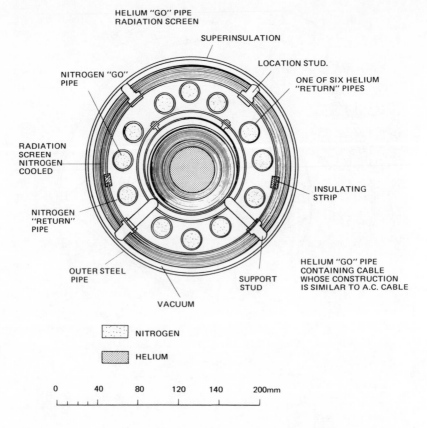

FIGURE 2.2.18.   A DC superconducting cable of transmission capacity of 3 GW at ± 110 kV.[66]

and relatively maintenance-free centrifugal compressors might be bulky and too expensive; screw machine might be compact but less efficient; quite efficient reciprocating compressors could require relatively high maintenance (see Table 2.2.7.). An optimization is necessary with respect to helium recycle flow, pressure levels, thermal locations of expanders, as well as refrigerant flow through the superconducting cable enclosure. Lower temperatures are successively achieved through a number of expansion machines, which are finally followed by Joule-Thomson expansion to obtain liquid helium.

Figure 2.2.24. shows a supercritical helium refrigerator with single-fluid flow for the temperature range 5.2 K to 6.7 K (probably an upper limit for operating Nb cables).[9,69] This design uses liquid nitrogen precooling, but one can design refrigerators for this temperature range without using liquid nitrogen. Figure 2.2.25. shows a refrigeration scheme for a Nb_3Sn cable proposed by BNL. It is designed to operate in the temperature range 6 K to 8 K.

It would be advantageous to return the coolant to the refrigerator through a duct within the cryogenic cable envelope, but if the thermal conductivity between the counterflowing coolant streams is too high, the coolant temperature will rise to an unacceptable level between refrigeration stations. BNL has proposed a modified version of the configuration of Figure 2.2.25[9] wherein the helium to cool the Nb_3Sb cable is returned to the refrigerator via the thermally isolated radiation shield. This approach, shown in Figures 2.2.26.[9] and 2.2.27.[62] could be very useful where a single-circuit installation is desired. There are three helium streams running through each heat exchanger. One stream is the refrigerant to be cooled; the second is a balanced counterflow of GHe just above liquefaction temperature; and the third is an unbalanced counter-flow stream.

Voltage	154 kV	110 kV	33 kV
Current	3.8 kA (1 GW)	45.5 kA (5 GW)	151 kA (5 GW)
Mode	a.c. (3 phase)	d.c.	d.c.
Conductor (inner)	Nb/NbTi/Cu (0.05/0.08/3 mm)	NbTi/Cu (0.3/5 mm)	NbTi/Cu (0.315 mm)
Inner dia./outer dia.	38/40 mm	115/120 mm	70/130 mm
Insulator	Mylar + LHe	LHe	Nylon
Thickness	11.87 mm	21.83 mm	0.5 mm × 2
Steel dia.	267.2 mm dia. (6.6 mm)	267.2 mm dia. (6.6 mm)	267.2 mm dia. (6.6 mm)
Maximum magnetic field	0.54 kG	1.52 kG	1.3 kG
Current density	4.3×10^2 A/cm	3.8×10^4 A/cm^2	5×10^4 A/cm^2
Electrical stress (eff.)	7.5 kV/mm	5.0 kV/mm	33 kV/mm
Estimated abnormal voltage	750 kV	330 kV	99 kV
Stress corresponding to above	36.5 kV/mm	15.1 kV/mm	99 kV/mm
LN$_2$ loss	1400 W/km	1500 W/m	1400 W/km
LHe loss	36 W/km	38 W/km	36 W/km

FIGURE 2.2.19.   AC and DC superconducting cable models cross sectionally photographed. (Courtesy of the Furukawa Electric Co. Ltd.)

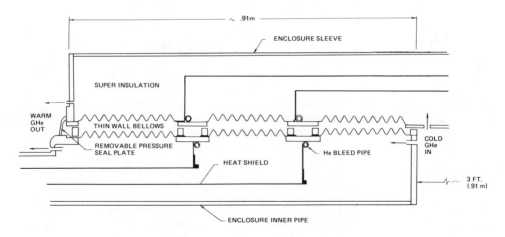

FIGURE 2.2.20.   Joint using bellows.[59]

2.3. OTHER ADVANCED CABLES

Several other new concepts have been under investigation with the principal objective of increasing transmission capacity. They are essentially forced-cooled cables. Water, air, and cryogenic refrigerants have all been proposed as cooling media. We shall discuss briefly some of these proposed schemes, in particular, evaporative-cooled cables, glass-insulated

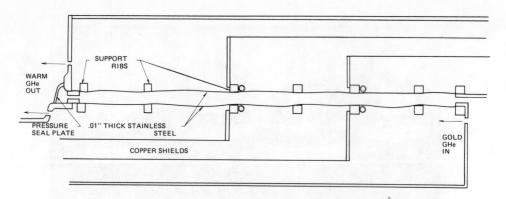

FIGURE 2.2.21.   Joint using ribbed membrane.[59]

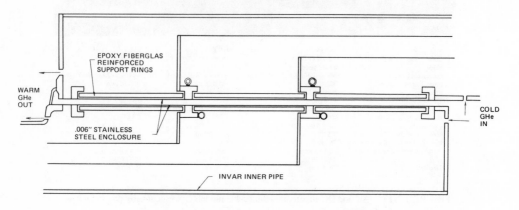

FIGURE 2.2.22.   Joint using supported membrane.[59]

Table 2.2.6.
JOINT CONCEPT COMPARISON[59]

Concept	Advantages	Disadvantages
Bellows	All metal system	Axially not loadable
	Allowance for cold wall contraction	Expensive
	Proven concept	Possibly unstable
		Possibility of thermal shorting across convolutions
Ribbed membrane	All metal system	Unproven concept
	Allowance for some contraction	Possibly unstable
	Low cost	Axially not loadable
	Axially loadable	Bi-material
	Stable	No allowance for cold wall contraction
Supported membrane	Low annulus heat leak	Possibly expensive
	Good chance for high reliability	

water-cooled cables, foam-insulated air-cooled cables, and oil-impregnated paper-insulated liquid nitrogen cables. All are intended for bulk power transmission. Somewhat different is the proposed 138-kV sodium conductor cable with extruded cross-linked polyethylene insulation. It is claimed that problems associated with differential thermal expansion and contraction between conductors and insulation, which may arise in conventional XLPE

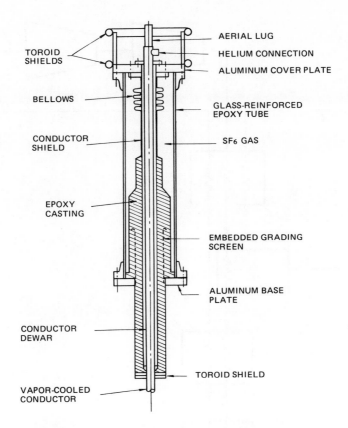

FIGURE 2.2.23.   Pothead design for 138-kV superconducting cable. (From Mausor, S. F., Burghardt, R. R., Fenger, M. L., Dakin, T. W., and Meyerhoff, R. W., *IEEE Trans. Power Appar. Syst.*, 95(3), 909, 1976. With permission.)

Table 2.2.7.
COMPARISON OF REFRIGERATION COMPRESSORS

	Reciprocating	Rotary (turbo)	Helical (screw)
Efficiency (%)	75—80	65—75	60—70
LHe flow rate[a](ℓ)/hr)	2—700	500—8000	100—2000
Pressure differential (psi relative to 1 atm)	140—2700[b]	150	140—300
Mean time to breakdown (hr)	2400—3000[c]	~8800	5000—6000
Best lubricant	Oil	Gas bearings	Dry or oil
Uses	Laboratories, small generators, low LHe production	Transmission cables, linear accelerators, commercial LHe production	Laboratories, transmission cables, generators

[a] Upper limit requires multistage compressors.
[b] Requires 5 compression stages.
[c] Piston rings must be replaced every 3000 hr. to ensure reliable operation.

cables, could be solved in cables with sodium conductors. (See also Volume I, Section 2.2.1.).[89]

2.3.1. Evaporatively Cooled Cables[12,90]

Heat generated in a cable can be removed by latent heat of vaporization. The evaporatively

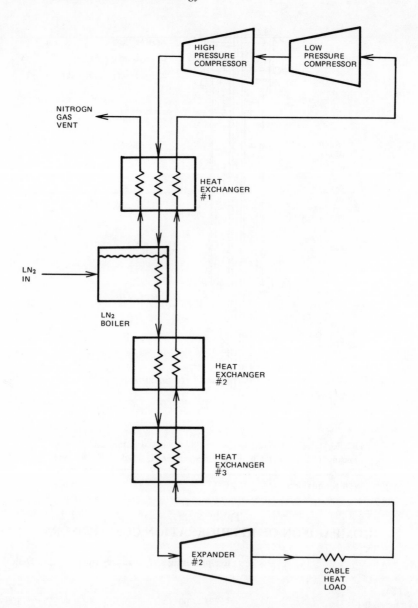

FIGURE 2.2.24.   Supercritical helium refrigerator for the temperature range 5.2 to 6.7 K.

cooled cable (EC cable) concept is based on this principle. Such cables are targeted on a probable carrying current of 2 to 12 kA. The coolant is selected for its ability to change from the liquid phase to the gas phase in the temperature range in which the cable operates. There are both internally cooled and externally cooled cables. The former type must be designed in such a way that the coolant does not affect the performance of the electrical insulation. External cooling is clearly less effective but it is less troublesome as regards insulation design. There are a number of options as far as the coolant is concerned. These will lead to different cable structures. Some typical cables will be briefly described.

A. Externally Evaporatively Cooled Cables

Water is the best choice as a coolant for externally evaporatively cooled cables because its latent heat of vaporization is adequate, it can change phase readily, and fortunately it is

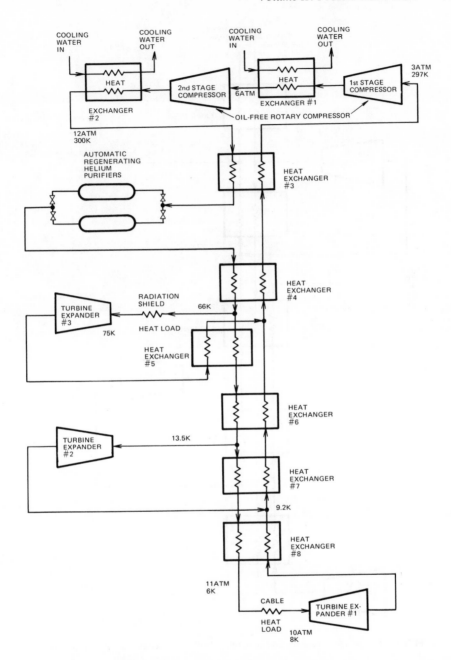

FIGURE 2.2.25. Refrigeration scheme for Nb₃Sn cable.[9]

abundant and inexpensive. Any existing cables can be used provided they are watertight. They are installed in a pipe containing water in the bottom half and water vapor in the upper half as schematically illustrated in Figure 2.3.1. Water can be kept at lower temperatures by reducing the pressure inside the pipe through constant evacuation. This technique is only suitable for flat cable installations. Evacuation up to 0.13 kg/cm² absolute will maintain the water temperature at 10°C over the entire length of the cable. However, for this purpose, vacuum pumps must be located at regular intervals along the cable route, or a large-radius water pipe must be selected. Some added considerations are necessary if this type of cable is to be installed on an uneven route.

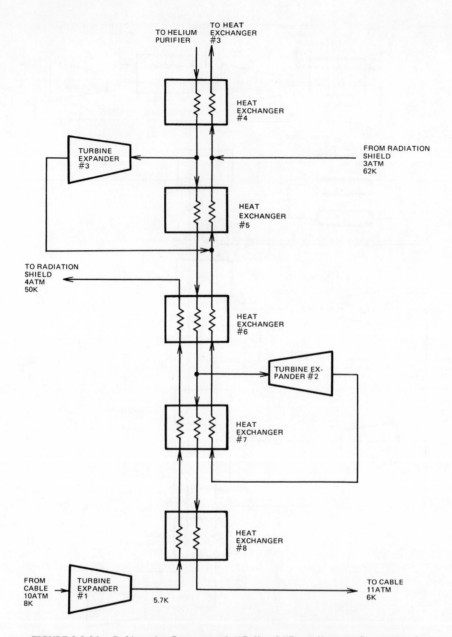

FIGURE 2.2.26.   Refrigeration flow to permit "Go" and "Return" counterflow cooling.[9]

Compared to externally water-cooled cables described in Volume II, Section 1.5., externally evaporative-cooled cables are more easily cooled to low temperature and inherently have no axial increase in temperature. On the other hand, cooling is indirect through the thermal barrier of cable insulation. Current capacity of this cable is therefore limited, being at most about 50% of the increase obtained with the externally water-cooled cable.

B. Internally Evaporatively Cooled Cables

Liquefied gases are utilized as a coolant for internally evaporatively cooled cables. The selected gas must have sufficient latent heat of vaporization, suitable vapor pressure characteristics, and excellent electrical insulating properties. Among these are Freon® gases: Freon® 12 (CCl_2F_2) and Freon® 22 ($CHClF_2$) are considered to be most suitable. Since the

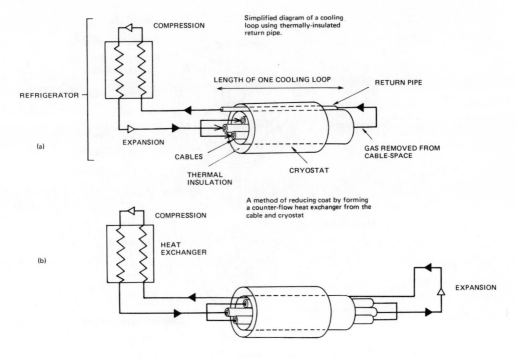

FIGURE 2.2.27. Cooling schemes for superconducting cables.[62]

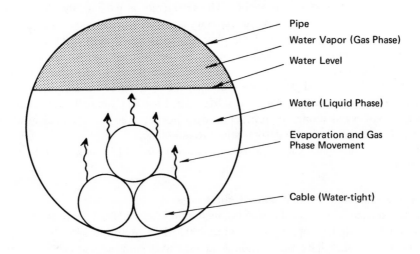

FIGURE 2.3.1. Externally evaporative cooled cable concept.

coolant is in direct contact with the cable conductor, the cooling is very effective. This results in a great increase in ampacity, the limit being determined by the burn-out phenomenon in heat transfer which accompanies boiling. This phenomenon is such that heat transfer increases up to a certain temperature as the temperature differential between the pipe-wall-surface and the boiling liquid increases, but decreases beyond this critical temperature. In other words, the heat transfer vs. the temperature difference has a maximum. In many instances the maximum in heat transfer is extremely large, in which case the expected cable ampacity will be limited instead by the pressure head produced in coolant circulation.

Cross sections of two types of internally evaporatively cooled cables are illustrated in Figure 2.3.2. The first is so designed as to have a wick structure inside each of the three

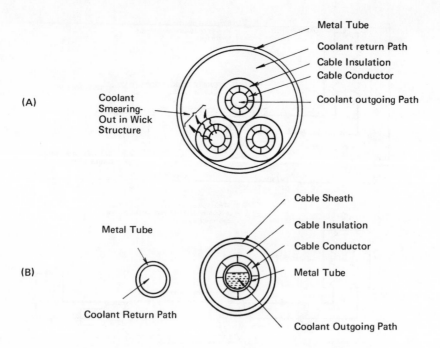

FIGURE 2.3.2.   Internally evaporative cooled cable concepts.

hollow aluminum and copper conductors. The refrigerant is fed into the three conductor bores, passes through the cable core via capillary action, returns in gaseous or mixed state through the remaining wide space of the pipe, and is recycled after being liquefied. The evaporated refrigerant acts as the electrical insulation while it removes heat to above-ground cooling units. For this reason, excellent insulation characteristics are required of it in this type of design. The second design, on the other hand, is based on a similar concept to conventional internally cooled cable. The refrigerant flows longitudinally in the cable bores with continuous evaporation. The refrigerant does not serve as the electrical insulation in this method, thus the insulation design is less troublesome.

Table 2.3.1. summarizes various kinds of internally-cooled EC cables and their features for comparison.

2.3.2. Glass-Insulated Water-Cooled Cables

A glass-insulated underground transmission system is now under investigation for a projected power transmission capacity of 1000 to 4000 MVA at 138 to 345 kV. It appears to offer potential for simplicity and economic savings over conventional cables of the same voltage and ampacity.

The design concept for this type cable is outlined in Figure 2.3.3.[91] The distinct advantages of glass insulation are that it permits a much higher operating temperature then organic insulation and that water can be used as the coolant. The specific heat of water is twice that of oil; water-cooled systems are less costly than oil-cooled systems and they are environmentally acceptable.

Each of the three conductors is installed inside a glass pipe with a thickness of 1/2 in. (12.7 mm). The pipe is coated internally with conducting materials to form a field-free region within. The glass, of course, also acts as the main electrical insulation. Thus, three glass pipes containing their respective cable conductor are placed inside a steel pipe; the whole is then filled with water. Glass pipes would be shipped in unit lenghts of 12 m and would be jointed in the field with solder glass. This would appear to be a nontrivial operation.

Table 2.3.1.
KINDS OF EVAPORATIVE COOLED CABLES

Item	Explanation	Characteristics
Evaporation temperature		
High temp.	To be evaporated at 40 ~ 80°C. Freon® 12 or 22	Efficient and economical
Ambient temp.	To be evaporated at 0 ~ 40°C. Freon® 12, 22 SF$_6$	Low thermal expansion and contraction Not heated inside tunnels and manholes
Low temp.	To be evaporated below ambient temp. CO$_2$	
Cryogenic temp.	To be evaporated below 100 K. N$_2$, H$_2$, He Technically CR and SC cables	Low conductor loss below 100 K
Cooling method		
Radial cooling	Coolant evaporates in conductor structure, moves radially and passes through the insulation	Long distance of feeding joint lengths
Longitudinal cooling	Coolant flows inside a hollow conductor in two-phase state	Either the coolant mixed type or the separate type can be applied
Spontaneous longitudinal cooling	Coolant circulates due to head and density difference without aid of pumps	No pumps needed Easy to keep temperature constant
Coolant and dielectrics		
Mixed type	Coolant acts as pressure medium to electrical insulation	Simple structure
Separate type	Coolant and electrical insulation are separated in structure	Applicable to any type of cables if covered water-tightly
Cable structure		
Pipe-type	Steel pipe sheath	High pressure, large size
Selfcontained	Aluminum or lead sheath	Single cores. In-pipe installation is possible
Others	Application of longitudinal method to I.P.B. bus ducts, CGI cables, etc	

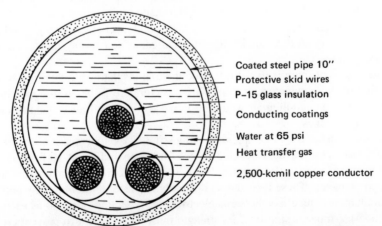

FIGURE 2.3.3.   Glass-insulated cable concept. (From *Electr. World,* October 1976, p. 53. With permission.)

A water-cooled system could operate at temperatures up to 250°C and still maintain good dielectric strength, but even at a conservative temperature rating of 100 to 125°C for the outer shield, the power capacity is very high. There is no need for any special cooling units

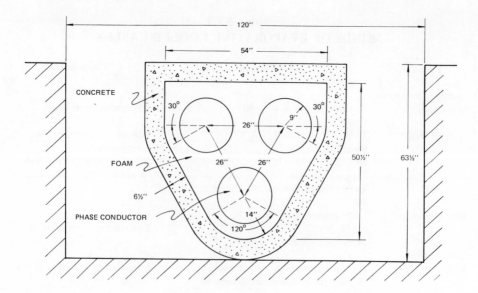

FIGURE 2.3.4.   A 345-kV foam-insulated air-cooled power transmission line.

for this system; it can utilize either ambient air or evaporative cooling. This is a distinct advantage compared with other bulk power transmission cables of the forced-cooled type.

2.3.3. Foam-Insulated Air-Cooled Cables[92]

A polyurethane-foam-insulated transmission line has been the subject of a feasibility and viability study. The design calls for no electrostatic screen around the foam insulation. In this respect it is more like an overhead transmission line buried underground rather than an underground cable. Conventional overhead conductors are replaced by thin-walled tubes of considerably greater diameter, great enough to keep their surface gradient below corona onset. These conductors would be laid in a trench and insulated from each other and from ground by a suitable dielectric foam which would also serve to support them mechanically. The whole would be enclosed in a moisture barrier made of combined plastic and aluminum foil for example. Foams of appropriate density have good mechanical strength and dielectric constant close to unity so that, unlike conventional cables, such a system would not draw an excessive charging current.

The disadvantage of foam is that it is a good thermal insulator, but this may be offset by forced cooling of the conductors. The phase conductors are to be so designed as to form air ducts through which cooling air would be flowed by modest blower facilities located at intervals of several miles.

Figure 2.3.4. illustrates an example of a proposed 345-kV transmission line of this type installed in a machine-excavated trench 5 ft (1.52 m) deep and about 10 ft (3.05 m) wide. Suspended in this trench are the 3 phases, each a hollow, thin-walled aluminum tube of 18 in. (45.7-cm) diameter. These tubes are held rigidly in place in the trench by polyurethane foam. Each aluminum tube has the same electrical cross section of metal as each phase of a standard 345-kV transmission line. Its strength must be sufficient to prevent crushing on short circuits. For mechanical protection, the foam would probably be placed in a concrete culvert. Field fabrication would be considered.

Conductor wall temperatures must be held to 100° C or below to keep the foam from deteriorating. Two possible schemes are air cooling and steam cooling. It is estimated that the heat disposal requirement would be about 25 kW per phase per mile (16 kW per phase per kilometer). This could be carried away axially through the tubing. A rigid aluminum

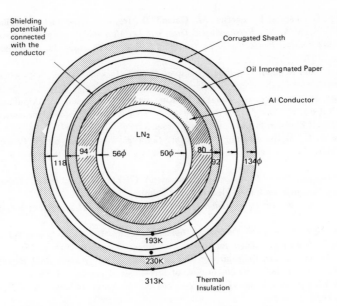

FIGURE 2.3.5. Liquid Nitrogen-cooled OF cable concept (DC ± 250 kV, 12,000 A, 6 GW).[93]

tubular conductor with expansion joints, a corrugated tubular aluminum conductor, and even a flexible all-aluminum center conductor surrounded by the flexible interlocked outer walls to form an air tunnel are currently under investigation as means to absorb thermal expansion and contraction.

2.3.4. Oil-Impregnated Paper-Insulated Liquid Nitrogen-Cooled Cables[93,94]

Dodecylbenzene (DDB) impregnated, laminated insulation made from either cellulose paper or polyethylene paper and polyethylene film exhibits formidable AC and impulse breakdown strengths at temperature down to −90 °C. A cable concept which would take advantage of this behavior, by using liquid nitrogen as a coolant, has been proposed. Both internally or externally cooled designs are conceivable.

Figure 2.3.5. shows a design with internal cooling intended for a 250-kV DC, 12,000-A, 6-GW oil-filled cable.[93] The conductor temperature of this cable would be about −193°C, and the temperature of electrical insulation about −80°C and −60°C on its inner and outer surfaces, respectively. It would operate under a working stress of about 36 kV/mm at the conductor surface. The outer diameter of the cable is small enough (about 134 mm) to be transported on a reel.

REFERENCES

1. **Anon.**, *Resistive Cryogenic Cable*, EEI Report SRD-74-042, Edison Electric Inst., New York, 1974, 1.
2. **Fox, G. R. and Burnstein, J. T.**, Underground cryogenic cable system, *Mech. Eng.*, August 1970, p. 7.
3. **Klaudy, P. A.**, Cryogenic Electrical Energy Transmission, *Electrotech. Cas.*, XXI 370, 1970.
4. **Arthur D. Little Inc.**, *Underground Power Transmission*, U.S. Department of Interior & EEI, ERC Publ., Electric Research Council, New York, 1971.
5. **Erb, J., Heinz, W., Hofmann, A., Köfler, H. J., Komarek, P., Maner, W., Nahar, A., and Heller, L.**, *Comparison of Advanced High Power Underground Cable Designs*, Gesselschaft für Kernforshung, M. B. H. Karlsruhe, West Germany, September 1975.

6. **Delile, G., Deschamps, L., Ferrier, M., Galand, D., Jegou, Y., Marquet, A., Pellier, P., Sabrie, J. L., and Schwab, A. M.,** *Perspectives Offertes par Les Materiels Electriques — Fonctionnant a Basse Temperature pour La Production et Le Transport D'Energie Electrique,* Electricite de France, Clamart, December, 1970.

7. **Bogner, G. and Schmidt, F.,** Energietransport durch Tiefgekühlte Kabel, *Naturwissenschaften,* 50(9), 414, 1970.

8. **Fukagawa, H., Okamoto, H., Tanaka, T., Kaminosono, H., Nagao, T., Akimoto, T., Hamamatsu, T., and Nakasa, H.,** *Investigation Research of Cryogenic Power Cables,* (in Japanese), CRIEPI Rep. No. 71021, Central Research Institute of Electric Power Industry, Tokyo, 1971, 1.

9. **Belanger, B. C.,** *Superconducting and Resistive Cryogenic Power Transmission Research in the U.S. — An Opportunity for Cryogenic Innovation,* 1973 Cryogenics Eng. Conf. Paper No. J-1, Boulder, Colo., 1973, p. 37.

10. **Anon.,** *Cryogenic Power Transmission,* (in Japanese), CRIEPI Advanced Power Transmission Series No. 17, Central Research Institute of Electric Power Industry, Tokyo, 1973, 1.

11. **Anon.,** *Advanced Power Transmission System,* (in Japanese), MITI Agency of Natural Resources and Energy, Tokyo, 1975, 1.

12. **Iizuka, K., Ed.,** *Power Cable Technology Handbook* (in Japanese), Denki-Shoin, Tokyo, 1974, 1.

13. **Graneau, P.,** Vacuum insulation for cryocables, *Cryogenics,* January 1975, p. 25.

14. **Lax, B.,** *Progress in the Application of Vacuum Insulation to Cryocables,* MIT Report to NSF (RANN Project) C670-No. 1, Massachusetts Institute of Technology, Cambridge, 1974, 1.

15. **Graneau, P.,** Vacuum insulation for cryocables and its resistance to discharges, *IEEE Trans. Electr. Insul.,* 9(2), 63, 1974.

16. **Graneau, P.,** *Magnetic Fields in Vacuum Insulated High Voltage Power Apparatus,* NSF-RANN Project, National Science Foundation, Washington, D.C., 1974, 1.

17. **Graneau, P.,** *Lichtenberg Figures Produced by High Voltage Discharges in Vacuum,* IEEE PES Meeting, Paper No. C73 251-6, IEEE, Piscataway, New Jersey, 1973, 1.

18. **Graneau, P. and Thompson, L. B., Jr.,** Three functions of vacuum in cryocables, *Cryogenics,* October 1972, p. 366.

19. **Graneau, P. and Jeanmonod, J.,** Voltage surge performance of vacuum-insulated cryo-cable, *IEEE Trans. Electr. Insul.,* 6(1), 39, 1971.

20. **Graneau, P., Parish, H. C., and Smith, J. L., Jr.,** Refrigeration requirements of the LN2 cryo-cable, *Trans. Am. Soc. Mech. Eng.,* 93-B(4), 1161, 1971.

21. **Afshartous, S. B. and Graneau, P.,** *Current Distribution in the LN2 Cryo-Cable,* IEEE PES Meeting Paper No. CP171-PWR, IEEE, Piscataway, New Jersey, 1970, 1.

22. **Graneau, P. and Jeanmonod, J.,** *Voltage Surge Performance of Vacuum Insulated Cryo-Cable,* IEEE Winter Meeting Paper No. 70 CP170-PWR, IEEE, Piscataway, New Jersey, 1970, 1.

23. **Graneau, P.,** Economics of underground transmission with cryogenic cables, *IEEE Trans. Power Appar. Syst.,* 89(1), 1, 1970.

24. **Graneau, P. and Jeanmonod, J.,** Economic assessment of a liquid-nitrogen-cooled cable, *IEEE Trans. Power Appar. Syst.,* 89(1), 8016, 1970.

25. **Anon.,** Cryogenic terminal tested at waltz mill, *Elecr. World,* December 1969, p. 28.

25a. **Graneau, P. and Thompson, L. B., Jr.,** Design and Testing of A 138 kV Prototype Termination for Vacuum Insulated LN2 Cryocable, unspecified, 1975, 1.

26. **Iwata, Z., Shiroto, K., Nakajima, M., Shinoda, S., and Kikuchi, K.,** *Development of High Voltage Cryoresistive Cable,* (in Japanese), Furukawa Elecr. Ind. Review, No. 53, Ichihara-shi, 1972, 37.

27. **Iwata, Z.,** *Electrical Properties of Insulating Materials at Cryogenic Temperature,* (in Japanese), Furukawa Elecr. Ind. Review No. 54, Ichihara—shi, 1973, 51.

28. **Iwata, Z., Nobuta, S., Nakajima, M., Matsumoto, K., and Kikuchi, K.,** *Development of A High Voltage Cryoresistive Cable,* Proc. 5th Int. Cryogenic Eng. Conf., Boulder, Colo., 1974, 206.

29. **Anon.,** *Resistive Cryogenic Cable — Phase A Report,* EEI Project RP 76-8, Edison Electric Inst., New York, 1969.

30. **Anon.,** *Resistive Cryogenic Cable — Phase B Report,* EEI Project RP 78-7, Edison Electric Inst., New York, 1970.

31. **Anon.,** *Resistive Cryogenic Cable — Phase C Report,* EEI Project RP 78-6, Edison Electric Inst., New York, 1971.

32. **Belanger, B. C.,** Dielectric Problems in the Development of Resistive Cryogenic and Superconducting Cables, Conf. Electr. Insulation and Dielectr. Phenomena, CEIDP Paper No. E-3, National Academy of Sciences, Washington, D.C., 1973, 7.

33. **Jefferies, M. J., Minnich, S. H., and Belanger, B. C.,** *High Voltage Testing of A High-Capacity, Liquid-Nitrogen Cooled Cable,* 1972 IEEE Underground Transmission Conf., IEEE, Piscataway, New Jersey, 1972.

34. **Anon.,** Underground aluminum cable for high power transmission successfully tested for operation at 320°F below zero, *Ind. Heating,* July 1972, p. 1266.
35. **Belanger, B. C.,** Lightning impulse and switching surge breakdown of liquid nitrogen-impregnated wrapped dielectrics for cryogenic cables, *IEEE Trans. Power Appar. Syst.,* 90(1971), 2616, 1971.
36. **Minnich, S. M. and Fox, G. R.,** *Comparative Costs of Cryogenic Cables,* IEEE PES Meeting Paper No. CP169-PWR, IEEE Piscataway, New Jersey, 1970, 1.
37. **Jefferies, M. J. and Mathes, K. N.,** *Insulation Systems for Cryogenic Cable,* IEEE PES Meeting Paper No. TP 44-PWR, IEEE, Piscataway, New Jersey, 1970, 1.
38. **Jefferies, M. J. and Mathes, K. N.,** Dielectric loss and voltage breakdown in liquid nitrogen and hydrogen, *IEEE Trans. Electr. Insul.,* 5(3), 83, 1970.
39. **Jefferies, M. J. and Mathes, K. N.,** *Progress in Cryogenic Dielectrics,* 1969 Annu. Rep. CEIDP, National Academy of Sciences, Washington, D.C., 1969, 153.
40. **Minnich, S. H. and Fox, G. R.,** Cryogenic power transmission, *Cryogenics,* June 1969, p. 165.
41. **Mathes, K. N.,** *Electrical Properties of Cryogenic Liquids and of Ice at Cryogenic Temperatures,* 1969 Annu. Rep. CEIDP, National Academy of Sciences, Washington, D.C., 1969, 20.
42. **Mathes, K. N.,** Cryogenic cable dielectrics, *IEEE Trans. Electr. Insul.,* 4(1), 2, 1969.
43. **Mathes, K. N.,** Dielectric properties of cryogenic liquids, *IEEE Trans. Electr. Insul.,* 207, 24, 1967.
44. **Glaser, P. E.,** Cryogenic insulations, *Machine Design,* August 1967, p. 146.
45. **Cockett, A. H. and Molnar, W.,** Recent improvements in insulants, *Cryogenics,* September 1960, p. 21.
46. **Tien, C. L. and Cunnington, G. R.,** Recent advances in high-performance cryogenic thermal insulation, *Cryogenics,* December 1972, p. 419.
47. **Barron, R.,** *Cryogenic Systems,* McGraw-Hill, New York, 1966.
48. **Nagano, H., Fukasawa, M., Kuma, S., and Sugiyama, K.,** Field test of liquid nitrogen cooled cryogenic power cable, *Cryogenics,* April 1973, p. 219.
49. **Maddock, B. J. and Male, J. C.,** *Superconducting Cables for AC Power Transmission,* CEGB Research No. 4 Central Electricity Generating Board, London, 1976, 11.
50. **McFee, R.,** Superconductivity — cryogenic key to low loss T and D ?, *Power Eng.,* 65, 80, 1961.
51. **McFee, R.,** Application of superconductivity to the generation and distribution of electric power, *Elecr. Eng.,* 81, 122, 1962.
52. **Forsyth, E. B.,** The Brookhaven Program to Develop A Helium-Cooled Power Transmission System, BNL-20444, Rec. Conf. Tech. Appl. Superconductivity, Alusha, U.S.S.R., September 16-19, 1975, 1.
53. **Forsyth, E. B.,** *Underground Power Transmission by Superconducting Cable,* Monograph BNL-50325, Brownhaven National Laboratory, Upton, N.Y., 1972, 1.
54. **Hammel, E. F.,** *A Proposal for DC Superconducting Power Transmission Line Prototype Development,* Monograph LASL P-94, 1972, 1; U.S. Atomic Energy Commission Contract W-7405-ENG. 36, Washington, D.C.
55. **Geballe, T. H. and Beasley, M. R.,** *Study of Critical Currents of Very Thin Multilayered Structures,* EPRI TD-193, Project 78-22, Final Rep. Electric Power Research Inst., Palo Alto, Calif., June 1976, 1.
56. **Beall, W. T., Jr.,** Development of A Niobium-Copper-Invar Composite Conductors for An AC Superconducting Power Transmission Cable, Unspecified 1974, 1.
57. **Bussière, J. F. and Suenage, M.,** Reduction of a.c. losses of Nb_3Sn by surface treatment, *J. Appl. Phys.,* 47(2), 707, 1976.
58. **UCC,** *Superconducting Cable System Program (Phase II),* Interim Rep., Part I, EPRI Project PR 7807-1, Electric Power Research Inst., Palo Alto, Calif., 1973, 1.
59. **UCC,** *Superconducting Cable System Program (Phase II),* Interim Rep. Part II, EPRI Project PR 7807-1, Electric Power Research Inst., Palo Alto, Calif., 1973, 138.
60. **Forsyth, E. B., Muller, A. C., and Rigby, S. J.,** *Some Theoretical Considerations Affecting the Design of Lapped Plastic Insulation for Superconducting Power Transmission Cables,* EPRI EL-269 (RP 7844-1), Electric Power Research Inst., Palo Alto, Calif., December 1976, 1.
61. **Furuto, Y., Miura, T., and Ikeda, M.,** *Electrification Test of Superconducting Model Cable,* Furukawa Electr. Rev. No. 54, Furakawa Electric Ind., Ltd., Ichihara-shi, May 1973.
62. **Forsyth, E. B.,** Superconducting power transmission: the perils and promise, paper submitted to the *Int. J. Energy,* 1976.
63. **Meyerhoff, R. W.,** *Testing of Subscale and Fullscale Single Phase Sections of A 3400 MVA AC Superconducting Transmission Line,* 1975 Cryogenic Eng. Conf., Paper R-8, Boulder, Colo., July 1975.
64. **Nurmepuu, K., Meyerhoff, R. W., Naatz, R. A., and Rooney, K. J.,** Feasibility of Incorporating Underground AC Superconducting Transmission in the Commonwealth Edison System in the 1990's, unspecified, 1976, 1.
65. **Croitoru, Z.,Dechamps, L., and Schwab, A. M.,** Transport d'énergie électrique par cryocâbles, *Entropie,* No. 56, Mars-Avril 1974, 47.
66. **Deschamps, L., Jegou, U., and Schwab, A. M.,** *Transport d'Énergie Électrique par Cryocâbles,* EdF Publ., Electricite de France, Clamart, April 1972, 1.

67. **Baylis, J. A.,** Superconducting cables for AC and DC power transmission, *Philos. Trans. R. Soc. London Ser. A:,* 275, 205, 1973.

68. **Mausor, S. F., Burghardt, R. R., Fenger, M. L., Dakin, T. W., and Meyerhoff, R. W.,** Development of a 138 kV superconducting cable termination, *IEEE Trans. Power Appar. Syst.,* 95(3), 909, 1976.

69. **Cairns, D. N. H., Swift, D. A., Edney, K., and Steel, A. J.,** Refrigeration and Circulation of Helium in Superconducting Power Cables, Proc. Int. Inst. Refrigeration Conf. Low Temp. Elec. Power, London, March 24—28, 1969, 155.

70. **Muller, A.,** *Recent Progress in Dielectric Materials Development for Superconducting Power Transmission Cables,* BNL Publ. PTP 55, Brownhaven National Lab., Upton, N.Y., March 1976, 1.

71. **Forsyth, E. B., McNerney, A. J., and Muller, A. C.,** Dielectric Design Consideration for a Flexible Superconducting Power Transmission Cable, paper submitted to the CDC Cryogenic Eng. Conf., Kingston, Ontario, July 22—25, 1975, 1.

72. **Garber, M., Bussiere, J. F., and Morgan, H. M.,** *Design of Double Helix Conductors for Superconducting AC Power Transmission,* BNL-21338, Brownhaven National Laboratory, Upton, N.Y., 1976, 1.

73. **Forsyth, E. B., Stewart, J. R., and Williams, J. A.,** *Long Distance Bulk Power Transmission Using Helium-Cooled Cables,* IEEE Conf. Rec. 1976 Underground T & D Conf., IEEE, Piscataway, New Jersey, 1976, 446, (76CH 119-7-PWR).

74. **Muller, A.,** Mechanical properties of insulating tapes at cryogenic temperatures, *Rev. Gen. Electr.,* 84, 568, 1975.

75. **Phillips, W. A., King, C. N., Thomas, R. A., and Norton, R. H.,** *Dielectric Insulation at Cryogenic Temperatures,* Final Rep. for EEI & DOI, RP78-11, Edison Electric Inst., N.Y., July 1973, 1.

76. **Schwenterly, S. W., Menon, M. M., Kernohan, R. H., and Long, M. H.,** AC Dielectric Performance of Helim Impregnated Multi-Layer Plastic Film Insulation, unspecified, 1976, 1.

77. **Meyerhoff, R. W.,** Development of the Dielectric System for An AC Superconducting Power Transmission Line, paper presented at the Underground T & D Conf., Dallas, Texas, April 1—5, 1974, 1.

78. **Meyerhoff, R. W.,** The Fault Recovery Performance of A Helium-Insulated Rigid AC Superconducting Cable, unspecified, 1975, 1.

79. **Bosack, D. J.,** Development of the Dielectric System for A Helium Cooled Superconducting Transmission Line, unspecified, 1975, 1.

80. **Meyerhoff, R. W.,** Use of Explosive Welding in the Manufacture and Installation of a Superconducting Electric Power Transmission Line, paper presented at the 5th Int. Conf. High Energy Rate Fabrication, Denver, Colo., June 24—26, 1975, 1.

81. **Beall, W. T., Jr.,** Development of a Niobium-Copper-Invar Composite Conductor for an AC Superconducting Power Transmission Cable, unspecified, 1974, 1.

82. **Meyerhoff, R. W.,** AC Superconducting Power Transmission, unspecified, 1972, 1.

83. **Bochenek, E., Voigt, H., Hildebrandt, U., Kuhman, H., and Scheffler, E.,** Supraleitendes Flexibles Hochleistrungs-Gleichstromkabel, *Elektrotech. Z. Ausg. B.,* 26, 215, 1974.

84. **Heuman, H.,** *Kabeltechische Probleme bei Projektierung und Bau von Übertragungsstrecken mit Supraleitern,* Publ. of Kabelwerke der AFG-Telefunken-Gruppe, 1972, 1.

85. **Deschamps, L., Giraud, P., Leroux, J. M., and Moisson, F.,** Problems Liés A La Réalisation de Cryocables, unspecified, 1970, 1.

86. **Aupoix, M., Giraud, P. J., and Carbonell, E.,** Les Cryoliasons, *Journees des Supraconducteurs,* 17—18 Mars, 1970, 1.

87. **Dubois, P., Eyraud, I., and Carbonell, E.,** Research and Power Transmission, Paper Presented at the 1972 Appl. Super. Conf., 1972, 1.

88. **Delile, G., Deschamps, L., Ferrier, M., Galand, D., Jegou, Y., Marquet, A., Pollier, P., Sabrie, J. L., and Schwab, A. M.,** *Perspectives Offertes par Les Materiels Electriques Fonctionnant a Basse Temperature pour La Production et Le transport D' Energie Electrique,* EdF Publication HM. 024-77, Electricite de France, Clamart, December 1970.

89. **Graneu, P.,** *Thermal Failure of High Voltage Solid Dielectric Cables,* IEEE PES Winter Meeting Rec. A76 037-2, IEEE, Piscataway, New Jersey, 1976, 1.

90. **Kubo, H. and Azuma, Y.,** *Evaporative-Cooling Type of Bulk Power Transmission Cables,* IEEE Rec. Electr. Wire Study Meeting Ec-71-4, IEEE, Piscataway, New Jersey, 1971, 1.

91. **Anon.,** Pipe-type systems still dominate, *Electr. World,* October 1976, p. 49.

92. **Greenwood, A. N., Anderson, J., and Singh, N.,** *Feasibility & Viability Study of an Air-Cooled, Foam-Insulated Transmission System,* EPRI Rep. EL-532, Electric Power Research Inst., Palo Alto, Calif., 1977, 1.

93. **Iwata, Z., Kikuchi, K., and Kawai, E.,** Low Temperature Properties of Oil Impregnated Paper Insulation, Rec. Int. Symp. High Voltage, Zürich, September 1975, 752.

94. **Hossam, A. A. and Salvage, B.,** *A Liquid-Nitrogen-Cooled, High Voltage, Direct Current Cable,* Int. Conf. High Voltage DC/AC Power Trans., IEEE, Piscataway, N.J., 1973.

Appendix

TABLE OF SCALE CONVERSION

Item			**SI Unit**
Length	1 mi	1,609 km	1.61×10^3 m
	1 ft	30.48 cm	3.05×10^{-2} m
	1 in.	25.4 mm	2.54×10^{-4} m
	1 mil = 10^{-3} in.	25.4 μm	2.54×10^{-7} m
Area	1 acre	4046.9 m^2	4.05×10^3 m^2
	1 in.^{2}	645 mm^2	6.45×10^{-4} m^2
	1000 kcmil	506.7 mm^2	5.07×10^{-4} m^2
Volume	1 gal	3785.43 cm^3	3.79×10^{-3} m^3
	1 British gal = 1.2 gal	4542.52 cm^3	4.54×10^{-3} m^3
Weight	1 lb	453.6 g	4.54×10^{-1} kg
Density	1 lb/ft^3	0.0160 g/cm^3	$1.60 \times$ kg/m^3
Electric field	100 V/mil	3.94 KV/mm, MV/m	3.94×10^6 V/m
Pressure	100 psi	7.03 kg/cm^2	
	14.22 psi	1 kg/cm^2	0.0981 MN/m^2
		1 atm, 760 mmHg	(MPa)
	145 psi	10.20 kg/cm^2	1 MN/m^2
Energy and heat	1 Btu	1054.866 J	1.05×10^3 J
	1 cal	4.186 J	4.186 J
Flow rate	100 gpm	6.31 ℓ/sec, 22.7 m^3/hr	6.31×10^{-3} m^3/sec
Thermal conductivity	1 Btu/hr·ft °F	1.7305×10^{-2} W/cm °C	1.73 W/m K
	1 Btu·in./hr·ft °F	1.4420×10^{-3} W/cm °C	1.44×10^{-1} W/m K
Temperature	°F	°C = (5/9)(°F-32)	K = C + 273.15

VISCOSITY CONVERSION TABLE

Poise   = c.g.s. unit of absolute viscosity = $\dfrac{gm}{sec \times cm}$

Stoke   = c.g.s. unit of kinematic viscosity = $\dfrac{gm}{sec \times cm \times density(to°F)}$

Centipoise   = 0.01 poise
Centistoke   = 0.01 stoke
Centipoises   = Centistokes × density (at given temperature)

To convert poises to $\dfrac{lb}{sec \times ft}$ or $\dfrac{lb}{hr \times ft}$ multiply by 0.0672 or 242 respectively.

	Saybolt seconds at			Redwood seconds at			Engler degrees at all temps.
Centistokes	100°F	130°F	210°F	70°F	140°F	200°F	
2.0	32.6	32.7	32.8	30.2	31.0	31.2	1.14
3.0	36.0	36.1	36.3	32.7	33.5	33.7	1.22
4.0	39.1	39.2	39.4	35.3	36.0	36.3	1.31
5.0	42.3	42.4	42.6	37.9	38.5	38.9	1.40
6.0	45.5	45.6	45.8	40.5	41.0	41.5	1.48
7.0	48.7	48.8	49.0	43.2	43.7	44.2	1.56
8.0	52.0	52.1	52.4	46.0	46.4	46.9	1.65
9.0	55.4	55.5	55.8	48.9	49.1	49.7	1.75
10.0	58.8	58.9	59.2	51.7	52.0	52.6	1.84
11.0	62.3	62.4	62.7	54.8	55.0	55.6	1.93
12.0	65.9	66.0	66.4	57.9	58.1	58.8	2.02
14.0	73.4	73.5	73.9	64.4	64.6	65.3	2.22
16.0	81.1	81.3	81.7	71.0	71.4	72.2	2.43
18.0	89.2	89.4	89.8	77.9	78.5	79.4	2.64
20.0	97.5	97.7	98.2	85.0	85.8	86.9	2.87
22.0	106.0	106.2	106.7	92.4	93.3	94.5	3.10
24.0	114.6	114.8	115.4	99.9	100.9	102.2	3.34
26.0	123.3	123.5	124.2	107.5	108.6	110.0	3.58
28.0	132.1	132.4	133.0	115.3	116.5	118.0	3.82
30.0	140.9	141.2	141.9	123.1	124.4	126.0	4.07
32.0	149.7	150.0	150.8	131.0	132.3	134.1	4.32
34.0	158.7	159.0	159.8	138.9	140.2	142.2	4.57
36.0	167.7	168.0	168.9	146.9	148.2	150.3	4.83
38.0	176.7	177.0	177.9	155.0	156.2	158.3	5.08
40.0	185.7	186.0	187.0	163.0	164.3	166.7	5.34
42.0	194.7	195.1	196.1	171.0	172.3	175.0	5.59
44.0	203.8	204.2	205.2	179.1	180.4	183.3	5.85
46.0	213.0	213.4	214.5	187.1	188.5	191.7	6.11
48.0	222.2	222.6	223.8	195.2	196.6	200.0	6.37
50.0	231.4	231.8	233.0	203.3	204.7	208.3	6.63
60.0	277.4	277.9	279.3	243.5	245.3	250.0	7.90
70.0	323.4	324.0	325.7	283.9	286.0	291.7	9.21
80.0	369.6	370.3	372.2	323.9	326.6	333.4	10.53
90.0	415.8	416.6	418.7	364.4	367.4	375.0	11.84
[a]100.0	462.0	462.9	465.2	404.9	408.2	416.7	13.16

Note: To obtain the Saybolt Universal viscosity equivalent to a kinematic viscosity determined at t°F, multiply the equivalent Saybolt Universal viscosity at 100°F by $1 + (t - 100)0.000064$; e.g., 10 centistokes at 210°F are equivalent to 58.8 × 1.0070, or 59.2 Saybolt Universal seconds.

[a]   At higher values use the same ratio as above for 100 centistokes: e.g., 100 centistokes = 110 × 4.620 Saybolt seconds at 100°F.

INDEX

E